BUCKLING OF SHELL STRUCTURES,
ON LAND, IN THE SEA AND IN THE AIR

This volume consists of papers presented at the International Colloquium on Buckling of shell structures, on land, in the sea and in the air, Lyon, FRANCE, 17–19 September 1991.

SCIENTIFIC COMMITTEE

Professor J. ARBOCZ	The Netherlands
Doctor A. COMBESCURE	France
Professor P. DOWLING	United Kingdom
Professor M. ESSLINGER	Germany
Professor G. GALLETLY	United Kingdom
Professor R. GREINER	Austria
Professor J. F. JULLIEN	France
Professor H.-R. MEYER-PIENING	Switzerland
Doctor J. ODLAND	The Netherlands
Doctor C. POGGI	Italy
Professor H. ŐRY	Germany
Professor M. POTIER-FERRY	France
Professor J. RATHÉ	Belgium
Doctor L. Å. SAMUELSON	Sweden
Professor H. SCHMIDT	Germany
Professor J. SINGER	Israel
Professor W. WUNDERLICH	Germany

ORGANIZING COMMITTEE

Professor J. ARBOCZ	The Netherlands
Doctor A. COMBESCURE	France
Professor P. DOWLING	United Kingdom
Professor G. GALLETLY	United Kingdom
Professor J. F. JULLIEN	France
Professor H. ŐRY	Germany
Professor J. RATHÉ	Belgium
Professor J. SINGER	Israel

BUCKLING OF SHELL STRUCTURES, ON LAND, IN THE SEA AND IN THE AIR

Edited by

J. F. JULLIEN

Civil Engineering Department
National Institute of Applied Science at Lyon
France

CRC Press
Taylor & Francis Group
Boca Raton London New York

CRC Press is an imprint of the
Taylor & Francis Group, an **informa** business

CRC Press
Taylor & Francis Group
6000 Broken Sound Parkway NW, Suite 300
Boca Raton, FL 33487-2742

First issued in paperback 2019

© 1991 by Taylor & Francis Group, LLC
CRC Press is an imprint of Taylor & Francis Group, an Informa business

No claim to original U.S. Government works

ISBN-13: 978-1-85166-716-1 (hbk)
ISBN-13: 978-0-367-86447-7 (pbk)

British Library Cataloguing in Publication Data
International Colloquium on Buckling of Shell
   Structures (1991: Lyon, France)
   Buckling of shell structures, on land, in the sea and
   in the air.
   I. Title   II. Jullien, J. F.
   624.17762

   ISBN 1-85166-716-4

Library of Congress CIP data applied for

Visit the Taylor & Francis Web site at
http://www.taylorandfrancis.com

and the CRC Press Web site at
http://www.crcpress.com

PREFACE

The theoretical formulation of the buckling load of thin slender structures of simple geometrical shape has been known for many years. Nevertheless, for engineers, buckling is often an obscure phenomenon and it is not widely known that the theoretical expressions can badly represent reality. Certain types of buckling lead to an instability whose behaviour at failure is not repeatable, a fact which in the past has often led to confusion. This situation is no longer true, but the domain has for too long remained the preserve of specialist engineers.

The results of laboratory studies have now shown that it is possible to obtain repeatability of the critical load and that the latter is very dependent on the nature of the imperfections of the shell, on the loading conditions, on the boundary conditions, etc. Many approaches, depending on the different sectors of application, have been made to assess these effects.

The regulations for dimensioning the structure are very varied; they depend on the complexity of the method of calculation to be used and on the degree of control over the construction phase. The calculation of the buckling load of a nuclear reactor vessel is not the same as that for a grain silo, an underwater shell, or for an aircraft fuselage. The analysis of the specialist calls upon the results of specific tests, on regulations, and upon his knowledge of the limitations of his calculations.

The science of the study of containers of various materials has, in the majority of cases, a uniformity of shape, of function and of loading. The grouping of the papers in this publication will, it is hoped, lift the "barriers" and give the "state of the art" for each sector.

More than 50 articles are grouped by theme: experimental methods and physical modelling; development of theoretical models; numerical methods and their implementation; codes, regulations and methods of dimensioning; comparison of different procedures.

Finally, I would like to thank all those who have helped with the Colloquium: the Scientific Committee, the Organising Committee and all those who have contributed their knowledge to the benefit of all.

J. F. JULLIEN

CONTENTS

Numerical Methods : General Applications

Numerical Methods : Specific Applications

Numerical Codes and Design

Round Table

Numerical Methods : Specific Applications

Experimental and Numerical Methods

POSTER SESSION

Experimental Methods

Theoretical Methods

Numerical Methods : General Applications

Numerical Codes and Design

Experimental and Numerical Methods

Experimental Methods

Numerical Codes and Design

Round Table

THE INFLUENCE OF INITIAL IMPERFECTIONS ON THE BUCKLING OF STIFFENED CYLINDRICAL SHELLS UNDER COMBINED LOADING[+]

HAIM ABRAMOVICH[*], JOSEF SINGER[**] AND TANCHUM WELLER[***]
Faculty of Aerospace Engineering
Technion, I.I.T., 32000 Haifa, Israel

ABSTRACT

The present experimental study aims at providing better inputs for improvement of the buckling load predictions of stiffened cylindrical shells subjected to combined loading. The work focuses on two main factors which considerably affect the combined buckling load of stiffened shells, namely initial geometric imperfections and boundary conditions. Six shells with nominal simple supports were tested under various combinations of combined loading. The Vibration Correlation Technique (VCT) is employed to define the real boundary conditions. The initial geometric imperfections of the integrally stiffened shells are measured in the present experiments *in situ* and are used as inputs to a multimode analysis which yields the appropriate "knockdown" factor for various combinations of axial compression and external pressure. Thus when employing the repeated buckling method for obtaining interaction curves, each point on the curve is adjusted (using the multimode analysis) for the measured degradation of the shell and this results in improved interaction curves and reduced experimental scatter. The measured geometrical imperfections of the preloaded shells can also serve as a contribution to International Imperfection Data Bank for future studies on the correlation between the manufacturing method of the shell and its geometric imperfections.

LIST OF SYMBOLS

A_{11}	=	cross-sectional area of stringer
b_1	=	distance between centers of stringers
c_1	=	width of stringer
d_1	=	height of stringer
E	=	Young's modulus of shell and stringer

[+]Sponsored in part by the Lena and Ben Fohrman Aerospace Structures Fund
 [*]Senior Lecturer
 [**]L. Shirley Tark Professor of Aircraft Structures
[***]Associate Professor

e_1	=	stringer eccentricity (distance from middle surface to centroid of stringer)
f	=	frequency
h	=	thickness of shell
I_{11}	=	moment of inertia of stringer cross section about its centroidal axis
L	=	length of shell
m, k	=	number of longitudinal half waves
n, l	=	circumferential wave number
P_{cr}, p_{cr}	=	theoretical buckling load, axial compression, and pressure, respectively
P_{const}, p_{const}	=	constant axial load and pressure, respectively
P_{exp}, p_{exp}	=	experimental buckling load, axial compression and pressure, respectively
P_{sp}, p_{sp}	=	calculated buckling load, axial compression and pressure, respectively, for shell with axial or rotational restraint
P_{ssx}, p_{ssx}	=	theoretical buckling load, axial compression and pressure, respectively, for SS type BCs
R	=	radius of middle surface of cylindrical shell
Z	=	$[1-\nu^2]^{1/2}(L^2/Rh)$, Batdorf shell parameter
ρ_{sp}	=	P_{exp}/P_{sp} or p_{exp}/p_{sp}
ρ_{th}	=	$P_{SS3_{imp}}/P_{SS3}$
ρ_{ssx}	=	P_{exp}/P_{ssx}, or p_{exp}/p_{ssx}
ξ	=	the imperfection amplitude

Notations for Boundary Conditions

SS-1	$w=M_x=N_x=N_{x\phi}=0$	C-1	$w=w_{,x}=N_x=N_{x\phi}=0$
SS-2	$w=M_x=u=N_{x\phi}=0$	C-2	$w=w_{,x}=u=N_{x\phi}=0$
SS-3	$w=M_x=N_x=v=0$	C-3	$w=w_{,x}=N_x=v=0$
SS-4	$w=M_x=u=v=0$	C-4	$w=w_{,x}=u=v=0$

INTRODUCTION

Stiffened cylindrical shells are widely used in many types of structures. In order to improve their buckling predictions by better definition of the boundary conditions, a nondestructive Vibration Correlation Technique (VCT) was developed at the Technion Aircraft Structures Laboratory (e.g., Refs. [1] or [2]). The VCT method was applied successfully to stringer stiffened cylindrical shells under axial compression, external pressure and their combination (Refs. [1]-[3]).

In the case of combined loading, there are two approaches to the construction of interaction curves: use of separate, nominally identical, shells for each point on the interaction curve, or use of repeated buckling of the same specimen. Singer (Ref. [4]) emphasized that with the use of separate shells the scatter in the experimental interaction curves is mainly due to minor differences among the specimens, whereas repeated buckling of the same specimen can provide a better experimental interaction curve. The previous Technion studies on stiffened cylindrical shells (Refs. [3], [5] and [6]) used both approaches for obtaining experimental interaction curves. It was concluded there that for stiffened shells the two approaches could be applied successfully, though the repeated buckling method seemed preferable. The initial results for the experimental interaction curves in Ref. [3], obtained by repeated buckling, were not conclusive, but those in Refs. [5] and [6] were already more consistent.

In Ref. [6] it was also shown that the sequence of loading, constant axial compression first and then increasing the external pressure until buckling occurs or reverse order of loading, does not influence the buckling loads significantly. This was somewhat surprising in view of the nonlinear behavior of the interaction curve. Since this observation was based on six specimens only, an additional test series was embarked upon to validate and reinforce this main conclusion. For more precision and more reliable interpretation of the data a new feature was incorporated in the present test series, the measurement of geometrical imperfections of the shell before during and after each buckling load combination. Although the initial geometric imperfections of the shell were already measured in the previous test series (Ref. [6]), they were there interpreted only qualitatively, without further processing by the special computer programs developed for axial compression (e.g., Refs. [7],[8]). Another improvement was measurement of these imperfections *in situ* and not in a separate rig as was done earlier (Refs. [5],[6],[8]). A detailed description of the new test system is presented in Ref. [9].

This approach of measuring the geometrical imperfections of shell prior to testing under combined loading is also applied in Ref. [10] where an optical method is used for measurement of the imperfections. Tennyson, Ref. [11] is using predetermined geometrical imperfection on a photoelastic epoxy plastic shell to study their influence on buckling under combined loading including axial compression and hydrostatic pressure. The effect of Z parameter on the buckling of isotropic shells under combined loading is discussed in Ref. [12], while Refs. [13]-[14] present studies performed on laminated composite shells under combined loading and prescribed geometrical imperfections.

It may be recalled, that geometrical imperfections are widely accepted as one of the main reasons for the low "knock-down" factors observed in cylindrical shells, in particular under axial compression. Extensive measurements of initial geometrical imperfections of stringer-stiffened cylindrical shells were therefore performed at the Technion, following those carried out at the California Institute of Technology (Caltech) in the late sixties (Ref. [15]). The measurements were performed in a special test rig (see Ref. [16]) which, however, could not be used for external pressure and combined loading and also could not accommodate the vibration scanning system employed for the VCT. Therefore the specimens were scanned only for initial imperfections and then transferred into another test rig for combined loading (Refs. [5], [6]).

This deficiency has now been removed and the present test rig can apply combined loading together with the ability of imperfection measurement as well as vibration scanning. A similar capability was incorporated in a different vibration scanning and imperfection measurement system developed for larger shells under axial compression (Ref. [17]).

Another feature of the present test series was the measurement of the growth of imperfections, for combined loading, similar to that performed at Caltech (Ref. [15]) for axial compression only. The imperfections scans are now recorded and with the aid of special computer programs harmonic components of two double Fourier series representations are calculated (for details see Refs. [8] and [9]). Then a multimode analysis program, MIUTAM (Ref. [7]) is employed to yield a theoretical "knock-down" factor

$$\rho_{th} = P_{SS3_{imp}} / P_{SS3} \tag{1}$$

In Eq. (1) $P_{SS3_{imp}}$ is the buckling load computed from the measured imperfections by MIUTAM and P_{SS3} is the classical-linear-theory axial buckling load for SS3 type boundary conditions. In order to calculate the influence of geometrical imperfections on the buckling of shells under external pressure and its combination with axial compression, MIUTAM was modified by Arbocz and Abramovich (Ref. [18]), yielding three new programs, PIIUTAM (axial compression and constant internal pressure), CPIUTAM (axial compression and constant external pressure) and CPIUTAMN (constant axial compression and variable external pressure).

The present test program includes six cylindrical stringer-stiffened shells on nominally simple supports. Each shell is subjected to various combinations of axial compression and external pressure. The repeated buckling approach is used, applying the VCT method to redefine the boundary conditions and the buckling loads. Initial geometrical imperfections, growth of imperfections and imperfections after buckling under external pressure and combined loading are measured *in situ*. Using the CPIUTAM and CPIUTAMN multimode analysis the influence of the geometrical imperfections is assessed. The repeatedly buckled points are then adjusted for the measured degradation, yielding improved interaction curves and reduced experimental scatter.

SPECIMENS, TEST SETUP AND PROCEDURES

The present test series included six stringer-stiffened aluminum-alloy shells whose dimensions are given in Table 1. The shells were integrally stiffened and were machined from thick-walled 7075-T6 aluminum-alloy drawn tubes, 254 mm outside diameter and 190.5 mm thick, by a special machining process developed at the Technion (see Ref. [19]). The shells were manufactured in pairs of nominally identical dimensions, cut from a single blank. This yielded three pairs of nominal "twin" shells, Y1A, Y2A, shells Y3A, Y4A and shells Y7B, Y8B (see Table 1). This allowed examination of the effect of the loading sequence by testing one shell with a given loading sequence and its "twin" under the reversed order of loading.

All the shells had nominally simple support boundary conditions (see Fig. 1b). They were experimentally found to be between SS3 and SS4. The test setup employed is shown in Fig. 1. It permits axial loading and external pressure to be applied separately or as combined loading. It also includes a system for measurement of the natural frequencies and their corresponding longitudinal and circumferential modes (for more details see Ref. [9]).

The vibrations of the shell are induced using a shaker and the response of the shell is recorded with a microphone. The natural frequencies of the shell are obtained using Lissajous figures and the corresponding modes are plotted on a X-YY plotter. The experimental test setup includes also a system for measurement *in situ* of geometrical imperfections of the shell, a rotating LVDT measuring the imperfections of the shell at a predetermined height. Then the LVDT is moved to another height, and is again rotated. These measurements yield a "topological map" of the geometrical imperfections of the shell. These imperfections are then stored on magnetic discs, using an A/D device and a PC computer, for further processing.

The test procedures were as follows: For each shell, after being mounted in the rig, the initial imperfections were recorded. Then the shell was loaded with a prescribed loading combination. For each loading, the shell was excited over a range of those natural frequencies, which yielded modes in the region of those predicted for buckling under the prescribed state of loading. The frequencies and the modes were monitored and recorded to be used later for the determination of the actual boundary conditions of the shell using the VCT.

After the shell buckled, the load was released and the imperfections of the shell were again recorded to measure the influence of the loading combination on the geometrical imperfections of the shell. Then the shell was loaded under another combination and the above procedure was repeated. One should note that the repeated buckling of the shell did not induce any substantial damage to the shell (which is continuously monitored by the measurement of the imperfections), except when buckling under axial compression, which usually is the last loading combination in a test.

TEST RESULTS AND DISCUSSION

This section contains descriptions of some of the tests conducted in the present series. More detailed test descriptions and complete results of the whole series of shell are reported in Ref. [9].

Table 2 summarizes the repeated buckling process employed in the present test series for the six shells tested. Throughout the tests the first loading combination was application of external pressure while the last combination was buckling under axial compression. After each buckling, measurement of the geometrical imperfections was carried out *in situ*. Vibration tests were also conducted for the application of the VCT method. Typical results for the vibration tests of shell Y1A under various loading combinations are presented in Fig. 2. The boundary conditions can be seen to be between SS3 and SS4. To emphasize the importance of repeated measurement of the geometrical imperfections of a shell some typical results are presented for shell Y2A (with more results reported in Ref. [9]). Figure 3 shows the measured imperfections at three stages of loading: a. initial, b. at external pressure p=8 [kPa], c. after buckling under external pressure. One can see some growth of imperfection in Figs. 3b and 3c compared to Fig. 3a. This growth is shown quantitatively in Figs. 4a-c where the circumferential variation of the predominant half wave sine Fourier representation are presented for the above three points of loading. As pointed out before, the double wave (sine and cosine) Fourier representation is used in the calculations of the ρ_{th} (see Refs. [8] and [9]).

The results of the present test series are presented in the usual form of interaction curves. Figure 5a shows the experimental results, while Fig. 5b depicts the same experimental results, after the assessment of the real boundary conditions (using the VCT method), which reduced the scatter and indicates a more realistic interaction curve. Using the measured geometrical imperfections one can obtain the "knockdown" factor ρ_{th} caused by them (see Table 3, for the shell Y3A). Knowing the ρ_{th} one can adjust the experimental results to take the geometrical imperfections into account yielding a better improved interaction curve, as can be seen in Fig. 5c.

CONCLUSIONS

The experimental studies on the six shells have extended the generality and applicability of the VCT as an adequate nondestructive tool for more reliable generation of buckling interaction curves of stiffened shells subjected to combined loading. The repeated buckling method to obtain reliable interaction curves appears to be very appropriate provided the "knockdown" factor of the shell due to the induced geometrical imperfections (caused by the repeated buckling process) and the real boundary conditions can be assessed and calculated.

It is shown again that measurement of the imperfections of the shell is a valuable tool. Using the measured geometrical imperfections as an input to two multimode analyses, CPIUTAM and CPIUTAMN on can calculate of the "knockdown" factor ρ_{th} for shells under external pressure, axial compression and a combination of constant external pressure and buckling under axial compression as well as a combination of constant axial compression and varying external pressure till buckling. This capability enables a complete evaluation of the effect of the growth of imperfections at all the intermediate points.

Another factor studied in the present test series was the sequence of loading and its influence on the buckling loads at the shell. It appears again, as in Ref. [6], that this effect is negligible inspite of the nonlinearity of the interaction.

REFERENCES

1. Singer, J., Vibration correlation techniques for improved buckling predictions of imperfect stiffened shells. In Buckling of Shells in Offshore Structures, Granada Publishing, London, 1982, 285-330.
2. Singer, J., Vibrations and buckling of imperfect stiffened shells - recent developments. In Collapse: The Buckling of Structures in Theory and Practice, Cambridge University Press, Cambridge, 1983, 443-481.
3. Abramovich, H. and Singer, J., Correlation between vibration and buckling of stiffened cylindrical shells under external pressure and combined loading. Israel J. of Technol., 1978, 16, 34-44.
4. Singer, J., On experimental technique for interaction curves of buckling and shells. Experimental Mechanics, 1964, 4, 279-280.
5. Abramovich, H., Singer, J. and Grunwald, A., Nondestructive determination of interaction curves for buckling of stiffened shells. TAE Report No. 341, Israel Inst. of Technology, Dept. of Aeronautical Eng., Haifa, Israel, Dec. 1981.
6. Abramovich, H., Weller, T. and Singer, J., Effect of sequence of combined loading on buckling of stiffened shells. Experimental Mechanics, 1988, 28, 1-13.

7. Arbocz, J. and Babcock, C.D., Prediction of buckling loads based on experimentally measured imperfections. In Buckling of Structures, Proc. of IUTAM Symposium, Budianski, B. (ed.) Springer-Verlag, New York, 1976, 291-311.
8. Abramovich, H., Yaffe, R. and Singer, J., Evaluation of shell characteristics from imperfection measurements. J. of Strain Analysis, 1987, 22, 17-32.
9. Abramovich, H. Singer, J. and Weller, T., Buckling of imperfect stiffened cylindrical shells under combined loading. TAE Report No. 653, Israel Inst. of Technology, Faculty of Aerospace Eng., Haifa, Israel, June 1991.
10. Waeckel, N., Jullien, J.F. and Kabore, A., Buckling of axially compressed imperfect cylinders and ring stiffened cylinders under external pressure. In Experimental Stress Analysis, Proc. of the 8th Int. Conf., Amsterdam, The Netherlands, May 12-16, 1986, Dordrecht Martinius Nijhoff Publishers, 1986, 123-132.
11. Tennyson, R.C., The effect of shape imperfections and stiffening on the buckling of circular cylinders. In Buckling of Structures, Proc. of IUTAM Symposium, Budiansky, B. (ed.), Springer-Verlag, New York, 1976, 251-273.
12. Tennyson, R.C., Booton, M. and Chan, K.H., Buckling of short cylinders under combined loading. J. of Applied Mechanics, 1978, 45, 574-578.
13. Tennyson, R.C., Buckling of laminated composite cylinders: a review. Composites (U.K.), 1975, 6, 17-24.
14. Booton, M. and Tennyson, R.C., Buckling of imperfect anisotropic circular cylinders under combined loading. AIAA J., 1979, 17, 278-287.
15. Singer, J., Arbocz, J. and Babcock, C.D., Buckling of imperfect stiffened cylindrical shells under axial compression. AIAA J., 1971, 9, 68-75.
16. Abramovich, H., Singer, J. and Yaffe, R., Initial imperfection measurements of stiffened shells and buckling predictions. Israel J. of Technol., 1979, 17, 324-338.
17. Rosen, A., Singer, J., Grunwald, A., Nachmani, S. and Singer, F., Unified noncontact measurement of vibrations and imperfections of cylindrical shells. In Proc. of the 7th International Conf. on Experimental Stress Analysis, Haifa, Israel, 23-27 August 1982, 524-538.
18. Arbocz, J., Personal Communication, Dept. of Aerospace Engineering, Delft University, Delft, The Netherlands, 1989.
19. Weller, T. and Singer, J., Experimental studies on the buckling under axial compression of integrally stringer-stiffened circular cylindrical shells. J. of Applied Mechanics, 1977, 44, 721-730.

Table 1. Dimensions and Geometrical Properties of the Shells

Shell	h(mm)	L(mm)	R/h	L/R	Z	d_1(mm)	C_1(mm)	b_1/h	$A_{11}/b_1 h$	$I_{11}/b,h^3$	e_1/h
Y1A	0.260	200.0	462	1.665	1222	1.750	0.90	34.23	0.681	2.57	-3.87
Y2A	0.259	200.0	464	1.665	1227	1.761	0.90	34.36	0.686	2.65	-3.90
Y3A	0.248	200.0	484	1.665	1280	1.762	0.90	35.89	0.718	3.02	-4.05
Y4A	0.246	200.0	488	1.665	1290	1.764	0.90	36.18	0.725	3.11	-4.09
Y7B	0.265	200.0	454	1.665	1198	1.755	0.90	33.58	0.700	2.45	-3.81
Y8B	0.267	200.0	450	1.665	1189	1.743	0.90	33.33	0.660	2.34	-3.76

$R = 120.22$ [mm]
$b_1 = 8.90$ [mm]
$E = 7.5 \cdot 10^4$ [MPa]
$\nu = 0.3$

All shells were manufactured from aluminum alloy 7075-T6 and stiffened by 84 integral external stringers.

Table 2. Repeated Buckling Sequence of the Tested Shells.

Loading combination / Shell	1st	2nd	3rd	4th	5th
Y1A	Ext. Pressure P_{exp}=16.1 [kPa] (10/1)*	P_{const}=7 [kN] P_{exp}=12.5 [kPa] (9/1)	P_{const}=12 [kN] P_{exp}=9.5 [kPa] (8/1)	Axial Compres. P_{exp}=32 [kN] (7/1)	—
Y2A	Ext. Pressure P_{exp}=16.6 [kPa] (10/1)	P_{const}=4 [kN] P_{exp}=14.3 [kPa] (10/1)	Axial Compres. P_{exp}=32.8 [kN] (7/1)	—	—
Y3A	Ext. Pressure P_{exp}=13.8 [kPa] (10/1)	P_{const}=11 [kPa] P_{exp}=6.2 [kN] (9/1)	P_{const}=8 [kPa] P_{exp}=13.5 [kN] (9/1)	Axial Compres. P_{exp}=30.8 [kN] (7/1)	—
Y4A	Ext. Pressure P_{exp}=13.7 [kPa] (10/1)	P_{const}=6.2 [kN] P_{exp}=11.3 [kPa] (9/1)	P_{const}=13.5 [kN] P_{exp}=8.5 [kPa] (9/1)	Axial Compres. P_{exp}=29 [kN] (7/1)	—
Y7B	Ext. Pressure P_{exp}=16.8 [kPa] (9/1)	P_{const}=12 [kPa] P_{exp}=10.7 [kN] (8/1)	P_{const}=6 [kPa] P_{exp}=22.9 [kN] (8/1)	Axial Compres. P_{exp}=28.1 [kN] (7/1)	—
Y8B	Ext. Pressure P_{exp}=17.5 [kPa] (9/1)	P_{const}=4 [kN] P_{exp}=14.3 [kPa] (9/1)	P_{const}=8 [kN] P_{exp}=12.3 [kPa] 9/1	P_{const}=12 [kN] P_{exp}=10.5 [kPa] (8/1)	Axial Compres. P_{exp}=23.2 [kN] 7/1

*(n/m)

Table 3. Theoretical and Experimental Buckling Loads and Modes - Shell Y3A.

Loading Run No.	1	2	3	4
Type of loading	External Pressure	Combined Loading P_{const}=11 [kPa]	Combined Loading P_{const}=8 [kPa]	Axial Compression
Experiment	13.8 [kPa] 10/1	6.2 [kN] 9/1	13.5 [kN] 9/1	30.8 [kN] 7/1
SS3	13.6 [kPa] 11/1	8.0 [kN] 11/1	16.1 [kN] 10/1	34.3 [kN] 9/1
SS4	16.9 [kPa] 13/1	22.9 [kN] 12/1	32.8 [kN] 11/1	55.4 [kN] 10/1
Equivalent Spring	15 [kPa] 12/1	8.5 [kN] 11/1	17.5 [kN] 10/1	41.0 [kN] 10/1
ρ_{SS3}	1.02	0.78	0.84	0.90
ρ_{SS4}	0.82	0.27	0.41	0.56
ρ_{sp}	0.92	0.73	0.77	0.75
ρ_{th}	0.92	0.85	0.80	0.82

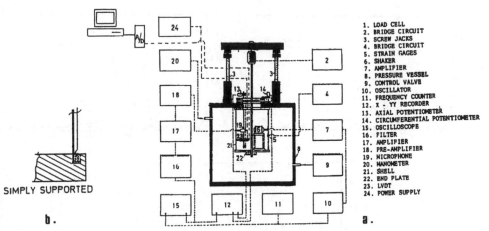

1. LOAD CELL
2. BRIDGE CIRCUIT
3. SCREW JACKS
4. BRIDGE CIRCUIT
5. STRAIN GAGES
6. SHAKER
7. AMPLIFIER
8. PRESSURE VESSEL
9. CONTROL VALVE
10. OSCILLATOR
11. FREQUENCY COUNTER
12. X - YY RECORDER
13. AXIAL POTENTIOMETER
14. CIRCUMFERENTIAL POTENTIOMETER
15. OSCILLOSCOPE
16. FILTER
17. AMPLIFIER
18. PRE-AMPLIFIER
19. MICROPHONE
20. MANOMETER
21. SHELL
22. END PLATE
23. LVDT
24. POWER SUPPLY

SIMPLY SUPPORTED

b.

a.

Fig. 1. Test Set-up a. Diagram of the test set-up.
b. Boundary condition detail.

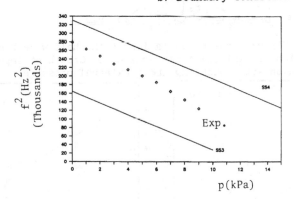

Fig. 2. Shell Y1A frequency squared versus applied loading P_{const}=12 [kN],
n=9, m=1.

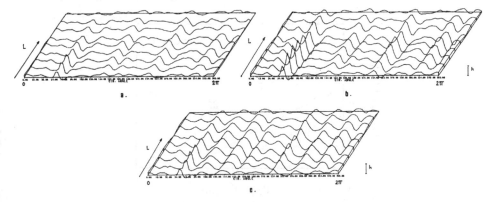

Fig. 3. Shell Y2A - Three dimensional plots. a. Initial imperfection.
b. Growth imperfection, p=8 [kPa].
c. Imperfection after buckling under external pressure.

10

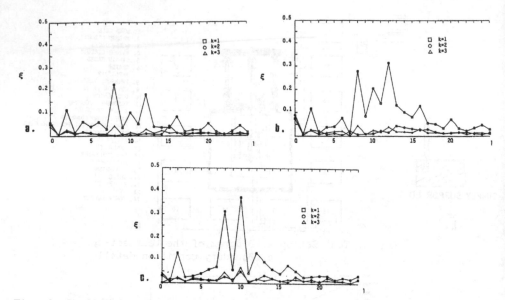

Fig. 4. Shell Y2A – Circumferential variation of the half wave sine Fourier
representation. a. Initial imperfection. b. Growth of imperfection.
c. Imperfection after buckling under external pressure.

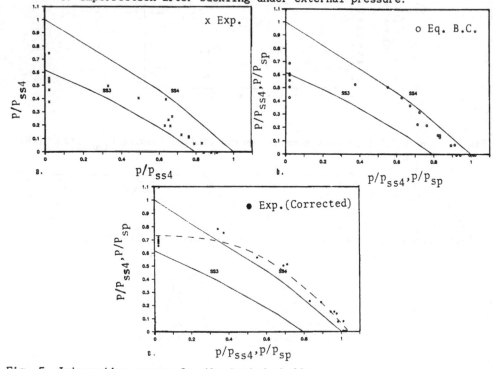

Fig. 5. Interaction curves for the tested shells.
a. Experimental results. b. Equivalent boundary conditions.
c. "Corrected" experimental results.

BUCKLING AND PLASTIC PUNCHING OF CIRCULAR CYLINDRICAL SHELL DUE TO SADDLE OR LUG LOADS

VLASTIMIL KŘUPKA

Institute of Applied Mechanics VÍTKOVICE, Brno 11, Box 32, Czechoslovakia

ABSTRACT

Theoretical solution of the compressive membrane stress due to the saddle load and its evaluation with respect to the buckling. Plastical punching of the saddle or lug into the cylindrical shell. Results of experiments and their comparison with the theoretical solution.

INTRODUCTION

The bifurcation buckling of a thin circular cylindrical shell subjected to a local contact load due to a saddle was studied.Besides peak stresses in the vicinity of the saddle horn, which can influence the lifetime of the structure through fatigue phenomena, the effect of stability is sub-stantial. It may cause buckling (Fig.1) of the shell in the vicinity of the saddle. The punching due to the saddle load through the shell can also happen (see Fig.2). The last phenomenom was observed in elasto-plastic and postbuckling state.

Figure 1.   Buckling due to a longitudinal membrane stress in the middle.

Figure 2.   Plastic punching of the shell due to the saddle·

Derivation of the limit load Q_p causing plastic punching was derived in [1] or [2] in the following form

$$Q_p = 2\ R_y\ .\ s\ .\ t\ .\sqrt{\frac{t}{r}} \qquad (1)$$

and experimentally checked in Tabs.1 and 2. The form of this deformation is very near to the footprint of the saddle (see Fig.2). Experiments are possible to find in [2], [3]. There is a very good agreement between experiments and our theoretical rel.(1) for loose saddle.

There are two kinds of membrane compressive stresses in the vicinity of the saddle which may lead to the loss of stability in buckling,i.e. the circumenferential membrane stress σ_s and longitudinal membrane stress σ_x.

The highest compressive circumferential membrane stress σ_s is concentrated only in a narrow zone around the horn of the saddle. Buckling in the form of a *kidney-shape* was observed there with our experiments. If a wear plate is welded to the vessel over the edges of the saddle, it can prevent very significantly the shell from this type of buckling. In addition our experiments show that the buckling of this type has a local self-constraited character.

On the contrary the effect of the longitudinal membrane stress σ_x cannot be neglected in any case. In the elastic state the distribution of the radial interface forces (acting between the shell and the saddle) is of a saddle form, with the maximum values at the both horns (see Fig.3a). Afterwards (by overcomming the yield point of local bending stress at the horns) it starts to change its form and the maximum interface forces will move to the middle (see Fig.3b and 3c). This fact evokes the increase in the longitudinal membrane stress which mostly causes the buckling at this place (Fig.1).The region with the compressive stress in the middle is comparatively large in respect of the compressive zone in the vicinity of horns. The buckling is of a *honey-shape* (see Fig. 1) which is characteristic for the loss of stability of thin shell subjected to a *pure compression*.

SOLUTION OF THE AXIAL MEMBRANE STRESS σ_x

As explained before the value of the compressive membrane stresses depens on the redistribution of the interface forces acting between the saddle and shell. At the beginning of the loading due to a rigid saddle the interface contact pressure is concentrated at both the horns. The redistribution has a saddle form as shown in Fig.3a. The solution of contact stresses due to the saddle was explained in [2] and [7] for instance. Increasing the load, the first local plastification (and high local radial deflection) occours over the small region around both the horns. Due to this high local radial deflections, the redistribution of the interface forces starts changing the form (see gradually from Fig.3a up to Fig.3c).

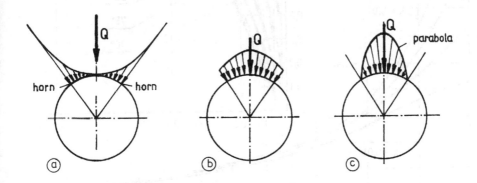

Figure 3.   Successive change of the redistribution of the radial interface forces after yielding at the region of the horn.

The parabolic form of the redistribution of the interface forces was used as the basic state for the calculation of the longitudinal compressive membrane stresses σ_x in the middle of the shell.

Simple solution [2] exists and respects the different width of the saddle 2b, the opening saddle angle 2ϑ and the saddle positions with respect to both the hinged ends (Fig.5). For the first estimation of stresses the graphs in Fig.4 are also at one's disposal (now they are icluded into the computer program - see [3]).

The compressive membrane stresses σ_x in the middle part of the saddle can be expressed in the following simple formula:

$$\sigma_x = f_{ox} \cdot f_{ax} \cdot \frac{Q}{t^2} \cdot \sqrt{\frac{t}{r}} \qquad (2)$$

where   Q - total force at the saddle;
        t - shell thickness;
        r - vessel mean radius;
        f_{ox} - a coefficient of the down graph in Fig.4 corresponds to the load in the middle of the shell being infinitely long (neglecting the bottom stiffening effect) depending on

$$k = \frac{b}{r} \cdot \sqrt{\frac{t}{r}} \qquad (3)$$

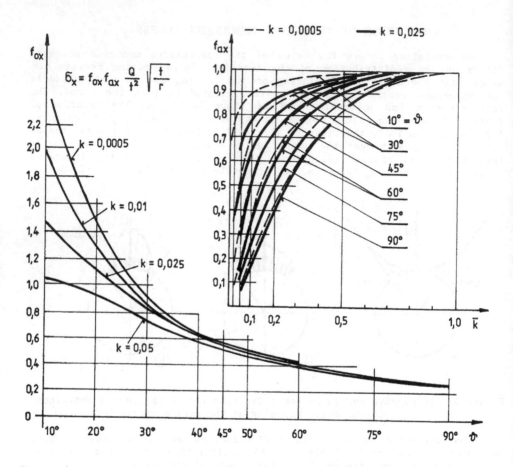

Figure 4.   Graphs for the determination of the maximum longitudinal membrane stress σ_x acc.to eq.(2).

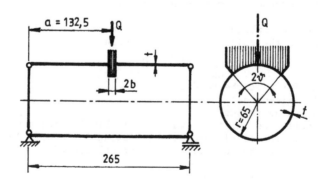

Figure 5.   Scheme of the loading of tested shells.

f_{ax} = coefficient that gives the stiffening effect of the bottom situated at the distance of

$$\bar{a} = \frac{2}{3} v + a$$

The coefficient is plotted in up graph in Fig.4 in terms of the width 2b of the saddle support, given by the angle ϑ and the coefficient

$$\bar{k} = \frac{\bar{a}}{r} \cdot \sqrt{\frac{t}{r}} \qquad\qquad (4)$$

ϑ - an angle - see Fig. 5, measured in the vessel-saddle contact;
b - half saddle width in the longitudinal direction;
v - formed head depth.

BUCKLING DUE TO AXIAL MEMBRANE STRESS σ_x

We shall solve the effect of buckling by means of Europien Recommendations ECCS, Buckling of Shells R 1.5.2 "Pure bending of the cylinder in the merional direction". The use of this relations will be in any case on the safe side, because the zone of compression is larger in all cases of a pure bending that in our case of local loading. This fact has been proved experimentally by our tests with many types of saddle loaded shells.

Tested shells of the radius 65 mm and lenght 265 mm are shown in the Fig.5. The combinations of three sheets of different thicknesses with the different yield point R_y of the material: t = 0,10 mm with R_y = 800 MPa; t = 0,22 mm with R_y = 360 MPa; t = 0,30 mm with R_y = 300 MPa and two saddle widths 2b = 5 mm and 20 mm with different angle of opening 2ϑ = 60°, 90°, 120°, 150°, 180° were chosen for tests.All results were recorded both for loose and welded saddle. 36 different types of shells and saddles were tested and their experimental results are compared with theoretical solution.

Comparisons between the calculated and measured loads Q in the case of buckling or postbuckling plastic punching are shown in the tables for the shells having their thicknesses t = 0,3 mm and t = 0,1 mm. Results for the shells with thicknesses t = 0,22 mm were similar to those with t = 0,3 mm.

In most cases tabulated in Tables and noted "buckling occurs first" the diagram of load vs. deflection was linear in the beginning (see phase "a" in Fig.6). In this first phase of loading, the kidney-shaped deformation area has been arising, step by step, around horns. By the end of this phase even the yield point was obviously exceeded in the vicinity of the both horns locally.This state however is self-constrained and in this way it effects very little the change of linear behaviour in the first phase of the loading.

When reaching the critical load bucklings of honeycomb shape arised in the upper part and, at the same time, the kidney-shaped deformation was enlarged. The most important fact, however, is that the load-carrying capacity has suddenly droped down (phase "b" in Fig.6), the stability has been lost and the shell starts buckling.Then the saddle starts squeezing (nearly by a constant load which corresponds to eq.(1)) into a shell cylinder in the shape as seen in Fig.2. This state corresponds with the phase "c" in the load-deflection diagram in Fig.6.

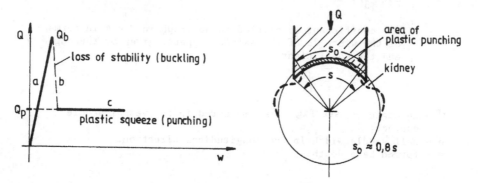

Figure 6. Diagram load vs.deflection for case when "buckling occurs first".

Figure 7. Tendency to deflect inwards in the vicinity of the horn and outward in the middle.

It remains to explain why the state of the plastic punching did not occur earlier and why the loss of stability due to buckling did not occur first. This question results from Tab.1 where the values for punching are lower than that for buckling. All the phenomenon is very complicated since it consists of different phases manifested by different kinds of equilibrium in elastic and elasto-plastic states and by an extensive deformation. The interface friction along the surface between the saddle and the shell, however, played a substantial role and produced considerable shearing forces that had not been considered in our theoretical solution. The presence of shear was demonstrated by crushed holes around the bolts in cases when we simulated a fix saddle-shell or with loose saddles which, in some cases of greater angles of ϑ, revealed apparent form of buckling as by combined shear and compression. It proves that the essential part of the load Q is transferred into the shell not only with radial interface forces but with tangential forces especially.

The relation (1) was derived under the conservative assumption that the contact resists only against radial forces. In reality even the interface shearing forces due to friction are produced there and the shell puts up resistance even by shearing. On considering the shear resistance the load-carrying capacity must be increased. When the critical state has been reached this contact friction, however, was suddenly released when buckling arose. At this moment (after lifting the friction) the state corresponds with the assumption used in teoretical solution because eq.(1) was derived on neglecting the effect of the shearing resistance. In this relation the contact circumference reduced to $s_o < s$ due to "the kidney effect" (see Fig.7). The new values of Q calculated with respect to the redused contact circumference s_o are signed in Tab.1 with square bracket.

Theoretical values for the buckling stress in Tabs.1 and 2 represent nearly doubled values than those from the experiments. They were used from the ECCS Recommendation [5] for the case of pure bending. The cause

TAB.1

t = 0,3 mm; R_y = 300 MPa;

Angle 2ψ	Width 2b	BUCKLING [N] Test	BUCKLING [N] Theory	Plastic PUNCHING [N] Test	Plastic PUNCHING [N] Theory	Remarks
60°	20w	900	417,2	690	831	buckling occurs first
	201	888		600	[505]	- " -
	5w	800	404,5	550		- " -
	51	720		500		- " -
90°	20w	1950	862,4	1100	1247	punching predominates
	201	1940		1000	[998]	buckling occurs first
	5w	1340	846,6	874		- " -
	51	1250		850		- " -
120°	20w	3750	1845	1700	1667	punching predominates
	201	3700		1650	[1334]	buckling occurs first
	5w	2350	1812	1600		- " -
	51	2030		1500		- " -
150°	20w	5200	3965	2075	2495	punching predominates
	201	4375		2000	[1996]	buckling occurs first
	5w	3550	3885	2000		- " -
	51	2300		1850		- " -
180°	20w	6500	5360	2800	2495	punching predominates
	201	5300		2600	[1996]	buckling occurs first
	5w	4500	5267,5	3000		shear effects buckling
	51	no buckling		2500		punching only

TAB.2

t = 0,1 mm; R_y = 800 MPa;

Angle 2ψ	Width 2b	BUCKLING [N] Test	BUCKLING [N] Theory	Plastic PUNCHING [N] Test	Plastic PUNCHING [N] Theory	Remarks
60°	20w	113	41,96	55		buckling occurs first
	201	105		45	(640)	- " -
	5w	100	40,96	65		- " -
	51	90		90		- " -
90°	20w	190	119,9	120		- " -
	201	220		110	(640)	- " -
	5w	180	109,2	110		- " -
	51	210		90		- " -
120°	20w	425	292,9	275		- " -
	201	440		230	(853)	- " -
	5w	380	248,9	250		- " -
	51	460		200		- " -
150°	20w	600	465,4	350		Buckling is influenced by shear
	201	600		290	(1066)	- " -
	5w	500	373,5	350		- " -
	51	475		215		- " -
180°	20w	640	653,2	buckling only 315		- " -
	201	820				
	5w	745	516,2	575	(1280)	Buckling starts in the zone which is inclined through 40°
	51	675		275		

lies partly in conservative assumption of distribution of radial interface forces shown in Fig.3c and partly in relation (4) of ECCS Recommendation [5], by that again follows from the conservative assumption that whole one half of the profile is in compression (like in pure bending). In real(namely with smaller angles of ϑ) the compression is concentrated merely on a much less area. For this reason the real theoretical critical stress has to be higher in our cases. The more precise theoretical solution accepting the local compression is not so far available.

In some cases the plastic punching can directly arise without preceding buckling.It was proved especially with narrow saddles (cases noted "punching only "or"punching predominates".
On the contrary, no plastic punching existed in cases of thinner shells given in Tab. 2. In all the cases the buckling predominates. It resulted mainly from the fact that material yield point had been considerably higher than that in previous cases. Even in these cases the load-carrying capacity has not been at once fully exhausted.The deformation,however, did not coincide with the punched saddle trace as seen in Fig.2. It possesed another shape and was developed over a larger region than in the first case of punching.

PLASTICAL PUNCHING OF THE LUG INTO THE SHELL.

The last part of our paper will deal with the effect of lug. The maximum momentous carrying capacity with the punching of the lug through the shell (see Fig.8) is possible to express by means of the eq.(1) in the following form

$$M_p = Q_p \cdot h = 2R_y \cdot t \cdot s \cdot h \cdot \sqrt{\frac{t}{r}} \qquad (5)$$

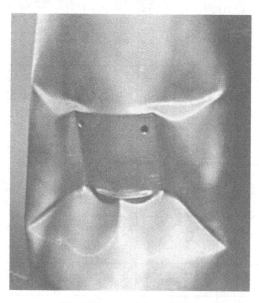

Figure 8.   Plastic punching due to the lug.

Plastic punching also usually limits the carrying capacity.Comparison between theoretical and experimental values is shown in Fig.9.

Under our considerations we neglected the effect of shear force F being produced in the contact of the lug with the shell. This effect can be neglected if e > h . With thin shells and short distance e ,however, it can come into effect and should be solved since this tangential force originates not only local stress concentration but could result even in buckling.

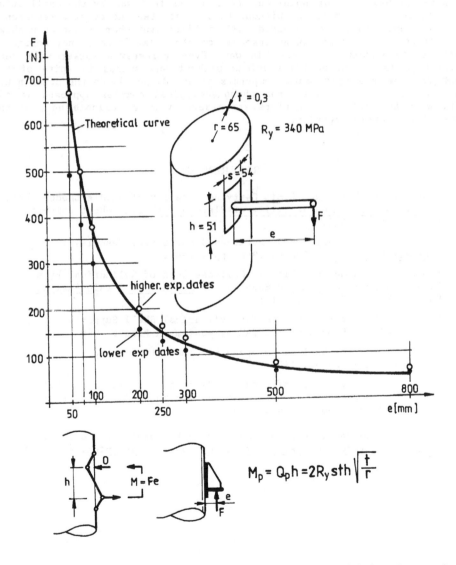

Figure 9. Comparison between the theoretical (acc.to eq.(5)) and tested results.

If e < h then shear force comes remarkably into effect and thus also the buckling resulting from it. The drop of carrying capacity due to buckling is a considerable one. It is of interest, however, that a postcritical recovery appears.

The problem remains rather complicated even when using the modern calculating methods since separate deformation processes take place in the region being geometrically and physically nonlinear and mainly they are complicated by different kinds of bifurcations being sometimes unforessen. Thus the introduced solution is only the first step necessary for the first estimation of dimensions and for the better understanding of physical priciple of solution. It must be followed by the application of computer finite methods. Without this first step, however, a researcher of our branch could be compared with a blind man whom is not even shown the direction of the way in an unknouwn countryside. If he is not correctly directed he surely will lose his way. Even a research worker will reel through a labyrinth of bifurcations without any succes in spite of the fact that he has up-to-date computers at his disposal. On the other hand, when we choose a correct and phenomenological complex approach we can get, with the help of computers, considerably more reliable and, at the same time, general results.

REFERENCES

1. Křupka,V.,Buckling and Limit Carrying Capacity of Saddle Loaded Shells In ECCS Colloquim on Stability of Plate and Shells, Ghent University, 1987, pp.617 - 622.

2. Křupka,V.,Saddle Supported Unstiffened Horizontal Vessels, in Acta Technica ČSAV, No 4, Prague 1988, pp.472-492

3. Tooth A.S. Jones N., Plastic Collapse Load of Cylindrical Pressure Vessels Supported by Rigid Saddles. In Journal of Strain Analysis, Vol 17, No 3, 1982 pp 187-198

4. Vokroj,P.,Expert Like System for Cylindrical Shell Support. In Preliminary Report from IUTAM Symposium in Prague 1990

5. Buckling of Steel Shells, European Recommendations ECCS, Fourth Edition, No 56, 1988 pp 9-15

6. Samuelson, Å.L., Effect of Local Loads on the Stability of Shells Subjected to Uniform Pressure Distribution in Preliminary Report from IUTAM Syposium in Prague 1990

7. Křupka,V.,The Background to a New Design Proposal for Saddle Supported Vessels. In Inter.Journal of Pressure Vessels and Piping, 46(1991)

TOWARDS AN OPTIMAL SHAPE OF CYLINDRICAL SHELL STRUCTURES UNDER EXTERNAL PRESSURE

J.F. JULLIEN, M. ARAAR
Concrete and Structures Laboratory, Bâtiment 304
INSA LYON, 20 avenue Albert Einstein
69621 VILLEURBANNE CEDEX France

INTRODUCTION

The use of thin shells is important in industrial construction. Examples are numerous : nuclear reactor vessels, supporting structures for marine platforms, under sea projects, reservoirs, etc. These structures are built in different materials, metal, composites, concrete, the choice depending on the dimensions, the loading, the objectives and the contents.

Cylindrical shells with or without stiffeners are used to carry a variety of loads such as external pressure, axial compression or various combinations of loading.

Under certain types of loading, the failure of thin shells is caused by buckling and the corresponding membrane stresses, calculated by classical methods, are below the strength of the material. Further more, under these loadings, the magnitude of the safety margin is important in order to allow for the imperfections of the shell. The design of circular cylindrical shells under external pressure embraces shells both with and without stiffeners. The sizing of the structure is, in general, simple and calls upon simplified methods or numerical methods of varying complexity.

Stiffening is a satisfactory solution for increasing the critical load of a shell. The cost depends on the construction material, the level of loading and the acceptable margins of safety.

A number of investigations into the buckling of circular cylindrical shells subject to external pressure have established a knowledge of the physical phenomenological behaviour which occurs before the onset of instability and they have brought an understanding of the way in which instability is initiated.

This state of knowledge now allows us to propose a surface profile which prevents the precritical geometry developing with its associated classical critical circular cylinder buckling load. Thus, with an identical quantity of materials, the strength of cylindrical shells is greatly increased by this novel geometry.

The work presented here is supported by experimental results and numerical analysis relating to the different phases of behaviour of various shells.

This novel form of more rigid geometry is also analysed under a combination of external pressure and axial compression.

THE EFFECT OF EXTERNAL PRESSURE ON CIRCULAR CYLINDRICAL SHELLS

The different theories concerning the behaviour of perfect thin circular cylindrical shells, subject to external pressure and having small or large displacements, predict an axisymmetrical precritical deformation. Instability is then characterised by the sudden change of this axisymmetric geometry by the appearence of a multimodal deformation in the circumferential and axial directions.

Geometrical defects are inevitable in the construction of thin shells both in practice and in the laboratory. Even with the numerous precautions undertaken during fabrication these imperfections appear and modify the effect of the boundary conditions of the shell. The imperfections create very small radial displacements at the point where they occur, the magnitude of this effect is comparable to that induced by an external pressure applied to the surface of the shell. This is why the shell often shows the initial state of circumferential multi-modal imperfections near or equal to the critical mode under external pressure as well as the nul axial mode.

Generally, in good quality construction, the maximum amplitude of these imperfections averages approximately between 20 and 30 % of the thickness of the shell which is defined by the relationship

R/t = 400 and Z = 1000. Analysis of laboratory tests show that the precritical deformation of the shell is not axisymmetrical but corresponds to the amplification of the imperfections on the critical mode with the creation of a geometry approximately that of the critical geometry (Fig. 1)

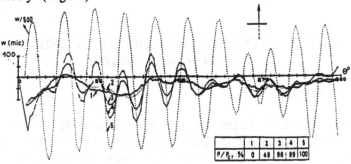

Figure 1. The radial geometry of the mid-height section of a near perfect cylindrical shell loaded by external pressure

At the time of loading, the variation in the amplitude of the imperfections is small but their magnitude grows asymptotically to that value which occurs at the critical load, there their size depends on the geometric parameters and the boundary conditions.

Similar behaviour has been observed [3] Fig. 2 under axial compression, where the sign of the curvature of a local imperfection determines its amplification. An inward imperfection is amplified and forms a deflection wave pointing towards the interior, whereas an outward imperfection remains constant.

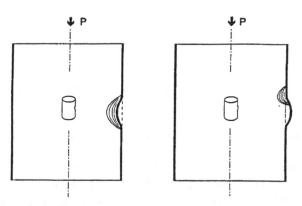

Figure 2. Precritical development of a localised geometrical imperfections in an axially loaded cylindrical shell

The experimental results and the calculated value of the critical loads under external pressure are very close for the dimensions investigated. These results confirm earlier work and the recorded differences are dependent principally on the quality of the boundary conditions.

THE EFFECT OF EXTERNAL PRESSURE ON CYLINDRICAL SHELLS OF OPTIMAL SHAPE

Basic principle

The new concept is based on the knowledge that the waves in the circumferential precritical geometry must be continuous and the observation that the precritical geometry is initiated by inward pointing imperfections.

We have sought to oppose the formation of the circumferential mode by creating a surface geometry of multiple meridian arches whose curvature is such that the arches always point outwards. The number of arches is equal to twice the number of the critical mode. The form of the resulting circular cylinder with multiple arches is shown in Fig 3 and is called an "ASTER" shell [1, 2]. The depth of the arches is another parameter to be considered.

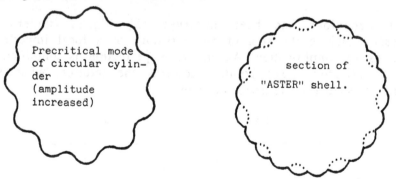

Figure 3. Section of cylindrical shell with multiple arches "ASTER"

Experimental and numerical validation of the concept

The behaviour of this novel from of "ASTER" shell is investigated experimentally and the result compared with that of a circular cylindrical shell, see references, and with another multiple arch shell whose number of arches is different from the Aster shell.

Details of specimens

The overall geometry of the three specimens is common : R = 75 mm, L = 120 mm, h = 150 to 160 microns. The three specimens are studied with encastre conditions at each end. These boundary conditions were obtained by fusing the ends of the shell into a base plate.

The number of arches in the Aster Shell was deduced from a linear analysis of the reference shell which gave the mode of buckling under external pressure. This modes is dependent on the quality of the boundary conditions. It is 12 when the shell is perfectly encastre at the two extremities and decreases to about 10 when the local axial displacements around the circumference at the assumed point of fixity are free. The experimental conditions are intermediate between these two and the observed mode is 11. This latter value is used in the following studies, hence the "Aster" shell has 22 arches. The number of arches chosen for the multi-arch shell used for comparison is arbitrary but should be smaller than that of the Aster shell ; the number chosen was 14, and the shell was designated VM14.

The form of the arches is deduced from the sinusoidal precritical geometry which developes on the circumference. For simplicity, and into order to facilitate fabrication, the arches are defined as an arc of that circle which approximates the half sine wave, Fig 4, (r = 20 mm for the "Aster" shell and r = 35 mm for shell VM14).

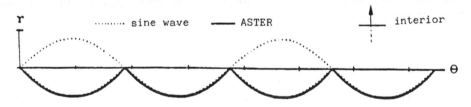

Figure 4. Approximation of half sine wave to an arc of a circle

The thickness of the arches is 2.3 mm for the "Aster" shell and 2.35 mm for the shell VM14. These values are a compromise between research requirments and the method of fabrication

For the given geometry of the Aster shell, the internal volume is 4.2 % greater and the weight of the constituent material is 5.3 % greater than that of the reference circular cylindrical shell of identical thickness.

The shells were constructed in nickel which was electrodeposited on a pattern having the geometrical form of the specimen.

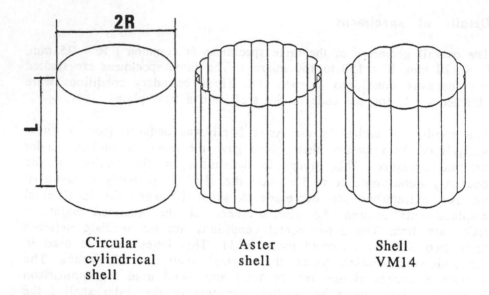

Figure 5. Geometry of the three types of shell studied

The production of the specimens was carefully controlled in all aspects in order to obtain the best experimental results. Dimensional accuracy, uniform mechanical characteristics, absence of residual stresses, minimisation of geometric imperfections, etc., are qualities that have the same level of importance as good test equipment and test procedure.

Test method

The shells were subjected to either external pressure or axial loading or a combination of both.

The experimental behaviour was entirely monotored by a computer which recorded and processed the desired readings.

At times, the load was kept constant to allow the radial and axial deformations to be measured. This was done by means of an axial transducer without contact with the surface.

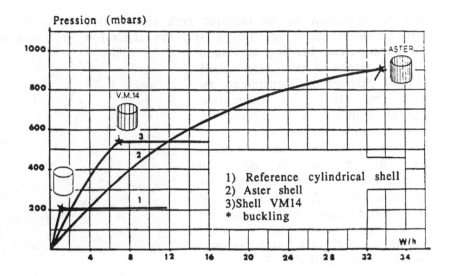

Figure 6. Comparison of the behaviour of the three types of shells under external pressure

Analysis of behaviour

The global behaviour of a shell under external pressure is characterised by the relationship between pressure and radial displacement. For the shells analysed the axial mode was the same. The measurement of radial displacement was taken at mid-height. The measurement points were defined after having observed that the precritical deformation of the "Aster" shell was totally axisymmetrical. Hence the characteristic point used was situated at the intersection of two arches for this shell and at the summit of a internal wave for the circular cylindrical shell.

The critical load under external pressure of the "Aster" shell was of the order of 4 times that of the reference circular cylindrical shell. The corresponding value for shell VM14 was 2.5 (Fig 6).

The results of the critical load and the precritical geometry show that it is possible to conceive forms of cylindrical shells having much improved loading capacity using a similar quantity of material. Certainly the radial stiffness of multiple arch shells in the precritical phase is greater than circular cylindrical shells ; of the order of 4 to 1 for the Aster shells. This result is because of the nature of the precritical deformation which is constrained to be axisymmetrical in the case of the Aster shell and which has a sinusoïdal mode for the circular shell. These two forms of deformation are coupled in the shell VM14. Compared to the circular cylindrical shell, the higher strength

28

and the stiffeness of the multiple arch shell⁻ follow from the lack of circumferntial membrane forces and the reduction in importance of the local radial bending curvature.

The 22 arches in the "Aster" shell should perhaps be considered an optimum since no geometric model reproduces axisymmetric deformation.

Only a variation in the curvature of the arches is observed corresponding to the constraints of each. The post critical state of the Aster shell under external load corresponds to a local instability of each arch at mid height of the shell (Fig 7).

Figure 7. Post-critical state of Aster shell under external pressure

This instability follows from local axial plastification on the shell surface. The behaviour is similar to a series of adjacent beams each having the section of a arch and of total breath equal to the circumference of the shell. Shell VM14 shows two types of failure (Fig 8) combining the form of the precritical mode with that of the failure of the arches if the number of the latter is not suffient to oppose it.

Figure 8 : Post-critical state of VM14 shell under external pressure

The numerical analysis of the multiple arch shells is carried out using linear three dimensional elasticity and small displacements, program Bilbo of the system Castem. These calculations emphasise the symmetries and the possibility of stress concentration at the intersection of the arches. The load at which bifurcation takes place is correctly forecast for the two multiple arch shells, Aster and VM14, as compared to the experimental critical load. In contrast, the axisymmetric mode of instability for the Aster shell is not forecast. The calculated critical mode for shell VM14 corresponds favourably with the test.

These comparisons have been made for an arch rise of 2.3 mm, which corresponds to 3 % of the radius and is 15 times the thickness. The effect of the rise of the arches was analysed numerically (table 1). The capacity of the shell is increased as the rise of the arches is increased. The critical dimension depending on the plastic criteria of the material.

Amplitude δ (mm)	δ/R %	δ/h	Radius of curvature r(mm)	Criticial pressure Pcr (mbars)	Pcr/Pref
1,63	2,2	10	25	415	1,7
2,32	3	15	20	1002	4,1
3,7	5	25	25	2140	8,7

Table 1. Influence of the rise of the arches on the critical pressure

EFFECT OF COMBINATION OF EXTERNAL PRESSURE AND AXIAL COMPRESSION ON ASTER SHELL

It is observed that the Aster shell has a precritical geometry under an axial compression load that is principally axisymmetric and which increases towards the exterior of the shell i.e. in the opposite direction to that observed under external pressure. The increase in the axial critical load, which corresponds to bifurcation, gives an experimental critical load which is 20 % greater than the corresponding load for the reference cylindrical shell tested under the same conditions (Figure 9).

The post critical geometry is a coupling of the zero mode (the barrel effect) and an axisymmetrical circumferential mode of 11 which corresponds to each pair of arches (Fig 10).

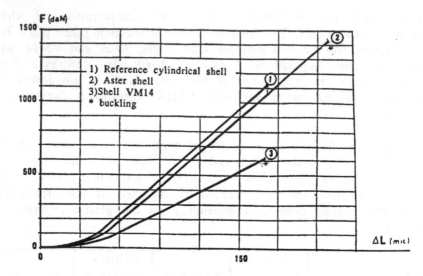

Figure 9 : Comparison of the behaviour of the three types of shells under axial compresion

Figure 10 : Failure state (ASTER) (unloaded)

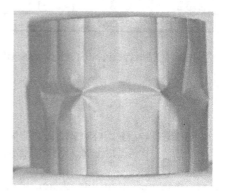

Figure 11 : Failure state (VMH, under load)

It should be noted that the behaviour of shell VM14 is not as good as that of the Aster Shell nor as good as the reference cylindrical shell (Fig11).

The Aster shell performs reasonably well under axial compression only and exceptionally well under external pressure when compared to the behaviour of the reference cylindrical shell. Moreover the axisymmetric precritical deformations under these two loadings are opposite to one another. We have made a series of tests under a combination of these two loads.

A diagram of the interaction of external pressure-compression for the Aster Shell is given in Fig12 together with that for the reference shell.

In this figure we give for the reference shell the experimental curve, a linear interaction curve according to classical theory and a curve based on linear axisymmetric calculations (INCA) ; for the Aster shell we only present the experimental curve. In the diagram unity corresponds to the theoretical critical load for the cylindrical shell under axial compression (and under external pressure).

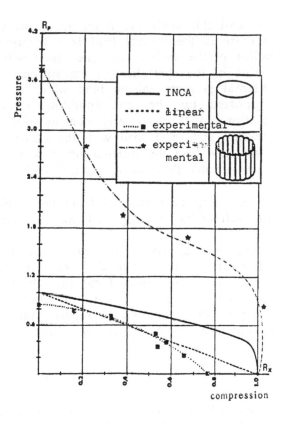

compression

Figure 12 : Interaction diagram of external pressure/axial compression for Aster and cylindrical shells

The experimental results for the two shells show that under external pressure a load of 3 to 4 times is sustained compared to that under uniaxial compression. This result opens a large field of application for shells under external pressure.

CONCLUSION

This research has concentrated on a particular type of geometric imperfection of cylindrical shells which has a serious effect on their critical load. Automatic stiffening is created by the inclusion of multiple arches on the circumference which are uniform over the whole height and whose number is optimised. This optimal number is equal to double the buckling circumferential mode under external pressure of the corresponding circular cylinder of the same thickness.

We call this shell the Aster shell and the principle is patented. The increase in the critical load is a function of the rise of arch and the radius of the shell . An Aster shell with an arch rise of 3 % of the radius of the shell has an autostiffening which gives an experimental external pressure failure load which is of the order of 4 times that supported by the corresponding circular cylindrical shell. This relationship is practically constant for all combinations of external pressure and axial load with the exception of axial load acting alone, under this load the relationship is less favourable but the load is still 20 % greater than the cylindrical shell.

These exceptional properties are the result of its stiff rigidity in local circumferential bending and weak circumferential membrane rigidity.

These results have been drawn principaly from the tests in the laboratory of the Aster shell and of the multiple arch shells. These shells were developed form the physical understanding of buckling and prebuckling behaviour of circular cylindrical shells and the analysis of geometrical imperfections, in particular, the average development of the geometry before and after buckling.

REFERENCES

1. ARAAR M., Contribution à l'autoraidissage des coques cylindriques vis-à-vis du flambage, Thèse de Doctorat, INSA de LYON, Mars 1990

2. JULLIEN J.F., ARAAR M., Coque à haute résistance à géométrie de révolution, Brevet n° 90.04277, INSA de LYON, Mars 1990

3. WAECKEL, N. et JULLIEN, J.F. Experimental studies of the instability of cylindrical shells with initial geometric imperfections. San Antonio (Texas), june 17-21, 1984, ASME : Special publications PVP, Vol. 89, p. 69-77.

BUCKLING OF AXIALLY COMPRESSED CYLINDRICAL SHELLS WITH LOCAL GEOMETRIC IMPERFECTIONS

S.KRISHNAKUMAR

University of Toronto Institute for Aerospace Studies
4925 Dufferin Street, Downsview, Ontario M3H5T6, CANADA

C.G.FOSTER

Department of Civil and Mechanical Engineering, University of Tasmania
GPO Box 252C, Hobart, Tasmania 7001, AUSTRALIA

ABSTRACT

The influence of large localized geometric imperfections on the stability and axial load carrying capacity of thin isotropic circular cylindrical shells was investigated experimentally. Diamond shaped 'Yoshimura facet type' dimples were introduced in otherwise near perfect cylinders and their behavior under compressive load monitored using a whole field grid reflection technique. Buckling tests were conducted to determine the effects of shell geometry and variations in the size and number of defects. An empirical formula was developed for estimating the degradation in buckling strength caused by single or multiple local facet type defects.

INTRODUCTION

The stability of axially loaded cylindrical shells is notorious for its sensitivity to initial imperfections. During the last four decades numerous studies (such as Refs.1-4) have been conducted on this subject proving that even minute variations in the geometry of the shell from the perfect cylindrical shape can result in drastic reductions in its load carrying capacity. The majority of these studies, however, deal with global imperfections, i.e. imperfections present over the entire circumference and/or length of the cylinder. Further most analyses are restricted to deviations of the same order of magnitude as the wall thickness. Thus the results of these studies are not applicable to large local deformations, such as the dents created by accidental lateral impact, wherein the central deviation of the shell wall from its initial geometry can easily extend over 20 to 30 times the wall thickness. The study described herein was aimed at investigating the effect of such localized geometric deviations or defects in the shape of the cylinder, upon its axial buckling strength. The role played by local lateral loads in causing premature buckling of axially loaded shells has been experimentally investigated earlier[5]. Other than some preliminary tests of the present type reported previously by Foster[6], major studies on local geometric defects in cylinders have been confined to those dealing with strength reductions caused by diamond-shaped dents in thick tubular columns[7,8].

It is significant that most geometric deformations generated 'spontaneously' by natural and man-made causes in real-life shell structures - whether they be offshore tubular columns suffering underwater collisions or beverage cans kicked around by ebullient youngsters - are diamond-shaped dents, which resemble the multi-faceted dimples of the 'Yoshimura buckle pattern' of the axially loaded cylinder[9]. It is possible that the resemblance is due to pure coincidence; however it appears more likely that the reason lies in the developable nature of the polyhedral surface formed by the Yoshimura pattern, which requires only a minimum amount of extensional energy in the shell for maintaining its deformed shape. On the basis of this observed resemblance, the local geometric imperfections employed in the current study were modeled by the diamond-shaped dimple of the Yoshimura pattern, as in the studies on tubular members[7,8].

Figure 1: Experimental Set-up.

TEST SPECIMENS AND APPARATUS

The tests were conducted on spun-cast epoxy cylinders with a mean radius of 77mm, and length L between 70 and 190mm. The thickness t ranged from 0.13mm to 0.39mm, with typical deviations of less than 2%. The geometric specifications for all the specimens are listed in Table 1. The modulus of elasticity E of the shell material (Araldite LC 261 mixed with Hardener LC 249) was measured at 3.05GPa. The thin-ness of the shells (R/t in the range of 200 to 600) and the relatively high elastic limit (approximately 70 MPa) of the epoxy mix ensured elastic buckling and hence repeatability of tests on each individual cylinder. This was also verified experimentally. The defects were introduced in the specimens by locally softening the shell wall with heat applied from a domestic hair dryer[6]. A cylindrical wooden mandrel with a 'V' shaped notch of the required dimensions was used as the template. The shell wall outside the region of the defect was protected during this procedure by a cardboard mask.

The compression tests were conducted in a rigid screw operated loading frame with a capacity of 20KN. A special purpose optical monitoring system[10], consisting of a cylindrical grid along the shell axis and a conical mirror at one end (Fig.1), was incorporated into the loading machine. The specimen cylinders were manufactured with a reflective inner surface, which reflected the grid lines from the central rod on to the conical mirror. The latter directed it to the plane mirror at the top. Thus a whole field two dimensional image of the inner shell surface reflecting the grid lines was generated in the top mirror, which was recorded by a camera. An axial grid, a circumferential grid and a helical grid were employed at different times at the shell axis (each being sensitive to different components of the shell's deflection to different degrees); these were respectively transformed into a radial, concentric and a spiral grid system at the image plane, when reflected off a perfect cylinder. Small deviations of the shell wall from its true cylindrical shape generated magnified distortions in the grid lines of the reflected image, thus providing a a sensitive deformation monitoring system.

DESCRIPTION OF THE TESTS

Defects of three different sizes were employed in this study. Their template specifications are shown in Fig.2. W_0 is the radial deviation (inward) from the cylindrical surface at

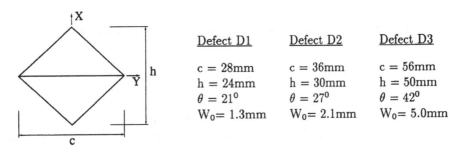

Defect D1	Defect D2	Defect D3
c = 28mm	c = 36mm	c = 56mm
h = 24mm	h = 30mm	h = 50mm
$\theta = 21^0$	$\theta = 27^0$	$\theta = 42^0$
W_0= 1.3mm	W_0= 2.1mm	W_0= 5.0mm

Figure 2: Template Dimensions of the Defects.

the center of the defect and θ the angle subtended by its circumferential width at the shell axis. The actual size of the imprinted imperfection was usually smaller than that of the template, this was later measured accurately from the specimen photographs. Each defect, D1, D2 and D3, was imposed on 12 shells. They were introduced at mid-section to avoid end-effects. The cylinders were subjected to several buckling tests: before introducing the defect, after the first defect was introduced and then with a second imposed on it. Multiple defects (up to 8 or 9 around the circumference, all at the same height) were imprinted on four selected specimens; these were tested in compression after the addition of each successive defect.

Using the optical system, images of the shell wall reflecting different grids were recorded at intervals during the loading of the cylinder to monitor the prebuckling behavior. Photographs were also taken immediately after buckling. The predominant effect of the imposed facet-shaped imperfection was to induce excessive deflection of the shell wall in its immediate vicinity. Figure 3a shows the prebuckling deformations (indicated by the distorted radial lines) on either side of the imposed defect in shell S19 at a load of 0.58 times the classical load. The growth of the prebuckling deformations resulted in premature collapse of the shell, marked by a sudden drop in load and the formation

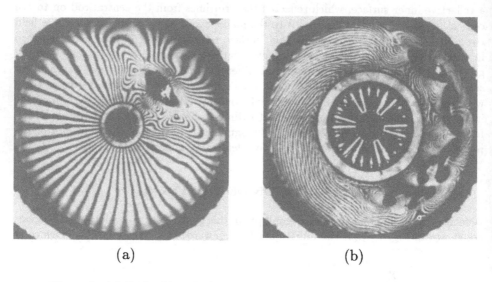

(a)   (b)

Figure 3: (a) Prebuckling Deformations   (b) Partially Collapsed Shell.

of facet-shaped buckles in the areas adjacent to the defect. The photograph of shell S16 (reflecting a spiral grid) in Fig.3b illustrates this partially collapsed condition. The defect in this case is the top-most facet at the 2 o'clock position. When partially buckled shells are loaded further, they usually develop additional facets, one after another, till the pattern extends all round the circumference. In some cases, the complete pattern developed instantaneously at the first incidence of collapse.

The load-displacement plots of shell S12, which was tested with multiple defects of size D3, are shown in Fig.4. The reduction in maximum load is obvious; so is the decrease in stiffness with increasing number of defects. The linearity of the prebuckling paths followed by the defective shell is noteworthy. For comparison, the load-displacement plots of two other shells, one with a circular hole of 30mm diameter, and another with an irregularly formed geometric imperfection, are shown in Fig.5. Not only are the maximum loads much lower in these cases, but the prebuckling paths are highly non-linear, when compared to those in Fig.4. (The abscissa of the plots in Figs.4 and 5 is

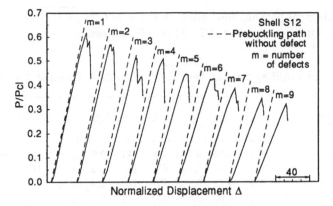

Figure 4: Load-Displacement Plots of Shell S12 with Multiple Defects.

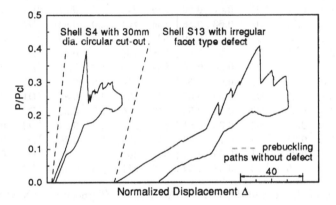

Figure 5: Load-displacement Plots of Shells with Other Types of Defects.

the non-dimensional value of end-shortening given by $\Delta = 10^4\delta\, L/R^2$, where δ is the actual relative displacement between the ends of the shell). It is felt that the linear prebuckling behavior, the relatively high stiffness and higher load bearing capacity of shells with diamond-shaped defects, are all attributable to the shape of the defect itself, which has straight ridge-like formations lining the edges of the facet, providing it with greater rigidity. It appears that while the interior of the facet remains ineffective in supporting compressive load, the facet is prevented from collapsing by its straight edges, which support a higher share of the load. At the mid-section nodes where the ridges meet, the stress flow is distributed outward. This gives rise to stress concentrations and excessive deformations which lead to the formation of the facets in the adjacent regions.

RESULTS AND DISCUSSION

The load data from the tests is presented in Table 1. The buckling loads η_0, η_1 and η_2, referring respectively to shells with none, one and two defects, are normalized with respect to the classical buckling load given by $Pcl = 2\pi Et^2/\sqrt{3(1 - \nu^2)}$, where ν is the Poisson's ratio. The type of defect (D1, D2 or D3) imposed on each shell is also listed.

TABLE 1

Geometric and Buckling Data of Test Shells With Zero, One and Two Defects

Shell	R/t	L/R	η_0	Defect	η_1	η_2	Shell	R/t	L/R	η_0	Defect	η_1	η_2
S8	559	2.2	0.88	[2]D2	0.75	0.71	S29	517	2.0	0.76	[2]D2	0.52	0 .54
S9	488	2.0	0.66	[3]D3	0.58	0.54	S30	598	1.2	0.85	[1]D1	0.55	0 .54
S11	461	2.0	0.90	[1]D3	0.62	0.52	S31	580	0.9	0.74	[1]D2	0.68	0 .60
S12	410	2.1	0.92	[1]D3	0.62	0.57	S32	511	2.0	0.72	[1]D2	0.70	0 .68
S14	485	1.3	0.84	[3]D3	0.58	0.61	S33	482	2.5	0.75	[3]D2	0.69	0 .65
S15	462	1.4	0.78	[3]D3	0.57	0.57	S34	268	1.8	0.90	[3]D2	0.68	0 .70
S16	426	1.5	0.82	[1]D1	0.67	0.56	S35	247	2.0	0.91	[3]D1	0.74	0 .68
S17	521	1.5	0.77	[1]D2	0.63	0.58	S36	201	2.0	0.94	[3]D1	0.71	0 .71
S18	317	2.5	0.96	[1]D3	0.71	0.50	S37	453	1.8	0.84	[3]D1	0.60	0 .59
S19	228	2.5	0.93	[1]D3	0.69	0.59	S38	406	1.6	0.79	[1]D1	0.61	0 .59
S20	408	2.0	0.87	[1]D2	0.66	0.56	S39	296	1.6	0.94	[3]D1	0.69	0 .66
S22	321	2.0	0.98	[2]D2	0.62	0.60	S40	354	1.0	0.91	[1]D1	0.76	0 .72
S23	410	2.1	0.89	[2]D2	0.65	0.60	S41	228	2.5	0.89	[2]D1	0.69	0 .58
S24	268	2.0	0.96	[2]D3	0.65	0.59	S42	356	1.0	0.96	[1]D2	0.75	0 .76
S25	198	1.1	0.87	[1]D3	0.62	0.61	S43	213	2.3	0.91	[1]D3	0.73	0 .64
S26	199	1.2	0.91	[1]D3	0.72	0.69	S44	436	1.2	0.90	[1]D1	0.69	0 .72
S27	436	1.3	0.90	[1]D1	0.65	0.57	S45	406	2.0	0.85	[1]D1	0.68	0 .59
S28	584	2.1	0.80	[1]D3	0.57	0.54	S46	211	1.8	0.93	[3]D2	0.71	0 .72

Location of the second defect :   [1]Adjacent   [2]Opposite   [3]Next tier

In majority of the cases the second defect was imposed either adjacent or diagonally opposite to the first defect, both located at the same axial height as the first one. In a third group of shells, the two defects were positioned at different heights, but near by each other, so that together they resembled adjacent facets in a two tier buckling pattern. It was found that this last produced considerably less load reduction than the first two. (Frequently shells with defects in separate tiers attained a higher load than they reached before introducing the second defect). Hence the third group has not been used for analyzing the data. The two locations at the same height, adjacent and diagonally opposite, appeared to affect the buckling strength to the same extent. The minimum load observed in these tests is about 50%, while the maximum is 76% of the classical load.

The ratios of the buckling loads obtained with the first defect to the loads recorded for the pristine shells are plotted against the radius to thickness ratio and the length to radius ratio of the shells in Figs.6 and 7 respectively. The regression lines in the two plots indicate that the shell geometry has virtually no influence on the degrading effect of the imposed defects. The normalized buckling loads recorded in tests with one and two defects are plotted against the circumferential size of the defects in Figs.8 and 9. The linear regression lines fitted to the two sets of data indicate a gradual reduction in

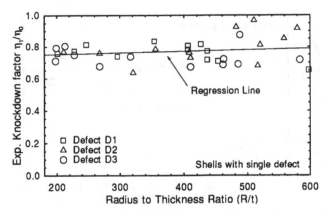

Figure 6: Effect of Shell Thickness on Buckling Load Reduction

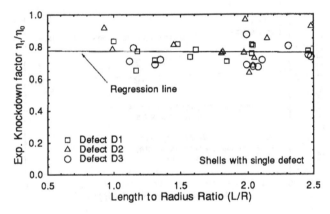

Figure 7: Effect of Shell Length on Buckling Load Reduction

buckling strength with increasing defect size (of about 5% over a range of 40 degrees), while the additional load reduction caused by the imposition of the second defect is between 5 and 6%. The results of the tests on the four shells with multiple defects of size D3 are plotted in Fig.10. Also plotted in the figure are the ratios of the effective stiffness moduli (given by the slopes of the prebuckling load-displacement paths) to the stiffness of the pristine shells. It may be noted that other than the drastic reduction in load (of over 20%) caused by introducing the first defect, the two relationships are more or less linear and parallel, showing approximately 5% reduction in magnitude with the addition of each successive defect. This suggests the possibility of an inter-relationship. Noting that the interior section of each defect is ineffective in supporting the compressive load, the reduction in shell stiffness may be attributed to the decrease in the overall cross-sectional area of the shell, which is directly proportional to the number of imposed defects. It follows that the load reductions associated with the subsequent defects are totally accountable by the reduction in stiffness caused by the loss in effective cross-sectional area.

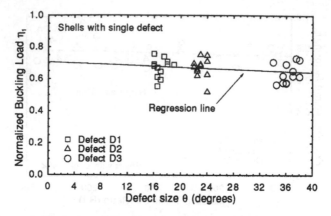

Figure 8: Variation of Buckling Load with Size for Single Defects

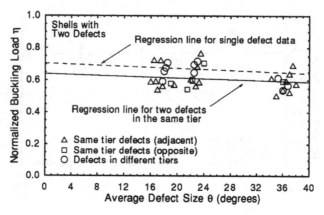

Figure 9: Variation of Buckling Load with Size for Shells with Two Defects

EMPIRICAL RELATION

The buckling data from the tests with one, two and multiple defects were used to develop a comprehensive empirical formula describing the effect of local faceted geometric defects on the axial buckling load of isotropic cylindrical shells. As seen in the foregoing section, the results are independent of the shell geometry. The main parameters affecting the degradation in buckling strength are the size and number of the imposed defects. The results indicate that the drop in load due to the presence of the first defect is drastic, but variations in the finite size of the defect has only a small influence. Hence an inverse exponential curve was fitted to the data, using multiple regression analysis, with the number of defects 'm' and the average circumferential size 'θ' as the two independent variables. The knockdown factor η, i.e., the ratio of the expected buckling load to classical buckling load, was obtained as :

$$\eta = 0.92 exp(-0.173.m^{0.461}.\theta^{0.187})$$

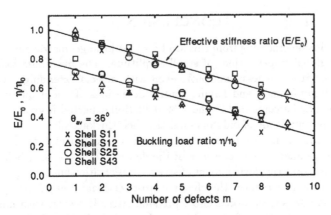

Figure 10: Variation of Buckling Load with Number of Defects

The multiple correlation coefficient for the above relation was computed to be 0.83. The constant 0.92 is based on the value predicted by non-linear theory for perfect shells with both ends clamped. Most of the specimens employed in this study were 'near-perfect' shells having initial knockdown factors (η_0 in Table 1) within 5 to 10% of this value. The foregoing equation is plotted against the average defect size for different values of m in Fig.11. These curves may be employed directly for a first estimate of the effect of

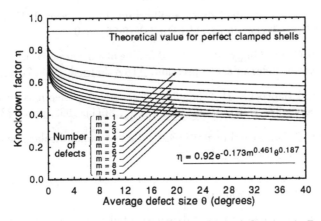

Figure 11: Empirical Relation for Effect of Local Geometric Defects

local defects in axially loaded shells. The applicability of the relation to defects of other types have not been assessed, albeit it is recognized that local geometric defects usually tend to develop the same shape as the diamond shaped facets employed herein.

CONCLUSION

One major conclusion emerging from this study is that local geometric defects are not as detrimental as global imperfections of small amplitude. The maximum load reduction registered in this study is less than 50%, even with defects extending over an eighth of the shell's circumference, and having central deviations of about 30 times the wall thickness. This may be compared to the 75% reduction expected from an axisymmetric dimple having an amplitude of only one-half the shell thickness[3].

The degradation in buckling strength caused by local facet type defects is independent of shell geometry. The presence of the defect affects the shell behavior in two ways: first, it promotes the development of stress concentrations in the regions adjoining its nodes; and second, it reduces the effective cross-sectional area of the cylinder. In a shell with a single defect, th former effect predominates. However, with number of defects the stress concentration effect remains unchanged, while the effective area reduces proportionally. Hence for shells with multiple defects both aspects have to be taken into account for determining the reduction in load bearing capacity.

REFERENCES

1. Donnell, L.H. and Wan,C.C., Effect of imperfections on buckling of thin cylinders and columns under axial compression. ASME J. of Appl. Mech. , 1950, 17(1), pp.73-83.

2. Koiter,W.T., The stability of elastic equilibrium. Ph.D Thesis, 1945, Delft. Transl. by Edward Riks, Techn. Rept. AFFDL TR-70-25, 1970.

3. Tennyson, R.C. and Muggeridge,D.B., Buckling of axisymmetric imperfect circular cylindrical shells under axial compression. AIAA Journal, 1969, 7(11), pp.2127-31.

4. Arbocz,J. and Babcock,C.D., The effect of general imperfections on the buckling of cylindrical shells. ASME J. of Appl. Mech. , 36(1), 1969, pp.28-38.

5. Ricardo,O.G.S., An experimental investigation of radial displacements of a thin-walled cylinder. NASA CR-934, 1967.

6. Foster,C.G., Axial compression buckling of conical and cylindrical shells. Exp. Mech. , 1987, 27(3), pp.255-61.

7. Ellinas, C.P., Ultimate strength of damaged tubular bracing members. ASCE J. of Struct. Engg. , 1984, 110(2), pp.245-59.

8. Ueda,Y. and Rashed,S.M.H., Behaviour of damaged tubular structural members. ASME J. of Energy Resources Techn., 1985, 107(3), pp.342-49.

9. Yoshimura,Y., On the mechanism of buckling of a circular cylindrical shell under axial compression. NACA TM 1390, 1955.

10. Krishnakumar,S. and Foster, C.G., Whole-field optical examination of cylindrical shell deformation. Exp. Mech. , 1989, 29(1), pp.16-22.

STATISTICAL IMPERFECTION MODELS FOR BUCKLING ANALYSIS OF COMPOSITE SHELLS

M K Chryssanthopoulos[*], V Giavotto[+] and C Poggi[Δ]

[*] Department of Civil Engineering, Imperial College, UK
[+] Department of Aerospace Engineering, Politecnico di Milano, Italy
[Δ] Department of Structural Engineering, Politecnico di Milano, Italy

ABSTRACT

Buckling design of cylinders under axial compression is sensitive to the assumptions made in the modelling of initial imperfections. Normally, imperfection modes are selected solely on the basis of buckling mode considerations and their amplitudes determined using existing tolerance specifications. Whilst this approach may be used for metal cylinders, it cannot be readily applied to the design of fibre-reinforced composite cylinders where the effects of manufacturing on imperfection characteristics have not yet been studied in any detail. This paper presents results from a statistical analysis on imperfections on two groups of composite cylinders manufactured by lay-up. Dominant features are quantified and the effect of fibre orientation on imperfections is examined. Simple models describing the random variability of imperfection modal amplitudes are presented in order to be used in buckling strength studies.

NOTATION

m	wavenumber along cylinder length
n	wavenumber along cylinder circumference
t	cylinder thickness
$w_0(x, \theta)$	cylinder initial imperfection function
w_i^{BF}	imperfection value at point i after 'best-fit' analysis
w_i^F	imperfection value at point i using Fourier representation
x	co-ordinate along cylinder length
E(A)	mean value of random variable A
N	total number of imperfection readings on cylinder surface
L	cylinder length
R	cylinder radius
θ	co-ordinate along cylinder circumference
ξ_{mn}	initial imperfection modal amplitude
$\sigma(A)$	standard deviation of random variable A

$\rho(A, B)$ correlation coefficient between random variables A and B

ϕ_{mn} phase angle associated with mode (m,n)

INTRODUCTION

Imperfection sensitivity has long been recognised as the main factor for discrepancies between experimental buckling loads and analytical predictions of shell structures, in general, and of cylindrical shells subject to meridional compression, in particular. In recent years, significant effort has been directed at detailed measurement of imperfections on cylindrical test specimens, as well as some full-scale components [1]. In most of these studies a standard method of data analysis has been adopted, based on the concept of the 'best-fit' cylinder [2]. Thus, the raw imperfections obtained from LVDT readings form the input data to a program that calculates a 'best-fit' cylinder through the entire grid of measured points and then re-computes the imperfections from this artificial perfect surface. This concept has enabled a unified datum to be established for shell imperfections and can be of particular use in comparative studies. These techniques have been used extensively in the study of metal shells [1, 3] but so far no systematic investigations have been carried out on composite cylinders, where several different manufacturing methods may be used with as yet unquantifiable effect on the resulting imperfection surfaces.

Following the 'best-fit' procedure, the resulting imperfections are analysed using two dimensional harmonic analysis to produce a set of Fourier coefficients. The advantage of this method is that, in analytical and numerical predictions of shells with measured imperfections, information on imperfection modes and amplitudes can be easily introduced and their effect studied parametrically. However, this method is particularly useful when imperfections are recorded on groups of similarly manufactured shells. This has prompted the creation of imperfection data banks [1]. The use of probabilistic methods in calculating the reliability of shells with random imperfections buckling in the elastic range has been extensively studied by Elishakoff and Arbocz [4, 5]. Recently, use of statistical techniques on Fourier coefficients of nominally identical cylinders has been made to arrive at characteristic imperfection models that are associated with a particular manufacturing method used in small-scale stringer stiffened cylinder models [6]. The results of the statistical analysis were subsequently incorporated into finite element models in order to quantify the effects of multi-mode imperfection patterns on elasto-plastic buckling strength [7].

The objective of this paper is to present the results of a statistical imperfection analysis on two groups of composite cylinders manufactured by lay-up. The first group (Series A) consists of sixteen nominally identical (i.e. with nominally identical geometric dimensions and material characteristics) cross-ply cylinders with symmetrical lay-up $(0/90)_s$, whilst the second group (Series B) consists of fourteen nominally identical angle-ply cylinders with symmetrical lay-up $(45/-45)_s$. These cylinders were manufactured as part of a research programme to investigate the buckling behaviour of composite cylinders under combined loading. The relevant geometric and material parameters, as well as the overall objectives and initial experimental results of this project, are presented in [8].

IMPERFECTION MEASUREMENTS

The imperfection surface of each cylinder has been recorded in a regular mesh with an interval of 1cm axially and 2cm circumferentially. This results in 110 measurements along any meridian and 46 measurements along any generator. At each point, measurements were taken on both the inside and the outside cylindrical surface. Following the 'best-fit' analysis, two dimensional Fourier analysis was undertaken using the following expression

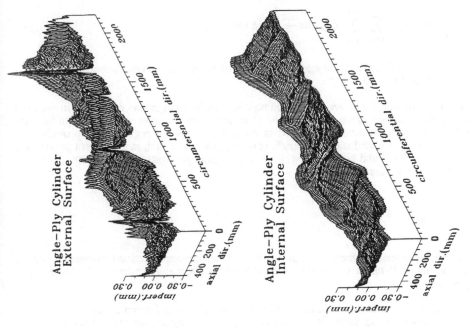

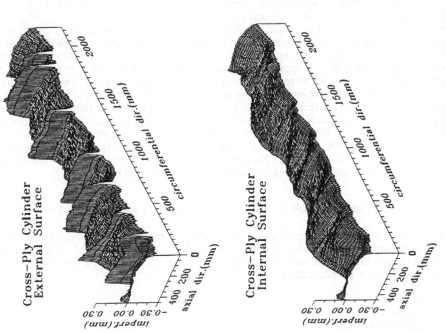

Fig. 1: Typical imperfection surfaces of composite cylinders

$$w_0(x, \theta) = \sum_{m=1}^{m_T} \sum_{n=0}^{n_T} \xi_{mn} \sin\frac{m\pi x}{L} \sin(n\theta + \phi_{mn}) \tag{1}$$

where $0 \leq x \leq L$ and $0 \leq \theta \leq 2\pi$. Each surface was described by a set of coefficients, ξ_{mn} and ϕ_{mn}, with $m_T = 20$ and $n_T = 40$.

It is worth noting that the above expression represents a half-range sine expansion in the axial direction, thus, imposing zero imperfection values at the two cylinder ends. Although this is is not strictly correct, the error introduced is confined to the end regions and, provided the number of terms calculated is not too small, is not significant. In fact, in the current programme both half-range (sine and cosine) as well as full-range expansions were evaluated using the following error function

$$e = \frac{1}{N} \sum_{i=1}^{N} (w_i^{BF} - w_i^F)^2 \tag{2}$$

In addition, comparisons where made at points of maximum imperfection (inwards/outwards). In general, the half-range sine series offered the best alternative in terms of accuracy and compactness.

Figure 1 shows imperfection surfaces (after 'best-fit') obtained for typical cylinders in both Series A (cross-ply) and Series B (angle-ply). It is interesting to note that, although internal surfaces appear to have similar characteristics (dominance of long imperfection waves in both axial and circumferential directions), the external surface is strongly influenced by the orientation of individual layers. In fact, the sharp peaks obtained on the external surfaces are the result of local thickness variations due to overlapping layers and, hence, should not be treated as initial geometric imperfections. In order to quantify this result, a detailed correlation analysis of Fourier modal amplitudes was undertaken. It was shown that $\rho(^{ext}\xi_{mn}, {}^{in}\xi_{mn})$, the correlation coefficient between modal amplitudes with identical wavenumbers obtained for external and internal surfaces for each group of cylinders, is positive and statistically significant for low wavenumbers in the circumferential direction. For $n \geq 15$, this correlation is not statistically significant, demonstrating that short wavelength modes are only present on the external surface due to localised thickness variations. As a result, internal surface measurements were used in the ensuing analysis to study the characteristics of imperfections in composite cylinders. However, the effect of local overlapping should be noted in the analysis of experimental results. This is clearly shown in Fig. 2(a), where the mid-height imperfection profiles corresponding to both external and internal surfaces have been plotted for a typical cross-ply cylinder. Fig. 2(b) schematically summarizes the correlation matrix between external and internal surfaces for cross-ply cylinders. The dark regions correspond to statistically significant correlation coefficients at a 99% confidence level using Fisher's test [9].

STATISTICAL ANALYSIS OF FOURIER COEFFICIENTS

A number of statistical parameters for both ξ_{mn} and ϕ_{mn} were calculated including univariate and bivariate statistics. In addition, fitting of appropriate probability distributions was examined by goodness-of-fit tests (Kolmogorov-Smirnov) and graphical methods. Several probability laws were tested, namely
(i)   uniform
(ii)   normal
(iii)   log-normal
(iv)   exponential
(v)   Weibull (two-parameter).

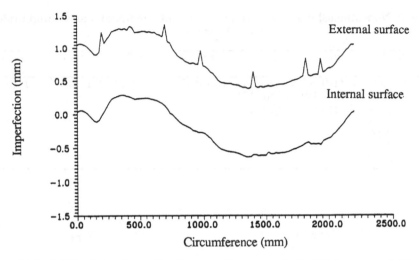

Fig. 2(a): Mid-height imperfection profiles at x=L/2 (thickness = 1.04 mm)

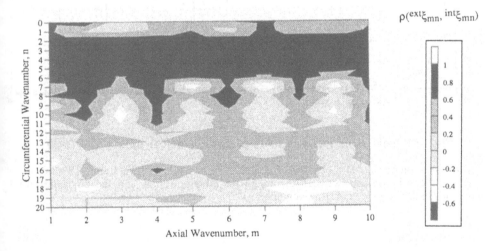

Fig. 2(b): Correlation matrix for $\rho(^{ext}\xi_{mn}, {}^{int}\xi_{mn})$

The results have been presented in detail in [10] and are summarized herein. For example, with reference to Figure 3, showing the results of mean value analysis, it is clear that dominant amplitudes are associated with long wavelengths in the circumferential direction. Modes with $n>15$ have a negligible influence on the imperfection profiles. The extreme values associated with each mode are in agreement with the mean value trends. Table 1 contains results of the mean value analysis for modes with mean value greater than 10% of the maximum mean value. It is worth noting that only about 2% of the total number of modes considered are included in this Table.

48

Table 1: Non-dimensional mean value matrix of dominant modal amplitudes
Series A (Series B)

	m=1	m=2	m=3	m=4	m=5	m=6	m=7
n=2	1.00(1.00)	0.31(0.34)	0.33(0.34)	0.15(0.17)	0.20(0.20)	0.10(0.11)	0.14(0.14)
n=3	0.44(0.51)	0.14(0.19)	0.15(0.17)				
n=4	0.36(0.44)	0.11(0.18)	0.12(0.14)				
n=5	0.14(0.26)						
n=6	0.12(0.16)						
n=7	0.15(0.19)						
n=8	- (0.16)						

Series A   Series B

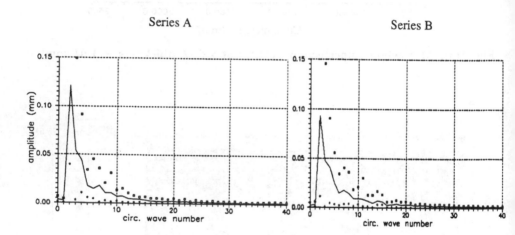

Fig. 3: Mean value analysis of imperfection modal amplitudes (m=1)

Due to the monotonic trend observed for $n \geq 2$, it is possible to fit a simple expression to describe the relationship between the mean modal amplitudes and the circumferential wavenumber for any given axial wavenumber. Thus,

$$E(\xi_{mn}) = e^{\alpha n^{\beta}} \tag{3}$$

where α and β are constants evaluated from sample mean values (Table 2). As demonstrated by Arbocz [1], the development of such simple expressions containing the important features of imperfection amplitudes enables comparison of characteristic models due to different manufacturing methods to be readily undertaken. The mean value of modes with shorter axial wavelength (m>2) can be approximately described by the following expressions

$$E\left(\frac{\xi_{mn}}{\xi_{1n}}\right) = \frac{1}{m} \quad \text{for m = 3, 5, ...} \qquad \text{and} \qquad E\left(\frac{\xi_{mn}}{\xi_{2n}}\right) = \frac{2}{m} \quad \text{for m = 4, 6, ...} \tag{4}$$

Table 2 : Constants for mean value models

Series	Modes	α	β
A	m=1, 2≤n≤20	-1.54	0.5
A	m=2, 2≤n≤20	-3.03	0.3
B	m=1, 2≤n≤20	-1.86	0.4
B	m=2, 2≤n≤20	-2.82	0.3

The next statistic to be examined was the standard deviation in order to quantify the dispersion of imperfection amplitudes. As shown in Figure 4, the coefficient of variation (= standard deviation/mean value) does not exhibit significant variability and this has prompted the idea of obtaining expressions that link the mean modal amplitude value to its standard deviation using regression techniques, i.e.

$$\sigma(\xi_{mn}) = \gamma E (\xi_{mn}) \tag{5}$$

where γ is a constant given in Table 3 for both cross- and angle-ply cylinders.

Table 3: Variability analysis results

m	γ (Series A)	γ (Series B)
1	0.65	0.72
2	0.69	0.56
3	0.66	0.71
4	0.68	0.56
5	0.66	0.71
6	0.69	0.57

Series A   Series B

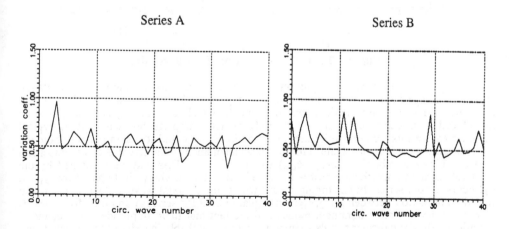

Fig. 4: Variability analysis of imperfection modal amplitudes (m=1)

Figure 5 shows typical results obtained from distribution fitting. Only the log-normal and Weibull models are shown, as these gave the best results for a large number of modal amplitudes investigated, together with sample data points obtained for two dominant modes of Series A and B. Following these graphical tests and the more formal goodness-of-fit tests, the parameters of the distributions have been obtained in order to describe fully random amplitude variability.

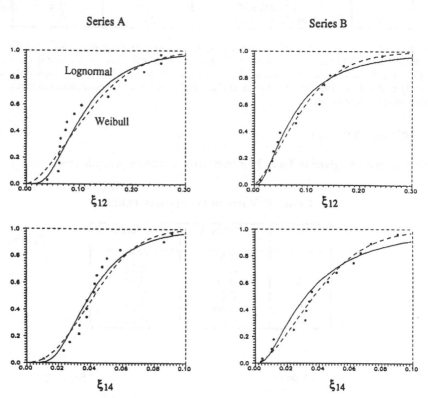

Fig. 5: Typical distribution fitting results

Further to the univariate statistics presented above, it was decided to explore the correlation structure of imperfection modal amplitudes. This type of analysis is useful in deciding how to combine modes in a multi-mode imperfection model which, as generally accepted, must be introduced in buckling studies in order to study the actual imperfection sensitivity of typical geometries. Results for both series are presented in Figure 6 and it is significant to note the high correlation coefficients obtained for modes with common circumferential wavenumber and odd axial wavenumber, i.e. $\rho(\xi_{1n}, \xi_{mn})$ for m = 3,5,... . This may be explained by the relatively smooth imperfection profile in the axial direction as shown in Figure 1 for typical cylinders in both series. In examining modes with common axial wavenumber, it was found that the correlation between $\rho(\xi_{12}, \xi_{13})$ was equal to 0.93 for series A and 0.61 for series B. These are the two most dominant modes and the fact that their amplitude is correlated is significant, since it indicates that these imperfections are probably introduced by a single factor in the manufacturing process.

Statistical analysis of phase angles has shown that these variables are described by a uniform distribution in the range (-π, π). The only significant correlation was found between angles with common circumferential wavenumber, similar to the results presented above for modal amplitudes. Thus, a characteristic imperfection model for these cylinders should include several modes of different axial wavelength for any given circumferential wavelength, with both amplitudes and phase angles of modes with m=2k+1 highly correlated to the corresponding values of the m=1 mode. Similarly, modes with m=2k should be related to the parameters of mode m=2.

On this basis, and considering the information contained in Table 1 regarding the amplitude of the dominant modes, the characteristic model for both series may be obtained from the following expression

$$w_0^c(x, \theta) = \sum_{m=1}^{7} \sum_{n=2}^{8} \xi_{mn} \sin\frac{m\pi x}{L} \sin(n\theta + \phi_{mn}) \tag{6}$$

where ξ_{mn} and ϕ_{mn} are random variables. Thus, simulation studies can be undertaken in order to describe the probabilistic properties of the characteristic model, as shown in [7] for steel cylinders. In particular, the mean value, $E(w_0^c)$, and the standard deviation, $\sigma(w_0^c)$, can be estimated at any point on the cylinder surface. In addition, extreme value properties can be determined either by considering a fixed threshold value, e.g. $\Delta=2t$, and calculating the probability of exceedance, $P(w_0^c>\Delta)$, or by specifying a desired probability level, e.g. $p_{crit}=0.05$, and calculating the corresponding imperfection value, Δ^*, that gives $P(w_0^c>\Delta^*)=p_{crit}$. The latter can be of more direct use in imperfection sensitivity studies since it gives rise to a characteristic shape that is associated with a constant exceedance probability at any point. Both these methods are currently investigated using the results of the statistical analysis and results will be reported at a later stage.

Finally, it is of interest to compare the results of the current study with those obtained in previous investigations. The dominance of ovalization has been a clear feature of imperfection distributions where the manufacturing does not involve joining (or welding) of a number of panels. Similarly, the inverse proportionality between amplitude and wavenumber (or wavelength) has also been observed in previous studies [1, 6].

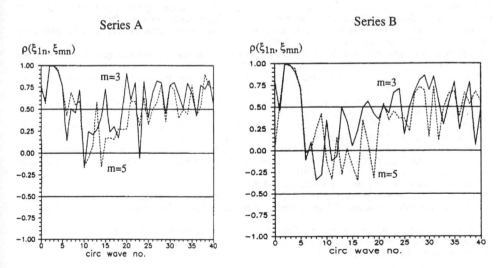

Fig. 6: Correlation analysis of imperfection modal amplitudes

Moreover, the significant correlation in modes with common circumferential wavenumber was also present in the imperfections of steel cylinders [6]. However, the correlation between ξ_{12}, ξ_{13} has not been previously recorded. The small influence of modes with n>15 is also much more evident in these cylinders compared to the stringer-stiffened cylinders, as might be expected.

The above discussion is based solely on the results of the statistical analysis. It is worth pointing out that in constructing a suitable characteristic imperfection model for use in buckling analysis, modes that are associated with high imperfection sensitivity should also be included, even though their mean amplitude and variability are expected to be small. In this context, results obtained in this study should be combined with the conclusions of theoretically based buckling analyses [11].

CONCLUSIONS

The results of the statistical analysis have revealed that, due to the common manufacturing process, several trends exist in the imperfection patterns. Various models have been developed that enable characteristic imperfection surfaces to be described. These can be used within numerical/analytical parametric studies to provide design recommendations for buckling of composite cylinders under combined loading. On the whole, the two series exhibit similar imperfection characteristics and it may be concluded that the same characteristic model could be adopted in buckling analysis. It is worth noting that this conclusion could not have been reached prior to the statistical analysis, especially considering the significant differences appearing on the external surfaces. In conjunction with similar studies on steel cylinders the results have quantified imperfection modelling assumptions and can be used to improve buckling strength predictions.

REFERENCES

1. Arbocz J, 'Shell stability analysis: theory and practice', in *Collapse: The Buckling of Structures in Theory and in Practice*, J M T Thompson and G W Hunt (eds), Cambridge University Press, 1983, pp. 43-74.
2. Arbocz J and Babcock C D, 'Prediction of buckling loads based on experimentally measured initial imperfections', in *Buckling of Structures*, B Budiansky (ed), Springer-Verlag, 1976, pp. 291-311.
3. Singer J, Abramovich H and Yaffe R, 'Evaluation of stiffened shell characteristics from inmperfection measurements', *J. Strain Analysis*, Vol. 22, 1, 1987, pp. 17-23.
4. Elishakoff I and Arbocz J, 'Reliability of axially compressed cylindrical shells with general nonsymmetric imperfections', *J. Appl. Mech.*, Vol. 52, 1985, pp. 122-128.
5. Elishakoff I, van Manen S, Vermeulen P G and Arbocz J, 'First-order second-moment analysis of the buckling of shells with random imperfections', *AIAA J.*, Vol. 25, 1987, pp. 1113-1117.
6. Chryssanthopoulos M K, Baker M J and Dowling P J 'Statistical analysis of imperfections in stiffened cylinders', *J. of Struct. Engg.*, ASCE, 117 (7), 1991, pp. 1979-1997.
7. Chryssanthopoulos M K, Baker M J and Dowling P J 'Imperfection modelling for buckling analysis of stiffened cylinders', *J. of Struct. Engg.*, ASCE, 117(7), 1991, pp. 1998-2017.
8. Giavotto V, Poggi C, Dowling P J and Chryssanthopoulos M K 'Buckling behaviour of composite shells under combined loading', Proceedings of *International Colloquium on Buckling of Shell Structures, on Land, in the Sea and in the Air*, Lyon, September 1991.
9. Hays W L, *Statistics*, 3rd edition, Holt-Saunders, 1981.
10. Ciavarella M, 'Modelli probabilistici di imperfezioni geometriche per lo studio dell instabilita di gusci cilindrici in materiale composito', Diploma Thesis, Dept. of Aerospace Engineering, Politecnico di Milano, 1991.
11. Poggi C, Taliercio A and Capsoni A, 'Fibre orientation effects on the buckling behaviour of imperfect composite cylinders', Proceedings of *International Colloquium on Buckling of Shell Structures, on Land, in the Sea and in the Air*, Lyon, September 1991.

BUCKLING BEHAVIOUR OF COMPOSITE SHELLS
UNDER COMBINED LOADING
by

Vittorio GIAVOTTO

Department of Aerospace Engineering

Politecnico di Milano - Italy

-

Carlo POGGI

Department of Structural Engineering

Politecnico di Milano - Italy

-

Marios CHRYSSANTHOPOULOS, Patrick DOWLING

Department of Civil Engineering

Imperial College - London - UK

ABSTRACT

The experimental programme presented in this paper is part of a research project aiming at improving the knowledge of the buckling behaviour of shell structures. The test results are undertaken on a series of nominally identical specimens. The geometric imperfections are recorded both in the inner and outer surfaces of the shells. The test programme covers the axial compression, torsion and combinations. The calibration of the numerical tools with the test results will allow wider parametric study to be performed useful for the definition of strength design criteria.

INTRODUCTION

The insufficient availability of design criteria for composite structures has largely restricted the efficient use of composite materials in civil, marine and aerospace industry. The gap is particularly severe in buckling strength prediction of composite shell structures. In the aerospace and sport car industry it has prevented the use of efficient stiffened shells in favour of sandwich panels. Although the latter are less prone to buckling they are generally heavier than stiffened shells.

It is known that even the buckling behaviour of isotropic shells depends on a wide range of parameters and the task becomes obviously more demanding in the case of composite shells. Some research in this field has been undertaken [1] but it has not led to systematic and widely applicable design criteria and is therefore of limited use to practising engineers. So far design data rely heavily on experimental testing. The use of numerical simulations for the analysis of different types of structures is commonly accepted as a replacement of expensive experimental programmes but some complex physical problems, such as buckling of shells, can only be solved by means of a combination of experimental, analytical and numerical activities.

The main objectives of the present research are to improve the knowledge of the behaviour of composite materials in shell structures and to provide scientific background for a better exploitation of the material properties together with control of the influence of processing conditions on product performance [2]. Furthermore, the results form suitable background material, through numerical and experimental studies, for the development of Eurocodes on composite shell structures and thin-walled components under combined loading.

In this paper the objectives of the general project are described and the first experimental results obtained at Politecnico di Milano are presented.

PROJECT DESCRIPTION

In order to achieve the main objective of producing reliable guidelines for the buckling design of composite shells the following intermediate goals are to be reached:

a) Assessment of the effect of different manufacturing methods on initial geometric imperfections. The statistical properties of geometric imperfections on several series of specimens made of composite materials with different lay-up configurations will be evaluated [3] and characteristic imperfection models for cylinders will be developed [4]. This will enable realistic tolerances on geometric imperfections to be established and incorporated in the guidelines. At the same time, this study will produce valuable information on the influence of manufacturing on product performance and will provide the necessary input for analytical models to predict buckling strength of imperfect shells [5].

b) Experimental assessment of the buckling response of thin wall composite cylinders under single or combined loading. A series of tests on cylindrical specimens will be undertaken to estimate the buckling and post-buckling behaviour. All the possible information regarding displacements, stresses and geometric imperfections will be recorded for a further calibration of the numerical tools. The possibility to test a series of nominally identical specimens will allow an assessment of the reliability of buckling test predictions.

c) Development and validation of a finite element package and of analytical procedures [6,7]. After a proper calibration of the numerical tools it will be possible to generate additional results to extend the applicability of the guidelines beyond the geometric, material and manufacturing range examined experimentally.

d) Development of strength design criteria for composite cylinders under axial compression, torsion and combined loading using both the experimental and numerical results.

e) Assessment of the suitability of the proposed guidelines in real design conditions and comparison with existing design methods in terms of weight and strength of typical applications.

In what follows the current status of the project is briefly summarized and the results of some of the aforementioned steps presented.

THE EXPERIMENTAL PROGRAMME

The experimental programme has been initiated at the Department of Aerospace Engineering of Politecnico di Milano on a group of composite cylindrical specimens.

The specimen characteristics

The cylindrical specimens are made with pre-preg fabric lay-up on a cylindrical mandrel and have the following geometric properties:

TABLE 1

Specimen geometry

t	1.04	mm
R	350	mm
L	700	mm

They present two thicker reinforced toes at the top and bottom to facilitate the fixing into the loading rig. (fig. 1). The actual shell length is therefore limited to the central part of the cylinder and equal to 550 mm.

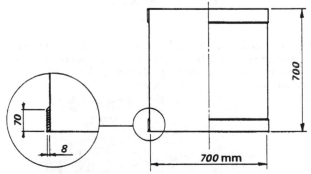

Fig. 1 Reinforced toes of the cylindrical specimen

The layers are made up using an orthogonal Kevlar fabric embedded into an epoxy resin matrix with the following properties (Table 2).

TABLE 2

Material properties of the lamina

E_{11}	$23450\,\mathrm{Nmm^{-2}}$
E_{22}	$23450\,\mathrm{Nmm^{-2}}$
G_{12}	$1520\ \mathrm{Nmm^{-2}}$
ν_{12}	0.20

being x_1 and x_2 the orthogonal in-plane axis directed as the fibres of the fabric in the lamina.

In Table 3 the lay-up orientation of the specimen in the current test series are summarised. The number of specimens of series A and B is sufficient to perform a statistical analysis of the geometric imperfections. Series C,D and E will allow to compare the different imperfection sensitivity of the quasi-isotropic lamination that is expected to be very high [7].

TABLE 3

Lamina characteristics of cylindrical specimens

Lay-up	No.	Thick. (mm)
A) $(0°/90°)_S$	16	1.04
B) $(45°/-45°)_S$	14	1.04
C) $(45°/0°/90°/45°)_S$	4	2.08
D) $(45°/-45°/0°/90°)_S$	4	2.08
E) $(0°/90°/45°/-45°)_S$	4	2.08

TEST APPARATUS

Loading rig

The loading rig is shown in Fig. 2. Axial force is provided by a hydraulic ram, but the actual load applied to the specimen is controlled by four adjustable screw stops, acting on the four corners of the loading platform. Thus, the loading machine is displacement controlled with a very good accuracy.

Figure 2. The loading rig

The lower end-plate is supported by three rollers laying on a horizontal surface (Fig. 3a), when torsion is to be applied by an independent mechanical system, or on sloped surfaces (Fig. 3b), for a fixed ratio of compression to torsion.

Figure 3a.   Figure 3b.

3a)   Flat supports of the rollers for the loading cases of independent torsion or compression.
3b)   Sloped supports of the rollers for the loading case of combined torsion and compression.

Compression load and axial shortening are measured, during the test, on three equally spaced points, corresponding to the three supporting rollers. This gives a measure of the accuracy of the loading process, in terms of load and displacement uniformity. Initially the specimens were fixed to the loading rig by pressing the reinforced toes of the cylinder against internally knurled steel rings fitted with two shoe-brakes. The provided grip was sufficient for the test but not uniform along the circumference. The system has been recently improved changing the shoe-brake type grip into a pressurized bag type system obtaining a uniform distribution of pressure. As a result a more regular buckling pattern was detected and the buckling load slightly increased.

Mapping of geometric imperfections

Shape imperfections have been recorded by an ad-hoc designed apparatus (Fig. 4) where the outer and the inner surface of the shell are scanned by two LVDT transducers. Data acquisition and surface scanning are computer controlled with an extreme flexibility on sampling pitch in both direction. Surface data are stored in a digital form suitable for subsequent computations. The imperfection surface of each cylinder has been recorded in a regular mesh with an interval of 1 cm axially and 2 cm circumferentially. This results in 110 measurements along any circumferential line and 46 measurements along any generator. At each point, measurements were taken on both the inside and outside cylindrical surface.

The main peculiarity of the recorded imperfection data consists in the fairly different aspect of the inner and outer surfaces of the cylinders. In the outer surface the thickness variation due to the overlapping of the layers was evident. This fact has been studied in detail to determine the common characteristic of the two surfaces and define a common reference system. The details of this analysis are reported in [2] where it is shown that the Fourier analysis of the inner surface is representative of the real geometric imperfections of the cylinder.

Fig.4- Apparatus for imperfection shape survey.

TEST RESULTS

The cylindrical models are tested under axial load, torsion and load combination. So far some cylinders of the A (cross-ply) and B (angle-ply) series have been tested under pure axial compression. The collapse loads of two cylinders of different series are reported in Table 4 and compared to the theoretical linear buckling load [7].

Typical plots of the axial load versus average axial displacement of one cylinder for each series are reported in fig. 5a,b.

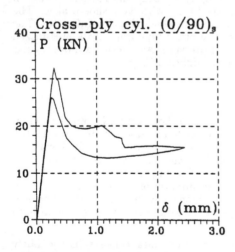
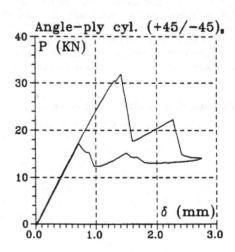

Figure 5a,b. Typical load-shortening curves

a) cross-ply cylinder $(0°/90°)_s$
b) angle-ply cylinder $(\pm 45°)_s$

TABLE 4

Test results		
Lay-up	P_u (KN)	Exp/Theory
A) $(0°/90°)_s$	32.5	0.88
B) $(45°/-45°)_s$	31.8	0.86

The pre-buckling displacements of the two cylinders confirm the expected theoretical values and differ approximately of a factor four. The buckling load has been reached in both cases without significant pre-buckling nonlinearities. The buckling load is very similar for the two different lamination as confirmed by the theoretical investigation [7]. In both cases the post-buckling curves reach approximately 50% of limit load. The unloading in both cases is completely elastic without any residual displacement.

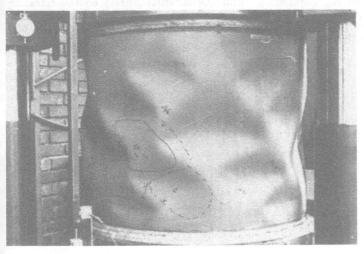

Figure 6. Typical buckling pattern for a cross-ply cylinder.

CONCLUSION

A research project aiming at improving the knowledge of the buckling behaviour of composite shell structures has been presented. It is based on an extensive experimental, numerical and analytical programme. The experimental equipment and the first test results obtained at Politecnico di Milano have been presented. The experimental programme will cover the axial compression, torsion and their combinations. The numerical tools, calibrated with the test results, will allow to perform a wider parametric study useful for the definitions of strength design criteria.

REFERENCES

[1]   Simitses G.J., Shaw D., Sheinman I., "Stability of impeerfect laminated cylinders: a comparison between theory and experiments" AIAA Journal, Vol. 23, July 1985, pp. 1086-1092.

[2]   Giavotto V., Poggi C., Chryssanthopoulos M. "Buckling of imperfect composite shells under compression and torsion", Proc. of Annual Forum of the American Helicopter Society, Atlanta, Georgia, March 1991.

[3]   Chryssanthopoulos M., Giavotto V., Poggi C. " Statistical imperfection models for buckling analysis of composite shells", Proc. of Int. Colloquium "Buckling of Shell Structures, on Land, in the Sea and in the Air", 1991, Lyon, France.

[4]   Chryssanthopoulos M K, Baker M J and Dowling P J " Imperfection modelling for buckling analysis of stiffened cylinders", J. of Struct. Eng., ASCE, July 1991.

[5]   Elishakoff I., Arbocz J. "Reliability of axially compressed cylindrical shells with random axisymmetric imperfections", J. Appl. Mech., Vol. 52, 1985, pp. 122-128.

[6]   Arbocz J., Hol J.M.A.M. "Koiter's Stability theory in a computer aided engineering environment", Int. J. Solids Structures, Vol. 26, No. 9/10, 1990, pp.945-973.

[7]   Poggi C., Taliercio A., Capsoni A. "Fibre Orientation Effects on the Buckling Behaviour of Imperfect Composite Cylinders" accepted for presentation at the International Colloquium "Buckling of Shells Structures, on land, in the sea and in the air", September 1991, Lyon, France.

EFFECT OF IMPACT DELAMINATIONS ON AXIAL BUCKLING OF COMPOSITE CYLINDERS

S.KRISHNAKUMAR and R.C.TENNYSON
University of Toronto Institute for Aerospace Studies
4925 Dufferin Street, Downsview, Ontario M3H5T6, CANADA

ABSTRACT

The development of delaminations induced by low velocity impact and their effect on the axial compressive buckling strength of thin laminated circular cylindrical shells was investigated experimentally. Image Enhanced Backlighting and Structurally Embedded Fiber Optic Sensors were employed for damage assessment. Buckling tests were also conducted on shells with implanted delaminations for comparison purposes and as a means for determining the influence of delamination area on the reduction in buckling strength.

INTRODUCTION

During recent years it has increasingly been recognized that a major weakness of lightweight fiber-reinforced laminate construction is its susceptibility to impact damage. In particular, attention has been focused on barely visible low velocity impact damage because of its significant detrimental effect on structural performance. Strength reductions of up to 40% have been observed in static and fatigue tests conducted on laminate coupons subjected to low velocity impact[1,2]. Damage due to low velocity impact is mainly internal in nature, occurring in the form of a delamination, which reduces the stiffness of the laminate considerably. A major consequence of delamination in panels subjected to compression, particularly when separation has occurred at an interface away from the mid-plane, is the development of delamination buckling[3], i.e. the local buckling of the thinner of the two separated layers. This can result in an unstable propagation of the 'debond', leading to a global reduction in stiffness and premature overall failure.

The effect of impact damage and delaminations in curved panels and cylindrical shells has received attention only recently. Experimental studies have been conducted to assess the degradation caused by impact damage on the hoop strength and burst pressure of complete circular cylinders[4,5]. Strength reductions of up to 40% were reported in the latter case. In tests performed with implanted mylar and teflon sheets simulating delaminations in curved cylindrical panels under axial loading[6], axial strength reductions of the order of 20 to 35% were observed, although no delamination buckling or debond growth was reported. An analytical investigation on delamination buckling in axially loaded cylindrical shells[7] has yielded results similar to those obtained for plane laminates, predicting the onset of instability at loads as low as 25% of that of the undamaged specimen. However, the occurrence of local buckling in cylinders with delaminations has yet to be observed experimentally. Interestingly, the only case where ply buckling (separation and failure of the outermost plies) has been identified as the major failure mode is in tests conducted on undamaged graphite/epoxy tubes subjected to axial compression[8]. Obviously, the incidence of a similar failure mode in shells which have pre-existing delaminations (caused by impact or other means) is a strong possibility.

This paper presents a study on the effect of delamination damage, in particular that arising from low velocity impact, on the load carrying capacity of thin composite cylindrical shells under axial compression. Experimentally the work involved three different aspects: (i) creation and assessment of delamination due to lateral impact on circular cylinders,(ii) estimation of degradation in buckling strength resulting from the imposed damage, and (iii) evaluating the influence of delamination size on the reduction in load bearing capacity of the shell. The last aspect was studied by conducting tests on cylinders with 'simulated' delaminations of controlled sizes, obtained by implanting thin teflon sheets between the plies of the cylinder during the manufacturing stage.

EXPERIMENTAL SET UP AND TEST SPECIMENS

The buckling tests were conducted on a Tinius Olsen hydraulic loading machine with a capacity of 250KN. A pendulum impactor, designed for low velocity impacts (less than 5 m/s) and capable of delivering impact energies of up to 150 Joules, was used for the impact tests. The impactor was designed to accommodate different hemispherical heads with radii of curvature ranging from 12.7 to 102mm (1/2 to 4 inches). The input and rebound velocities of the impactor were measured using a reflective type Object Sensor (TRW OPB 125A) mounted at the location of maximum velocity of the pendulum and connected to an Apple Computer whose internal clock (with a frequency 1.0 Mhz) was used for the time measurement. Calibration tests showed that the measured input velocity and the estimated impact energy were within 1% of the theoretical values computed from energy considerations of the compound pendulum.

The test specimens were manufactured with glass fiber reinforced epoxy (3M SP-1003) unidirectional material, using a belt wrapping machine to facilitate rolling of the prepreg on to an aluminum cylindrical mandrel. The shell was then bagged and cured at 175^0C with a vacuum and external pressure of 500 KPa for two hours. The cylindrical specimens had a mean radius of 100 mm and average thickness of 0.19 mm per ply. They were mounted into 25 mm thick aluminum end-rings, using a Hysol TE6175/HD3561 epoxy system. The average length of the mounted specimens was 150 mm. The tests reported here were conducted on 8 ply shells with a lay up of $(+45_2, -45_2)_s$.

IMPACT TEST DAMAGE

Preliminary tests performed on shells with different thicknesses and ply orientations, employing impactor heads with different radii of curvature, established that the curvature of the projectile head played a critical role in determining the extent of delamination. It was found that there is a limiting value for the radius of curvature of the hemispherical head; when this was exceeded, no delamination was produced, even when energies were high enough to cause severe transverse cracking and fiber failure in the cylinder. The limiting radius was determined to be 25mm and 51mm for 6 and 8 ply shells, respectively. Although all impactors with smaller radii induced delaminations, the least amount of transverse cracking on the front and rear surfaces of the shell occurred when the curvature was the least, hence the impactor with the limiting radius of curvature was employed in the majority of the tests.

The particular lay-up employed for the 8 ply shells had only two distinct interfaces between plies of dissimilar orientations, where delaminations could occur: one between the second and the third plies, and the other between the sixth and the seventh plies. The response of the shells being predominantly in the flexural mode, the damage was primarily induced at the farther (inner) interface, while the delamination at the outer

(a) (b)

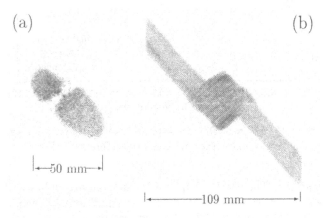

|←—50 mm—→|

|←————109 mm————→|

Figure 1: (a)"Peanut" and (b) Extended Delaminations.

interface, when present, was restricted to a very small area. The geometry of the delaminations generated belonged to two distinct categories, depending upon the magnitude of the impact energy. At energy levels below a threshold value, the induced delamination had the classical "peanut" shape[9], whereas at energies above the threshold level, it had an extended "Zee" shape, as seen in Figs. 1a and 1b, respectively. The latter shape, not mentioned hitherto in the literature, was generated repeatedly and consistently in a number of cylinders, and hence identified as a characteristic response specific to cylindrical laminates subjected to impact energies above the threshold level. The straight edges of this delamination, which are typical of this mode, are caused by the presence of transverse cracks in the two innermost plies of the shell wall, which restrain the inter-laminar crack from propagating sideways. It may be noted that the limbs of the "Zee" shape as well as the major axis of the "peanut" delamination, are oriented at 45^0 to the shell axis, i.e. parallel to the fibers in the lamina immediately below the interface, as predicted by the Bending Stiffness Mismatch[9] and Peel Separation[10] models for delamination. The extended shape in Fig.1b is clearly due to the action of peel forces described in the latter model; its peculiar asymmetry appears to be caused by the difference in stiffnesses in the axial and circumferential directions of the shell.

Apart from geometry, the major difference between the two modes of damage is in the size of the delamination area. The peanut shaped ones are much smaller by comparison. In Fig.2 impact energy is plotted against delamination area for a number of 8 ply shells having a length of about 150 mm. It can be seen that the two modes are

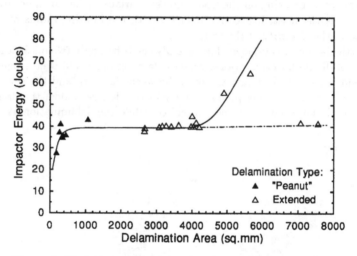

Figure 2: Variation of Delamination Area with Impact Energy
in Tests Conducted on Fourteen Different Shells.

separated by a large gap. The peanut shaped delaminations are all below 1000 sq.mm in size while the Zee shaped ones exceed 2600 sq.mm. The threshold energy is seen to be just under 40 Joules. The direct relationship between impact energy and damage size established earlier for peanut shaped delaminations[9] appears to hold good even in the present case, until the threshold level is reached. At this point there is a drastic increase in the delamination growth, indicating a sudden instability in crack propagation. This instability is associated with the change in delamination mode. Above the threshold level the delamination size once again appears to be proportional to the magnitude of energy as indicated by the rising curve in Fig.2. However, in this region, the shells suffered a considerable amount of extraneous damage, in the form of intra-laminar cracks, which also increased in proportion to the energy. This limited the possibility of creating 'clean' delaminations of larger size by increasing the impact energy.

The results plotted in Fig.2 were obtained from tests performed on different shells. Subsequently, more exact results were obtained by conducting a series of impacts at different locations on the same shell, which are plotted in Fig.3. This shell had a length of 145mm. Again, the sharp transition from one mode to another can be observed at a threshold level of about 38 Joules. The similarity between the lower and upper portions

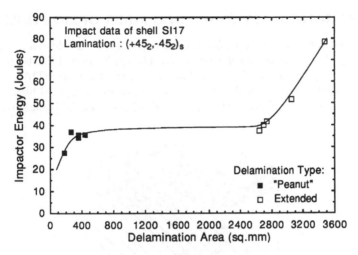

Figure 3: Variation of Delamination Area with Impact Energy
in Tests Conducted on a Single Shell (SI17).

of the curve suggests that the rates of growth of the Peanut and Zee delaminations are about the same, although at higher energies there is additional intra-laminar damage. The threshold for transition from one mode of delamination to the other appeared to be directly related to the shell length and the lamination sequence. For instance, it was found that the threshold for 8 ply $(0_2,90_2)_s$ of 150 mm length was about 90 Joules.

DAMAGE DETECTION WITH IMAGE ENHANCED BACKLIGHTING

The delaminations shown in Figs.1a and 1b were mapped using a simple but highly effective optical damage assessment system, known as the Image Enhanced Backlighting Technique[11]. The technique is specifically suited to mapping of delaminations, but is applicable only to transparent or translucent materials. It basically involves illuminating the specimen from behind, and viewing it in transmitted light. The information pertaining to the delamination is extracted from changes caused by its presence in the transmitted intensity. Briefly, the process consists of recording the images of the specimen taken before and after the impact, subtracting them digitally and then selectively enhancing the grey scale values to provide the desired definition and sharpness in the image of the delamination. Although the backlighting method does not provide data regarding the depth at which delaminations occur, its sensitivity makes it possible to distinguish between delaminations at different interfaces. For instance, the darker shade in the square central portion in Fig.1b denotes the presence of a smaller delamination at the outer interface overlying the "Zee" shaped one at the inner interface. The use of this technique permitted multiple impact tests to be performed on a cylinder, by ensuring that each delamination zone was confined to a local area in the shell wall.

THE STRUCTURALLY EMBEDDED FIBER OPTIC SYSTEM

In some of the test specimens a second optical technique, using embedded fiber optic sensors, was employed for mapping the delamination. The application of this technique was made possible mainly by the development of an etching treatment at UTIAS[12], which facilitates controlled alteration of the cross-section properties of optical fibers to tailor their strength to match the damage threshold levels of the material being investigated.

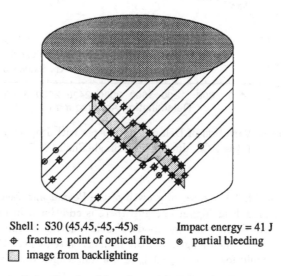

Shell : S30 (45,45,-45,-45)s   Impact energy = 41 J
⊕ fracture point of optical fibers   ⊚ partial bleeding
▨ image from backlighting

Figure 4: Delamination Mapping with Fiber Optic Sensors

By interrogating the fibers using a laser beam, their fracture can be readily monitored, since it causes a sudden drop (of over 90%) in the in the intensity of transmitted light. In translucent materials, such as the glass/epoxy system employed here, the profuse bleeding of light from the broken fibers is visible to the naked eye, hence the fracture points can be easily located. With their strength appropriately tailored, the fibers fracture along the edges of the delamination occurring at the adjacent interface, providing a discrete mapping of the boundary of the inter-laminar fracture.

Application of the fiber optic sensor system in the cylindrical specimens presented some technological difficulties, such as design of the shell's mounting rings to provide safe passage for the optical fibers through them, embedding the fibers between the prepreg layers while rolling them on to a cylindrical mandrel in the belt wrapper, and ensuring their survival through the cure process in the autoclave. These were overcome by repeated experimentation. Figure 4 shows the results obtained in a specimen with embedded optical fibers. The impact was made with a 51mm dia. impactor using 41 Joules of energy. The fracture points of the optical fibers are seen to match well with the mapping of the delamination area provided by the backlighting technique. Similar results, though not as spectacular, were obtained in other cases with extended as well as peanut shaped delaminations, attesting to the applicability of optical fiber sensors for delamination monitoring.

BUCKLING TESTS ON SHELLS WITH IMPLANTED DELAMINATIONS

Buckling tests were conducted on several shells with simulated delaminations, created by embedding thin sheets of teflon[6] (thickness less than 0.025mm) between the second and the third inner plies during the manufacturing stage. Two groups of specimens were tested: one having delaminations along the full length of the cylinder, and the other with inserts going around the entire circumference of the shell. The benchmark was established by testing control specimens without any delaminations. The specifications, delamination size, impact as well as buckling data for all the specimens are listed in Table 1. The knockdown factors (ratio of buckling loads with delamination to that without delamination) recorded for all shells with implanted teflon sheets are plotted in Fig.5. The abscissa is the ratio of the delamination area to the surface area of the shell. The

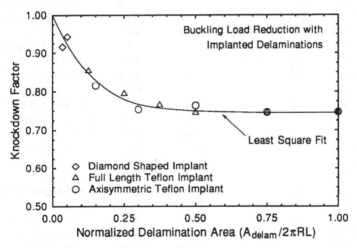

Figure 5: Knockdown Factors of Shells with Implanted Delaminations

buckling load drops steeply in the initial stage, but soon levels off and asymptotes to a value of about 75% of the pristine load. Data belonging to both groups of shells follow the same trend, indicating that the reduction in load is independent of the orientation of the delamination. Also plotted in the figure are the values obtained for two shells which had diamond shaped teflon inserts, providing a closer approximation of the actual delamination created by impact. These had approximately the same area as the Zee shaped delaminations observed in the impact tests.

In every test with implanted delaminations, the two debonded layers invariably buckled together, developing, in all but one case, the classic asymmetric (checkerboard) buckling pattern normally observed in thin cylinders under axial compression. Only the shell with the narrowest axisymmetric teflon insert (22mm wide), buckled into an axisymmetric mode. (The half-wavelength predicted by theory for axisymmetric buckling in this shell is 24mm). Even in this case, the layers on either side of the implant appeared to deform together, bulging outward forming an axisymmetric ripple with the same half-wavelength as the width of the delamination. Thus no evidence of delamination buckling was observed in any of these tests, nor was there any indication of propagation of the delamination before the onset of buckling. It appears that the occurrence of delamination buckling in cylindrical shells is restricted by geometry, at least

TABLE 1

Specifications and impact data for buckling test specimens

Shell No.	Thickness (mm)	Shell Length (mm)	Impact Energy (Joules)	Delamination Area (sq.mm)	Buckling Load (KN)	Remark
SI02	1.600	265			103.6	Control specimen
SI03	1.600	127			104.7	Control specimen
SI08	1.575	149	39.5	4203	90.3	Impact test
SI09	1.575	149			100.1	Control specimen
SI10	1.575	149	39.7	3964	90.1	Impact test
SI11	1.575	152		4925	94.3	Diamond-shaped Insert
SI12	1.575	138			100.1	Control specimen
SI13	1.575	140		22080	79.7	Full-length implant
SI14	1.575	140		3087	91.7	Diamond-shaped insert
SI15	1.600	140		88316	78.1	Axisymmetric implant
SI16	1.600	145		45735	77.9	Full-length implant
SI17	1.600	145			104.6	Control specimen
SI18	1.549	140	40.3	3618	83.3	Impact test
SI20	1.549	147		69480	72.3	Full-length implant
SI21	1.549	145		34270	74.1	Full-length implant
SI22	1.549	65			96.9	Control specimen
SI24	1.549	143	39.9	3157	88.1	Impact test
SI25	1.549	144		68060	72.3	Axisymmetric implant
SI26	1.549	146		11500	82.8	Full-length implant
SI27	1.524	145		13710	76.5	Axisymmetric implant
SI28	1.524	145		27420	70.7	Axisymmetric implant
SI29	1.524	145		45690	71.6	Axisymmetric implant
SI31	1.524	120			93.7	Control specimen
SI39	1.524	155			93.9	Control specimen
SI40	1.524	160	41.3	7069	81.4	Impact test
SI41	1.524	156	41.0	7555	77.4	Impact test

Radius = 100mm.   Lamination sequence : 8 Ply $(45,45,-45,-45)_s$ for all shells

where the thinner layer is on the inside of the shell, as in the present case. The curvature of the cylinder prevents inward axisymmetric deformation of the inner layer, while its buckling into an asymmetric pattern would require radially outward movement which is restricted by the outer layer. In either case, the thinner layer is constrained, and its buckling delayed to match that of the thicker segment. The suppression of delamination buckling in this manner is significant, for it allows the cylinder to carry a much higher load. This is evidenced by the results in Fig.5, wherein the maximum registered load reduction is only about 25%.

BUCKLING LOAD REDUCTION IN IMPACTED SHELLS

The "Zee" delaminations generated in the impact tests, although they appear large, occupy only a small extent of the total surface area of the shell (less than 10%). The peanut shaped delaminations are even smaller and did not appear to cause any degradation in the buckling strength of the shells. The load reductions observed in the case of shells with "Zee" delaminations are plotted in Fig.6. It may be observed that in all cases

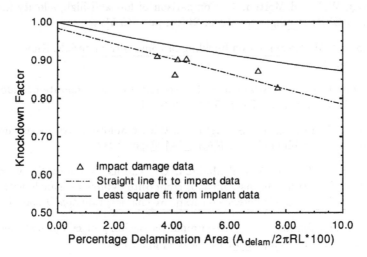

Figure 6: Knockdown Factors of Shells with Impact Damage

the data points of the impacted specimens lie below the segment of the curve obtained from implant data (Fig.5). This can be attributed to the extraneous intra-laminar damage accompanying the impact induced delaminations. Considering that the curve from Fig.5 represents the effect of pure delaminations, the agreement between the two results is rather good.

CONCLUSION

The development of delaminations due to low velocity lateral impact on thin laminated cylindrical shells has been investigated. The shape of the delamination produced was mapped accurately using two different optical techniques. Two distinct modes of delamination have been identified, and a threshold energy level for changing from one mode to another was determined experimentally for a specific shell geometry. The effect of inter-ply debonding on the buckling strength of axially loaded cylindrical shells was determined by a series of tests on shells with implanted teflon sheets and the empirical curve generated. No delamination buckling was observed in these tests. The results of buckling tests conducted on shells with impact damage were compared with those obtained with simulated delamination. It was found that the specimens damaged by impact have lower buckling loads than those with simulated delaminations due to the presence of additional matrix damage generated by the impact.

REFERENCES

1. Starnes, Jr., J.H., Rhodes, M.D. and Williams, J.G., The effect of impact damage and circular holes on the compressive strength of a graphite epoxy laminate. NASA TM 78796, 1978.

2. Cantwell, W.J. and Morton, J., Comparison of low and high velocity impact response of CFRP. Composites, 1989, 20(6), pp. 545-51.

3. Kachanov, L.M., Delamination buckling of composite materials, Kluwer Academic Publishers, 1988.

4. Christoforou, A.P. and Swanson, S.R., Strength loss in composite cylinders under impact. Trans. ASME, 1988, 110 (2), pp.180-4.

5. Poe, Jr., C.C. and Illg, W., Strength of a thick graphite/epoxy rocket motor case after impact by a blunt object. NASA TM 89099, 1987.

6. Horban, B., Palazotto, A. and Maddux, G., The use of stereo X-ray and deply techniques for evaluating instability of composite cylindrical panels with delaminations. SEM Spring Conference on Exp. Mechanics, Houston, Texas, June 1987.

7. Sallam, S and Simitses, G. J., Delamination buckling of cylindrical shells under axial compression. Comp. Struct., 1987, 7, pp. 83-101.

8. Vizzini, A,J. and Lagace, P. A., The role of ply buckling in the compression failure of graphite epoxy tubes. 25th Structures, Struct. Dynamics and Mtrls. Conf., Proc.AIAA, 1984, pp.342-50.

9. Liu, D., Impact-induced delamination - a view of bending stiffness mismatching. J. Comp. Mtrls., 1988, 22, pp.674-92.

10. Clark, G., Modelling of impact damage in composite laminates. Composites, 1989, 20,(3), pp. 209-14.

11. Glossop, N.D.W, Tsaw, W., Measures, R.M. and Tennyson, R.C., Image-enhanced backlighting: a new method of NDE for translucent composites. J. of NDE, 8(3), 1989, pp.181-193.

12. Measures, R.M., Glossop, N.D.W, Lymer, J., LeBlanc, M., West, J., Dubois, S., Tsaw, W., and Tennyson, R.C., Structurally integrated fiber optic impact damage assessment system for composite materials. Applied Optics, 28(13), July 1989, pp.2626-33.

Stability-Analysis of Elastic Shells of Revolution with the Transfer-Matrix-Method
- A fast and reliable approach based on the exact solution of the shell equations -

Rittweger, A. / Öry, H.
Institut für Leichtbau
Technical University of Aachen

Introduction

In aerospace structures shells of revolution are used as main construction elements, for instance as aeroplane fuselage, rocket structure, fuel-tank, habitation modul of space-station etc.
For this light weight structures sophisticated calculation-methods are needed not only for static and dynamic investigations but also for buckling analysis.

Mass saving principles require to find the optimum design. For this purpose often a large number of parametric calculation will be needed.
Therefore, reliable but also fast and economic methods are requested. In many cases, the Transfer-Matrix-Method (TMM), representing the true solution of the differential equation, fulfills these requirements.

The partial differential equation system of an axisymmetric shell with arbitrary meridian has been developed basing on a complete and consistent second order theory using the normal-hypothesis. Provided that the displacements are small and the material linear elastic the second order theory is in many cases a good tool for stability analysis.

By developing the physical states in circumferential direction in a Fourier-series it is possible to eliminate the circumferential coordinate. This way the partial differential equations are transformed into an ordinary differential equation system of first order in meridional direction. In the case of axisymmetric prestress the equations of each wave-number can be evaluated seperately.

The integration of the differential matrix (the differential equation system in matrix form) for arbitrary boundary conditions leads to the transfer-matrix (the mathematicians call it integral-matrix) which represents the exact solution of the problem.
Changing the transfer-matrix into an element-stiffness-matrix makes it possible to compute easily complex branched shells in the same procedure as the most standard Finite-Element codes: by building up a total-stiffness-matrix.
The difference between the Finite-Element-Method (FEM), where the element-stiffness-matrix is based mostly on approximation functions which are only valid for small elements, is that you have here the exact solution independent of the size of an element. Therefore the best possible quality of the analysis and radical reduction of degrees of freedom is reached. This means reliable analysis, less storage in the computer and in many cases faster calculations.

If the axisymmetric shell becomes irregular (large cut outs or single discret stiffeners in meridional direction) a hybrid calculation is possible (TMM for the regular shell, the rest of the structure with FEM), which procedure was demonstrated by Dieker /1/. Therefore the element-stiffness-matrix of the axisymmetric shell basing on the transfer-matrix is not only a special solution for a special problem, it is also an improvement for standard FE codes.

The here presented transfer-matrix was not only applied for stability analysis, but also for static (arbitrary distributed loads, including thermal and hygroscopic strain) and dynamic analysisses (including axisymmetric prestress).
In this paper the application of the TMM for stability analysis of shells of revolution is demonstrated in the case of axisymmetric loads (further developments will take into account non- axisymmetric prestresses) for shells with:

- arbitrary meridian (including cylinders and circular ring plates)
- arbitrary boundary conditions
- complex branched structure
- stepwise variable stiffness in meridional direction
- anisotropic wall-stiffness-matrix
- combined loads

In the case of axisymmetric loads non-linearities (geometrical and physical) can be taken into account by incremental approach.

The Shell Equations

Definitions

The shell geometry is described by the radius r (Fig.1) and two geometric data (angle α and radius of the meridian's curvature ρ), which are depending on the radius function $r(x)$ or $r(s)$. By this kind of description all shells of revolution with arbitrary meridian are represented provided that the axis of rotation is a straight line.
The shell coordinates are the meridian coordinate s and the circumferential coordinate, represented by the circumferential angle ϕ, in the middle of the wall and the normal coordinate z. The displacements u,v,w of a point in the middle of the wall are related to the s,ϕ,z-coordinates (Fig.2).

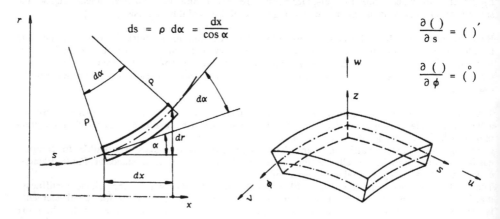

$$ds = \rho \, d\alpha = \frac{dx}{\cos \alpha}$$

$$\frac{\partial \, (\,)}{\partial s} = (\,)'$$

$$\frac{\partial \, (\,)}{\partial \phi} = \overset{\circ}{(\,)}$$

Fig.1   Fig.2

The internal forces and moments related to the length of a shell element (ds or rdφ) result from the integration of the stresses over the wall's sectional area. Fig.3 shows the postitive directions of the internal and external forces and moments (without their derivatives). Their definitions (1a-f, 2a-d) take into account the trapezoidal form of the sectional area.

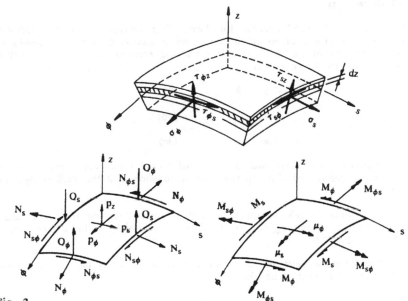

Fig. 3

$$N_s = \int_{-t/2}^{t/2} \sigma_s \left(1 + \frac{z}{r} \cos\alpha\right) dz \qquad (1a)$$

$$M_s = -\int_{-t/2}^{t/2} \sigma_s \left(1 + \frac{z}{r} \cos\alpha\right) z\, dz \qquad (2a)$$

$$N_\phi = \int_{-t/2}^{t/2} \sigma_\phi \left(1 - \frac{z}{\rho}\right) dz \qquad (1b)$$

$$M_\phi = \int_{-t/2}^{t/2} \sigma_\phi \left(1 - \frac{z}{\rho}\right) z\, dz \qquad (2b)$$

$$N_{s\phi} = \int_{-t/2}^{t/2} \tau_{s\phi} \left(1 + \frac{z}{r} \cos\alpha\right) dz \qquad (1c)$$

$$M_{s\phi} = \int_{-t/2}^{t/2} \tau_{s\phi}\left(1 + \frac{z}{r} \cos\alpha\right) z\, dz \qquad (2c)$$

$$N_{\phi s} = \int_{-t/2}^{t/2} \tau_{\phi s} \left(1 - \frac{z}{\rho}\right) dz \qquad (1d)$$

$$M_{\phi s} = -\int_{-t/2}^{t/2} \tau_{\phi s} \left(1 - \frac{z}{\rho}\right) z\, dz \qquad (2d)$$

$$Q_s = \int_{-t/2}^{t/2} \tau_{sz} \left(1 + \frac{z}{r} \cos\alpha\right) dz \qquad (1e)$$

$$Q_\phi = \int_{-t/2}^{t/2} \tau_{\phi z} \left(1 - \frac{z}{\rho}\right) dz \qquad (1f)$$

Kinematic Relations

The relations between strain and displacement are here needed in two kinds:

A) valid for large displacements to determinate the equilibrium equations analytically (principle of virtual work)

Basing on the Novozhilov-strain for large displacements (including quadratic terms) the kinematic relations of the three-dimensional continuum are formulated only in functions of the displacements u,v,w and rotations β_s, β_ϕ of the shell's middle-surface by the constrains:

$$u_z = u + \beta_s z \qquad (3a)$$

$$v_z = v + \beta_\phi z \qquad (3b)$$

$$w_z = w \qquad (3c)$$

The constrains (3a,b) mean that the sectional areas of the wall element remain plain under loading and (3c) expresses the assumption, that the shell is rigid in z-direction .

B) valid for small displacements (without quadratic terms) provided that the transverse shear strain is neglectable ($\gamma_{sz}=0$ and $\gamma_{\phi z}=0$) to introduce the material law for a plain stress state into the shell equations.

Using the kinematic relations of A) without quadratic terms the two equations for the transverse shear strain give:

$$\gamma_{sz} = 0 \quad \Rightarrow \quad \beta_s = -(w' + \frac{u}{\rho}) \qquad (4a)$$

$$\gamma_{\phi z} = 0 \quad \Rightarrow \quad \beta_\phi = -\frac{(\overset{\circ}{w} - v\cos\alpha)}{r} \qquad (4b)$$

The rotations β_s and β_ϕ can be expressed by the displacements, the formulation (4a,b) is equal to the normalhypothesis.

The relations between strain and displacements, valid for small displacements and negligible transverse shear strain, are with the new definition (5):

$$\Psi = w' + \frac{u}{\rho} \qquad (5)$$

$$\varepsilon_{s_z} = \frac{1}{1 - \frac{z}{\rho}} [u' - \frac{w}{\rho} + z (-\Psi')] \qquad (6a)$$

$$\varepsilon_{\phi_z} = \frac{1}{1 + z\frac{\cos\alpha}{r}} \left\{ \frac{1}{r}(\overset{\circ}{v} + u\sin\alpha + w\cos\alpha) + z [\frac{\cos\alpha}{r^2}\overset{\circ}{v} - \frac{\overset{\circ\circ}{w}}{r^2} - \frac{\sin\alpha}{r}\Psi] \right\} \qquad (6b)$$

$$\gamma_{s\phi_z} = \frac{1}{1 - \frac{z}{\rho}} \left\{ v' + z [\frac{\cos\alpha}{r}v' - \frac{\sin\alpha}{r}(\frac{\cos\alpha}{r} + \frac{1}{\rho})v - \frac{\overset{\circ}{\Psi}}{r} + \frac{\overset{\circ}{u}}{r\rho} + \frac{\sin\alpha}{r^2}\overset{\circ}{w}] \right\}$$

$$+ \frac{1}{1 + z\frac{\cos\alpha}{r}} \left\{ \frac{\overset{\circ}{u} - v\sin\alpha}{r} + z [-\frac{\overset{\circ}{\Psi}}{r} - \frac{\sin\alpha\cos\alpha}{r^2}v + \frac{\sin\alpha}{r^2}\overset{\circ}{w}] \right\} \qquad (6c)$$

Equilibrium

The equilibrium equations were determined by the principle of virtual work (7a–d) /2/. Provided that the system is conservativ five equations for equilibrium are found by using the variational calculus /3/, /4/. The kinematic relations valid for large displacements (A) were taken into account.

$$\delta W = \delta W_e + \delta W_i = 0 \tag{7a}$$

$$\delta W_i = - \iiint\limits_{(V)} (\sigma_s \delta\varepsilon_s + \sigma_\phi \delta\varepsilon_\phi + \tau_{s\phi} \delta\gamma_{s\phi} + \tau_{sz} \delta\gamma_{sz} + \tau_{\phi z} \delta\gamma_{\phi z}) \, dV \tag{7b}$$

$$\delta W_e = \int\limits_{s_0}^{s_1} \int\limits_0^{2\pi} (p_s \delta u + p_\phi \delta v + p_z \delta w - \mu_s \beta_s + \mu_\phi \beta_\phi) \, r \, d\phi \, ds \, +$$

$$+ \int\limits_0^{2\pi} [\bar{N}_s \delta u + \bar{N}_{s\phi} \delta v + \bar{Q}_s \delta w - \bar{M}_s \delta\beta_s + \bar{M}_{s\phi} \delta\beta_\phi] \, r \, d\phi \tag{7c}$$

$$dV = (1 - \frac{z}{\rho})(1 + z \frac{\cos\alpha}{r}) \, r \, ds \, d\phi \, dz \tag{7d}$$

The non-linear terms in the equilibrium equations were only taken into account in connection with the membran-forces (N_s, N_ϕ and $N_{s\phi}$). The non-linear terms were linearized by interpreting the membran-forces in these terms as "prestress", which results out of a first-order-theory (linear equilibrium). Provided that the displacements are small this very first iteration is good enough.

The sixth equilibrium equation (moments arround z) was found by observing:

$$N_{s\phi} - N_{\phi s} = \frac{M_{s\phi}}{\rho} - \frac{\cos\alpha}{r} M_{\phi s} \tag{8}$$

Indeed this equation (8) is automatically fulfilled by the definition of the internal forces (1c,d and 2c,d).

Boundary Conditions

The boundary conditions for negligible transverse shear strain were found by variational calculation:

$$N^* = N_s + (u' - \frac{w}{\rho}) N_s + \frac{\overset{\circ}{u} - v \sin\alpha}{r} N_{s\phi} \qquad \text{or} \quad u = 0 \tag{9a}$$

$$N_{s\phi}{}^* = N_{s\phi} + \frac{\cos\alpha}{r} M_{s\phi} + v' N_s + \frac{1}{r}(\overset{\circ}{v} + u \sin\alpha + w \cos\alpha) N_{s\phi} \qquad \text{or} \quad v = 0 \tag{9b}$$

$$Q_s{}^* = Q_s + \frac{1}{r} \overset{\circ}{M}_{s\phi} + \Psi N_s + \frac{1}{r} (\overset{\circ}{w} - v \cos\alpha) N_{s\phi} \qquad \text{or} \quad w = 0 \tag{9c}$$

$$M_s{}^* = M_s \qquad \text{or} \quad \Psi = 0 \tag{9d}$$

Elastomechanical relations

The stress state is assumed as two dimensional and the material behaviour as linear elastic. The Hooke's law for anisotropic material (without thermal or hygroscopic strain)

$$
\left\{
\begin{array}{c}
\sigma_s \\
\sigma_\phi \\
\tau_{s\phi}
\end{array}
\right\}
=
\left[
\begin{array}{ccc}
c_{11} & c_{12} & c_{13} \\
c_{21} & c_{22} & c_{23} \\
c_{31} & c_{32} & c_{33}
\end{array}
\right]
\left\{
\begin{array}{c}
\varepsilon_{s_z} \\
\varepsilon_{\phi_z} \\
\gamma_{s\phi_z}
\end{array}
\right\}
\tag{10}
$$

together with the kinematic relations for small displacements and negligible transverse shear strain (6a–c), which are in short form

$$
\varepsilon_{s_z} = \cfrac{1}{1 - \cfrac{z}{\rho}} \left[\, \varepsilon_s + z\, \kappa_s \,\right]
\tag{11a}
$$

$$
\varepsilon_{\phi_z} = \cfrac{1}{1 + z\, \cfrac{\cos\alpha}{r}} \left[\, \varepsilon_\phi + z\, \kappa_\phi \,\right]
\tag{11b}
$$

$$
\gamma_{s\phi_z} = \cfrac{1}{1 - \cfrac{z}{\rho}} \left[\, \gamma_{s\phi 1} + z\, \kappa_{s\phi 1} \,\right] + \cfrac{1}{1 + z\, \cfrac{\cos\alpha}{r}} \left[\, \gamma_{s\phi 2} + z\, \kappa_{s\phi 2} \right]
\tag{11c}
$$

give the elastomechanical relations:

$$
\left\{
\begin{array}{c}
N_s \\
N_\phi \\
N_{s\phi} \\
N_{\phi s} \\
-M_s \\
M_\phi \\
M_{s\phi} \\
-M_{\phi s}
\end{array}
\right\}
=
\left[
\begin{array}{ccc}
k_{11} \; k_{12} \; \cdots \; k_{18} \\
\cdot \qquad\qquad\qquad \cdot \\
\cdot \quad \cdot \qquad\qquad \cdot \\
\cdot \qquad \cdot \qquad\quad \cdot \\
\cdot \qquad\quad \cdot \qquad \cdot \\
\cdot \qquad\qquad \cdot \quad \cdot \\
\cdot \qquad\qquad\qquad \cdot \\
k_{81} \; \cdots \cdots \; k_{88}
\end{array}
\right]
\left\{
\begin{array}{c}
\varepsilon_s \\
\varepsilon_\phi \\
\gamma_{s\phi 1} \\
\gamma_{s\phi 2} \\
\kappa_s \\
\kappa_\phi \\
\kappa_{s\phi 1} \\
\kappa_{s\phi 2}
\end{array}
\right\}
\tag{12}
$$

Here ε_s, ε_ϕ, $\gamma_{s\phi 1}$, $\gamma_{s\phi 2}$, κ_s, κ_ϕ, $\kappa_{s\phi 1}$, $\kappa_{s\phi 2}$ are the terms in (11a–c) or (6a–c) respectively, which are independent of z.

In this paper a difference between the shear forces $N_{s\phi}$ and $N_{\phi s}$ and the torque $M_{s\phi}$ and $M_{\phi s}$ is made in order to fulfill the sixth equilibrium equation (8), necessary for consistent shell equations.

The coefficients k_{ij} of the wall-stiffness-matrix [K] result from the integration of the material stiffness over the z-coordinate. The trapezoidal form of the sectional area or the curvature of the coordinates respectively has been taken into account because of a consistent formulation.

One consequence of this accuracy is that even if the wall-stiffness-matrix is orthotropic (the axis of orthotropy are the meridional and circumferential direction) and if the wall is built up symmetrically to the shell's middle surface (symmetric orthotropic) there is still a coupling between membran forces and κ or moments and ε respectively.

Differential-Matrix

The five equilibrium-equations and the elastomechanical-relations (12) give together with (5) the partial differential equation system of the form (13) provided that the prestress is axisymmetric (otherwise the matrices $[A_i]$ are depending on the circumferential coordinate too)

$$\frac{\partial}{\partial s} \left\{ y_{(s,\phi)} \right\} = \sum_{i=0}^{4} [A_i(s)] \frac{\partial^i}{\partial \phi^i} \left\{ y_{(s,\phi)} \right\} + \left\{ L_{(s,\phi)} \right\} \qquad (13)$$

$$\left\{ y_{(s,\phi)} \right\} = \left\{ u \quad v \quad w \quad \Psi \quad N_s^* \quad N_{s\phi}^* \quad Q_s^* \quad M_s^* \right\}^T \quad : \text{state vector}$$

$$\left\{ L_{(s,\phi)} \right\} \quad : \text{load vector (external loads)}$$

As already stated, by developing $\left\{ y \right\}$ and $\left\{ L \right\}$ in circumferential direction in a Fourier-series (14), the partial differential equations are transformed into a ordinary differential equation system and in the case of axisymmetric prestress the equations of each wave number m can be evaluated seperately (otherwise there would be a coupling between the wave numbers). That leads to an ordinary differential equation system of first order in meridional direction with 16 equations (15).

$$\left\{ y_{(s,\phi)} \right\} = \sum_{m=0}^{\infty} \left(\left\{ y_{c_m(s)} \right\} \cos m\phi + \left\{ y_{s_m(s)} \right\} \sin m\phi \right) \qquad (14)$$

$$\frac{\partial}{\partial s} \left\{ \frac{\left\{ y_{c_m} \right\}}{\left\{ y_{s_m} \right\}} \right\} = [A_{m(s)}] \left\{ \frac{\left\{ y_{c_m} \right\}}{\left\{ y_{s_m} \right\}} \right\} + \left\{ \frac{\left\{ L_{c_m} \right\}}{\left\{ L_{s_m} \right\}} \right\} \qquad (15)$$

For the special case of symmetric loading (no shear prestress $N_{s\phi}$) and orthotropic wall-stiffness-matrix (including eccentricity) a reduction of (15) to 8 equations is possible.

Transfer-Matrix

The integration of the differential-matrix for arbitrary boundary conditions gives the transfer-matrix $[B_m]$, which describes the behaviour between two state vektors at $s=s_0$ and $s=s_1$:

$$\left\{ \frac{\left\{ y_{c_m} \right\}}{\left\{ y_{s_m} \right\}} \right\}_{s_1} = [B_m] \left\{ \frac{\left\{ y_{c_m} \right\}}{\left\{ y_{s_m} \right\}} \right\}_{s_0} + \left\{ \frac{\left\{ P_{c_m} \right\}}{\left\{ P_{s_m} \right\}} \right\} \qquad (16)$$

If the differential-matrix is integral-exchangeable (that is the case for constant coefficients like for a cylindrical shell), solutions with semi-analytical integration methods (Lee-Magnus-series, spectral analysis /5/) are well known.
In general (for arbitrary meridians) numerical standart procedures like Runge-Kutta or extrapolation algorithms can be used.

Element-Stiffness-Matrix

To analyse a complex, branched shell structure composed of different shell elements it is practicable to build up a total-stiffness-matrix. The element-stiffness-matrix is found by arranging the transfer-matrix (16) in that way, that the forces of the state vector at s_0 and s_1 are collected in one vector, just as the displacements. Also a transformation to a global coordinate system has to be carried out.

The here used transfer-matrix leads to an element-stiffness-matrix, which is symmetrical and invariant against rigid body movements (the reward of the consistent shell equations' formulation).

Linear-Stability-Analysis

The bifurcation loads of each wave number were investigated by analyzing the eigenvalues of the total-stiffness-matrix (searching the load for eigenvalue equal zero).

The minimal critical load (at the critical wave number) is often called Euler-buckling-load.

In seek of simplicity we did not take into account the pre-buckling deformations of the shell due to the internal pressure or to the Poisson's effect, although it would be possible.

Tank under Internal Pressure

Fig (4) showes a satellite tank made of titanium with a rouhgly elliptical outer shape and a common bulkhead (conical and spherical part) with a small screen in the center.

The tank has partly different wall-thickness in meridional direction and is fixed by the small conical shell outside the tank, a slotted skirt which was modellized as an orthotropic shell (boundary conditions: clamped).

In mission the tank is pressurized with the internal pressure p=33 bar and has been investigated here for stability.

The calculation with the TMM gave as result an Euler-buckling-load of $p_{crit}=56.684$ bar internal pressure at the wave number m=57 (Fig 5), compared to a linear analysis with the Finite-Difference code BOSOR-4 a relativ difference of 0.4% ($p_{crit}=56.9$bar at m=53).

With the TMM a modellization of the structure was made with 23 shell elements compared to 335 shell elements needed in BOSOR-4 for convergency (nearly 15 times more).

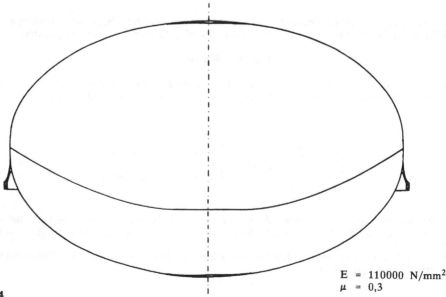

$$E = 110000 \ N/mm^2$$
$$\mu = 0,3$$

Fig. 4

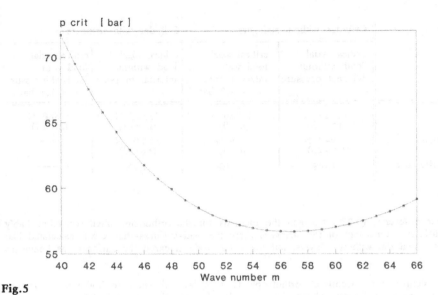

Fig.5

Orthotropic Cylindrical Shell under Axial Load and Internal Pressure

A stringer stiffened cylindrical shell model (Fig6), produced by galvanic deposit and used for experimental research at Laboratoire Betons Et Structures of INSA in Lyon (France), which represents a part of the ARIANE V rocket stage, was investigated for Euler-buckling-load with the TMM and compared with the french FE code INCA.

In the table (1) the results of the eccentric orthotropic (stringer stiffened) shell compared to the same shell without stringers (isotropic shell) are listed.

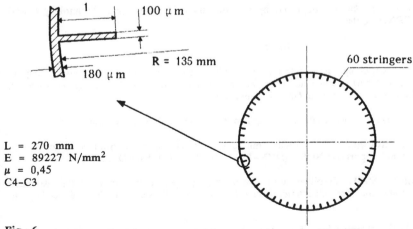

L = 270 mm
E = 89227 N/mm^2
μ = 0,45
C4-C3

Fig. 6

	Cylinder without stringers		Cylinder with stringers	
	critical axial load without internal pressure	critical axial load with internal pressure p = 1,5 bar	critical axial load without internal pressure	critical axial load with internal pressure p = 1,5 bar
TMM	11727 N (m=23)	11780 N (m=0)	13547 N (m=18)	27660 N (m=14)
FEM (INCA)	11942 N (m=23)	12084N (m=0)	12640 N (m=18)	———
relative difference	1,8%	2,5%	7,2%	———

Table 1

The higher difference (7,2%) between the methods for the orthotropic shell results probably from a different "smearing" of the stringers (here the wall-stiffness-matrix was calculated like a composite material with two layers, but the INCA code smeared the stiffeners as beams to the skin).

Another example for excentrical orthotropy is a cylindrical shell (r=250mm, l=510mm, t=1.25mm) under axial load, whose skin is made of composite material (12 layers carbonepoxy each with the same thickness).
The critical axial load of two composite shells with different layers' orientation compared to the method of Tennyson are for the boundary condition SS4-SS3 (table 2):

	Composite 1 $[90,0]_{3 S}$	Composite 2 $[0_6, 90_6]$	Material data:
TMM	N_{crit} = 99.00 N/mm (m=11)	N_{crit} = 71.80 N/mm (m=13)	$E_{\shortparallel}$ = 123550 N/mm^2
			$E_{\perp}$ = 8707 N/mm^2
Tennyson	N_{crit} = 99.96 N/mm (m=11)	N_{crit} = 72.46 N/mm (m=13)	$G_{\#}$ = 5695 N/mm^2
			μ = 0,319

Table 2

Acknowledgement

This research has been kindly supported by the Ministry of Research and Technology (FRG) within DARA/ERNO contract.

Literature

/1/ S. Dieker: Kopplung der Methode der finiten Elemente mit der Randwertintegralmethode und dem Übertragungsmatrizenverfahren; Baden-Baden, 1987, Congress FEM'87, published by IKO Software Service GmbH (Stuttgart)

/2/ H.L. Langhaar: Energy Methods in Apllied Mechanics ; J.Wiley & Sons, 1962

/3/ H.Öry, A.Rittweger: Die Schale mit einer durch kubische Splines angenäherten Meridiankurve; Final report DYNOST (DYN-TB-120) of DARA/ERNO contract, 1991

/4/ A.Rittweger: Statik, Stabilität und Eigenschwingungen anisotroper Rotationsschalen beliebigen Meridians mit der Übertragungsmatrizenmethode; Dissertation RWTH Aachen (to be published)

/5/ H. Öry, H.G. Reimerdes, S.Dieker: Berechnung von interlaminaren Spannungen in mehrschichtigen Faserverbundwerkstoffen mit Hilfe von Übertragungsmatrizen; Z.Flugwiss. Weltraumforsch., Bd.8 Heft 6, 1984

SIMPLIFIED ANALYSES FOR THE BUCKLING OF SHELLS OF REVOLUTION

by R. Abdelmoula*, N. Damil**, K. Hatem, M. Potier-Ferry and L. Siad
Laboratoire de Physique et Mécanique des Matériaux,
URA CNRS n° 1215
Université de Metz, Ile du Saulcy
57045 Metz

* EDF, Département MMN, Clamart
** Université Hassan II, Casablanca, Maroc.

In many cases, the buckling mode of shells of revolution has a typical shape, with more rapid derivatives in the azimutal direction than in the axial one. This occurs for instance, with cylindrical shells under external pressure or in torsion, but generally not in axial compression. In the same cases, the buckling load converges to same asymptotic values, for large values of the BATDORF parameter Z. For example, YAMAKI (1984) established the following formula for the buckling pressure of an isotropic cylinder

$$P_c/E = 0.927 \ (t/R)^{5/2} \quad (R/L)$$

with free boundaries in the axial direction ($N_x = 0$) and the same multiplied by 1.48 with clamped boundaries in the axial direction (u=0). So the buckling load is mainly influenced by the axial boundary conditions, which had been early recognized. (SOBEL (1964)). The difference between these two buckling loads may much greater than 1.5, as for corrugated shells (DEBBANEH (1988)).

The first result presented here is an explanation of the role of boundary conditions. Indeed, we establish that, for large values of Z, there exists a boundary layer. This permits to distinguish between a local and a global form of the mode, where the second one is valid away from the boundaries and leads to an approximation of the buckling load. Some boundary conditions influence only the "local mode" (for instance clamping) while the axial boundary condition leads to a boundary condition for the "global mode", that is different from the original one. Moreover the "global mode" satisfies simplified equations, that are more or less equivalent to the assumption of a quasi-inextensional mode. In the case of cylinder under pressure, the simplified differential equations can be solved analytically , which reestablishes YAMAKI's asymptotic formulae. The range of validity of the approximation depends on boundary conditions, it is about $Z > 100$ in the case

$N_x = 0$ and $Z > 1000$ for the case $u = 0$.

This simplified analysis can be applied to a lot of buckling problems as well as vibration problems. The second interesting result is the buckling of anisotropic shells under external pressure. In that case, the buckling load depends only an two constitutive coefficients, that are respectively the azimutal bending stiffness and the axial membrane flexibility. An application under study is the influence of mouldings on the strength of cans. A slightly different application is the plastic buckling of torispherical shells under internal pressure, where the anisotropy is induced by plasticity. In that case, an analytical formula for the buckling pressure can be obtained, that depends on the two similar constitutive coefficients. Here, because the simplified equations have not constant coefficients, a Ritz procedure completes the quasi inextensional approximation. In addition to its own interest, this analytical formula permits an easier discussion of the influence of the plasticity model on buckling.

Lastly, this simplified analysis has been applied to nonlinear buckling and to the imperfection sensitivity of an elastic cylindrical shell under external pressure. Roughly, the idea is to insert the approximate mode in the classical KOITER's postbuckling analysis. So we get a reduction of critical load that is proportional to

$$(a_0 / t \sqrt{Z})^{2/3}$$

a_0 being the amplitude of the initial deflection. The same has been done in the case of a high wavenumber $(n > 8)$ and in the presence of an initial imperfection that is localized on one or two buckles, by using a general theory that we recently established for this purpose. This yields a reduction of critical load that is proportional to

$$a_1 / t \sqrt{Z}$$

The interaction between the two types of imperfections has also been considered.

N. DEBANNEH, doctoral dissertation, INSA, Lyon 1988
L.H. SOBEL, AIAA Journal (1964)
N. YAMAKI, Elastic stability of circular cylindrical shells (1984)

AN ENERGY BASED CONCEPT FOR DYNAMIC STABILITY OF ELASTIC STRUCTURES

Dieter Dinkler and Bernd Kröplin
Institut für Statik und Dynamik der Luft- und Raumfahrtkonstruktionen
Universität Stuttgart
Pfaffenwaldring 27, D–7000 Stuttgart 80, Germany

ABSTRACT

The paper deals with the development of stability estimates for structures under time depending loads. The estimates are based on energy norms and can be used for the investigation of safety against loss of stability by buckling. The static analysis is used to compute the critical strain energy which is necessary to effect the transition from the pre- to the postbuckling range. Starting from a stable state of equilibrium the presented procedure allows to decide whether the structure stays for a certain load history within the critical bounds, which separate the motion around the prebuckling state from the motion in a postbuckling region. For an arbitrary load history Galerkin's procedure is applied to the equations of motion using a representation of the deformations from static modal analysis. The stability is proved by comparing the load induced energy and the critical strain energy.

INTRODUCTION

For elastic limit load analysis of thin–walled structures under static loads a number of well–known methods exist. In case of linear structural behaviour they are mostly based on eigenvalue analysis and in case of nonlinear structural behaviour on accompanying eigenvalue calculations along a load path. In /10/ an energy–based concept of structural stability is proposed, that works with the strain energy and its first and second variations in order to describe stability properties of elastic structures.

For time–dependent loads in structures with linear behaviour the load deformation dependence often is represented in frequency domain. Nonlinear structures require more

detailed treatment as far as buckling is concerned. Because of the lack of reliable estimates the equations of motion including all nonlinearities are often solved by a direct
integration in time. The disadvantage of the computation of the deformation behaviour
in time is the fact that the solution only gives an information about the motion in the
considered time interval, but not about the further motion in the future. Furthermore
it is not possible to decide whether the motion is stable or not. However, the numerical
effort seems to be often inadequate, particularly as the results are hard to interprete
because of the numerous interacting modes.

The most important point in the judgement of motions of elastic systems is the statement about stability. Fundamental works considering the theory of stability of elastic
structures are for example Leipholz /1/ and Thompson/Hunt /2/. These works treat
mostly questions of stability of equilibrium states, while the stability of nonlinear dynamic systems is discussed e.g. in Hayashi /3/ and Thompson/Stewart /4/.

The fundamental investigations concerning stability of motions and a definition of stability in case of perturbed initial conditions date back to Liapunov, see /6/. The procedures
that allow a decision about stability or not are given by Liapunov as his first and second
method. The first method considers the time dependence of the deformations and needs
a direct integration of the equations of motion as it is mentioned above. The second
method is based on investigations to a Liapunov function V that is given by Hamiltons
function H. H contains the cinetic energy T and the strain energy Π. However, this
method is only valid, if we take into account infinitesimal perturbations of a given state
of motion. An information about the distance to a critical state is not available.

In the following a motion is defined as a stable motion, if the deformations of the
structure remain in the prebuckling region without a transition to the postbuckling
range. The transition to a postbuckling region may be indicated by an increasing of a
local or an overall deformation mode. Normally a proper measurement for the state of
deformation may be a *Norm* of the deformation field. This paper proposes the Liapunov
function V as a measurement for the judgement of the motion. Assume elastic material
behaviour the difference in V of a given stable state of motion to an indifferent one may
be computed and is defined as the critical energy ΔV_{crit}. This critical energy is of finite
size and is a scalar measurement for the distance to a critical state of motion as well.

In the design of a structure one may choose the critical energy for investigations
to the safety, if the stability of a stable state of motion against some additional time
dependent perturbation loads is taken into account. In this case a comparison of the load
induced energy T_e and the energy V_0 related to initial conditions with the critical energy
allows the proof of the structures safety against buckling without a direct integration of
the equations of motion.

The proof of stability is given by (1).

$$
\eta = \frac{\Delta V_{crit} - T_e - V_0}{\Delta V_{crit}}
\quad
\begin{array}{ll}
> 0 : & \text{stable} \\
= 0 : & \text{indifferent} \\
< 0 : & \text{unstable,}
\end{array}
\tag{1}
$$

where η is called the *degree of stability*.

The main topics of the proposed method are the following three steps:

- Identification of limit states between prebuckling and postbuckling and calculation of the energy related to these critical states.

• Calculation of the energy related to the time-dependent load influence.

• Comparison of the critical energy and the load induced energy.

The method presented here are based on elastic material behaviour. The load functions are short time impulse loads, load level changes are considered in /9/.

BASIC ASSUMPTIONS

Basis of the theory of stability of elastic continuous systems is the energy balance in rateform (2). It contains the rates of strain energy, kinetic energy and the external loads. Temperature effects /9/ are neglected.

$$\dot{V}_v = \int \{\dot{\gamma}_{ij}\, e_{ijkl}\, \gamma_{kl} + \rho\, \dot{v}_i\, \ddot{v}_i\}\ dR^3 - \int \dot{v}_i\, \bar{s}_i\, dO = 0 \tag{2}$$

with the initial condition $V_{v_0} = c$ at time $t = t_0$.

$(\dot{\ })$: derivative with respect to time
e_{ijkl} : elasticity of the material
ρ : material density
v_i : displacements in direction of the space coordinates x_i, $i = 1, 2, 3$
$\bar{s}_i$: loads

The Green strains γ_{ij} are defined as

$$\gamma_{ij} = \frac{1}{2}\left(v_{i,j} + v_{j,i} + v_{k,i}\, v_{k,j}\right). \tag{3}$$

Alternatively the potential energy may be formulated in stresses and strains as a mixed formulation. Equation (4) contains the nonlinearity one order lower than (2), which leads to some advantages in numerical investigations.

$$\dot{V}_{\sigma,v} = \int \{\dot{\gamma}_{ij}\, \sigma_{ij} + \gamma_{ij}\, \dot{\sigma}_{ij} - \dot{\sigma}_{ij}\, f_{ijkl}\, \sigma_{kl} + \rho\, \dot{v}_i\, \ddot{v}_i\}\ dR^3$$
$$- \int \dot{v}_i\, \bar{s}_i\, dO_\sigma - \int \dot{s}_i\, (v_i - \bar{v}_i)\, dO_v = 0 \tag{4}$$

f_{ijkl} : flexibility of the material
σ_{ij} : 2. Piola–Kirchhoff stress–tensor
s_i : stress–vector

Thin–walled shell structures are treated by means of the Kirchhoff–Love–Hypothesis and by introducing a reference surface. The mixed formulations with moderately large rotations describe the rate of the potential dependent on the displacements u_α, u^3, the velocities $\dot{u}_\alpha$, $\dot{u}^3$, the stress resultant axial forces $n^{\alpha\beta}$ and bending moments $m^{\alpha\beta}$ and the loads p^α, p^3 of the reference surface

$$\dot{V} = \dot{V}(u_\alpha, u^3, \dot{u}_\alpha, \dot{u}^3, n^{\alpha\beta}, m^{\alpha\beta}, p^\alpha, p^3). \tag{5}$$

The discretisation of the reference surface with finite elements leads after the integration over the reference surface to the matrix notation (6)

$$\dot{V}_z = \dot{z}^T \left\{ A^L(z_0)\, z + A^N(\tfrac{1}{2}z)\, z + M\, \ddot{z} - p \right\} = 0. \tag{6}$$

z :   vector of the unknowns $z^T = \{ u^T,\, n^T,\, m^T \}$
$A^L(z_0)$:   linear matrix, dependent on the initial state z_0
$A^N(\tfrac{1}{2}z)$:   geometric nonlinear matrix
M :   mass matrix
p :   vector of loads

THE CRITCAL ENERGY

The motion describes a trajectory in the phase–plane of displacements and velocities, see figure 1. The distance of the trajectory to the static equilibrium state is a measure for the induced energy by initial conditions. Critical states of motion are indicated by a critical trajectory, which is called *separatrix*. This trajectory is the limit–curve for all *locally stable* motions in the pre- and post–buckling range, respectively. For investigations to stability the energy level ΔV_{crit} that indicates the *separatrix* is of interest. In case of conservation of energy this energy level holds for all states of motion on the *separatrix*. The criterium of a stable motion now is satisfied, if the state of motion at the end of the perturbation time is inside of the *separatrix*.

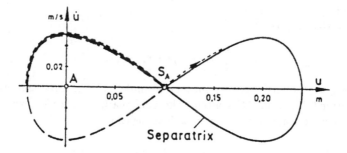

Figure 1 :   Phase plane of motions

A condition for computing the critical energy level may be derived, if the static axis is taken into account. The critical deformations on the static axis ($\dot{u} = 0$) are given as the distance from the state of equilibrium to the saddle–point S. The condition for computing the distance is obviously, if the strain energy projection orthogonal to the phase–plane is considered for the special case of $\dot{u} = 0$.

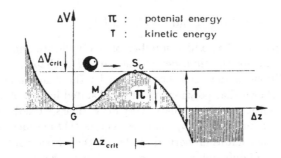

Figure 2 : Energy profile for $\dot{u} = 0$

The stable equilibrium state is given by the minimum G of the potential energy. A perturbation of this equilibrium state leads to a *locally stable* motion of the structure with the energy level ΔV. An increase from ΔV to ΔV_{crit} shifts the structure to the saddle–point S_G. S_G gives the critical deformation state of the structure, since two *locally stable* motions in the pre and the postbuckling range meet here. A further increase of the energy leads to a motion in the whole pre– and postbuckling state. Hence the decisive factor for the stability of the equilibrium state G is the energy level, which is given by the maximum of the potential energy. Therefore the first variation of the potential energy has to vanish in case of critical deformations

$$\delta\Pi = 0. \tag{7}$$

(7) leads to a system of nonlinear algebraic equations (8) for the critical state vector Δz_{crit}. The nontrivial solution describes the unstable equilibrium state and may calculated iteratively as a nonlinear *eigenvalue problem*.

$$\delta z^T \left\{ A^L(z_G) + A^N(\tfrac{1}{2}\Delta z_{crit}) \right\} \Delta z_{crit} = 0. \tag{8}$$

(8) means a vanashing secant stiffness matrix. The critical energy, which will be needed for the stability criteria may be calculated now for the equilibrium state G by means of integration of (6) for the static case

$$\Delta V_{crit} = \Delta z_{crit}^T \left\{ \tfrac{1}{2} A^L(z_G) + \tfrac{1}{6} A^N(\Delta z_{crit}) \right\} \Delta z_{crit}. \tag{9}$$

CALCULATION OF THE LOAD ENERGY

The aim of the investigations is to compare the critical energy with the load energy which is induced by perturbations, see (1). The external energy depends on the function of the perturbation in time. Hence the energy balance (6) has to be integrated over the time of perturbation t_s

$$V_{t_0} = \int_{t_0}^{t_s} \dot{z}^T \left\{ A^L(z_0)\, z + A^N(\tfrac{1}{2}\Delta z)\, z + M\,\ddot{z} - p \right\} dt = 0. \tag{10}$$

Since the exact solution is not known, a Ritz–like method is chosen. The approximation of the space–time–function of the state–vectors

$$z = \alpha_{ij} \Delta z_i \, \phi_{j(t)} \,, \quad t_0 \leq t \leq t_s \tag{11}$$

consists of the spatial pattern Δz_i and a number of freely chosen shape functions $\phi_{j(t)}$ with unknown scalers α_{ij} for the time–dependent part. The shape functions have to satisfy the initial conditions of the perturbed motion for the deformations and the velocities. Experience has shown that polynomials are well suited in case of impulsively loaded structures since the initial motion oscillates very little. The spatial distribution of the unknown state–vector is approximated by Lanczos–vectors /11/ in order to represent the effects of the load and some additinal normal modes for the autonomous part of the oscillation, see /8,9/. In case of different critical modes each mode has to be considered.

With the approximation (11) the energy balance (10) is first calculated for spatial coordinates. The calculation of the quadratic forms can be performed with standard finite element programs under static loading. The integration in time can be performed analytically or numerically for example with Gauss. After integration a nonlinear algebraic system of equations has to be solved for the unknown factors α_{ij}.

$$\{V_{AL} + V_{AN}(\alpha_{ij}) + V_M\}_{klij} \, \alpha_{ij} - V_{pkl} = 0 \tag{12}$$

With little effort (12) can be solved iteratively. With the calculated α_{ij} the function of the motion is known and the induced energy T_e follows

$$T_e = \alpha_{ij} V_{pij} \,. \tag{13}$$

With less effort the energy V_0 may be calculated. V_0 is the energy, that is induced by initial conditions z_0, $\dot{z}_0$ related to the equilibrium state z_G. With

$$\begin{aligned} \Delta z &= z_0 - z_G \\ \Delta \dot{z} &= \dot{z}_0 \end{aligned} \tag{14}$$

V_0 follows using (6) and (9)

$$V_0 = \Delta z^T \{ \tfrac{1}{2} A^L(z_G) + \tfrac{1}{6} A^N(\Delta z) \} \Delta z + \tfrac{1}{2} \Delta \dot{z}^T M \, \Delta \dot{z} \,.$$

NUMERICAL EXAMPLE

The application represent the method for impulsively loadings. For clarity the investigations are limited for a shallow arch. Further examples are given in /9/. For complex shell structures the method is totally analog if the discretization leads to a form similar to (6). If the strain energy describes the nonlinearities exactly, the method leads to an illimit exact judgement of the stability.

The geometry is given in figure 3. The load deflection behaviour for static loads is described in /7/. The stable state of equilibrium G is calculated for a symmetric load of $P = 200\,N$. For this state first the critical energy ΔV_{crit} is calculated with (8) and

(9). A sufficiently large perturbation of an antisymmetric load pattern, e.g. with the time distribution of figure 4 accelerates the structure beyond the separatrix such that loss of stability occurs. For small perturbations the structure moves in the prebuckling range around the equilibrium state G. The level of the initial load and the perturbation pattern in the time are arbitrary and do not influence the accuracy of the result.

The approximation of the state–vector (11) is done with the state vector Δz which describes the nontrivial solution of (8). The motion in time caused by the related energy T_e are calculated with (10). The values T_e is computed with a polynomial with eight terms. The comparison of critical energy and external energy shows that the error can be kept arbitrarily small. The convergence of the external energy T_e for different approximations in time is very well in comparison with a direct integration of the equations of motion.

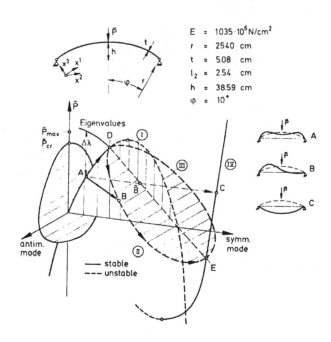

Fig. 3 :   Load deflection behaviour for static loading /7/

CONCLUSIONS

A method is given, which allows a simplified stability investigation of snapping and buckling structures under time–dependent loads. The degree of stability is calculated from the comparison of the load–induced energy and the critical energy stored in the structure using energy norms. The deformation state related to the critical energy is calculated from the condition of vanishing secant stiffness with respect to the initial load state of the structure. For different limit criteria it is possible to use other deformation state vectors without changing the method.

The results show that the loss of stability is not related to the induced impulse but to the work of the perturbating loads, in this case the induced energy.

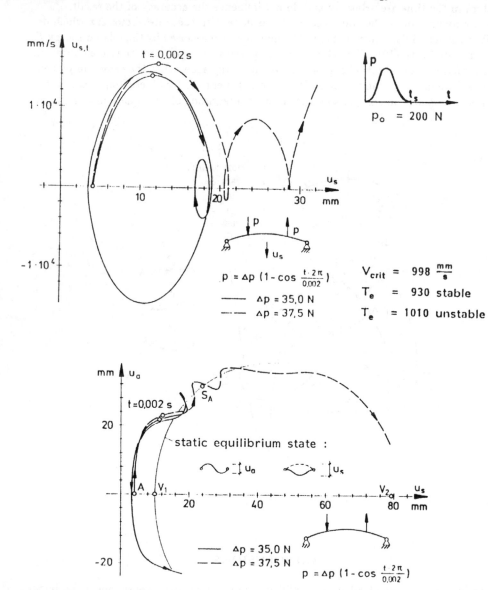

Fig. 4 :   Motion of a flat arch under antisymmetric impact loading

REFERENCES

/1/ Leipholz, H.: Stabilität elastischer Systeme. Karlsruhe: Verlag G. Braun 1980

/2/ Thompson, J. M. T.; Hunt, G. W.: A General Theory of Elastic Stability. London: John Wiley & Sons 1973

/3/ Hayashi, C.: Nonlinear Oscillations in Physical Systems. New York: Mc Graw–Hill Book Company 1964

/4/ Thompson, J. M. T.; Stewart, H. B.: Nonlinear Dynamics and Chaos. Chichester: John Wiley & Sons 1986

/5/ Koiter, W. T.: On the Thermodynamic Background of Elastic Stability Theory. Problems of Hydrodynamics and Continuum Mechanics. L. I. Sedov Anniversary Vol. SIAM. Philadelphia: 1969

/6/ Malkin, J. G.: Theorie der Stabilität einer Bewegung. München: Verlag R. Oldenburg 1959

/7/ Kröplin, B.H., Dinkler, D., Hillmann, J.: An energy perturbation applied to nonlinear structural analysis, in: Computer Methods in Applied Mechanics and Engineering 52 (1984) 885-897

/8/ Dinkler, D.: Stability of thin–walled elastic structures against finite perturbations, in : Structural Dynamics, W.Krätzig e.a. edtrs., A.A.Balkema, Rotterdam (1991)

/9/ Dinkler, D. and Kröplin, B.: Stability of Dynamically loaded Structures, in: Computational Mechanics of Nonlinear Behaviour of Shell Structures. Atluri Ed. (1989). Springer Verlag

/10/ Kröplin, B. and Dinkler, D.: An energy based concept of structural stability, in : European Conference on New Advances in Computational Structural Mechanics, P.Ladeveze, O.Zienkiewicz edtrs., Giens - France (1991)

/11/ Lanczos, C.: An Iteration Method for the Solution of the Eigenvalue Problem of linear Differential and Integral Operators. Journal res. nath. bur. Standards 45 (1950) 255 - 282

STABILITY ANALYSIS OF COMPOSITE SANDWICH SHELLS

László P. Kollár
Department of Reinforced Concrete Structures
Technical University of Budapest
Budapest, Hungary 1521

ABSTRACT

The system of differential equations that describes the buckling of shallow generally anisotopic composite sandwich shells is presented. First, the Donnell-equation of buckling is derived; then, neglecting the higher order terms, a system of linear differential equations on the basis of which the bifurcation load can be calculated. A closed form solution for the calculating the buckling load is also derived if the mid-surface of the shell can be described by a second order function.

INTRODUCTION

In the last two decades, many papers were published on the buckling of anisotropic plates and shells. Many authors took into account the effect of shearing deformation while others dealt with sandwich structures [1-11]. They investigated different shapes, boundary conditions, and made different restrictions regarding anisotropy.

This paper can be considered as the generalization of [7], where the author investigated the effect of shearing deformation on the buckling load of anisotropic shells. Our aim is to generalize these results for multi-layer sandwich shells, which are widely used in the aerospace industry. An important result of this work is that the shell can be generally anisotropic. The author will derive the Donnell-type equation of shallow sandwich shells, and also closed-form solutions for the determination of the bifurcation loads.

BASIC ASSUMPTIONS AND NOTATIONS

In the derivation, we assume that the structure behaves in a linearly elastic manner, and we use the assumptions of the shallow shell theory.

The investigated structure is a "sandwich with thick faces" [1], consisting of three layers: the two faces and the core (Fig.1). A perfect bonding between the layers is assumed. The faces are made of composite materials. Each can contain several fiber reinforced plies, which may be ordered in any sequence. The core, generally, is much less stiff than the faces. It can be honeycomb, foam etc. [1].

In the faces, the Kirchhoff-Love hypothesis is assumed (but not in the whole cross-section). On each face, the plies are replaced by one generally anisotropic layer, the stiffness of which can be calculated on the basis of the layup and on the material characteristics of the plies [10]. In the faces, we neglect the transverse shear deformation, and assume that the faces are incompressible perpendicular to the surface.

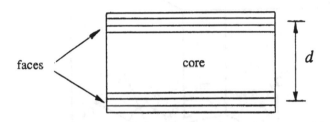

Figure 1. The build-up of the composite shell

The core is "antiplane" [1]: it is soft in its plane, and its transverse shear stiffness is finite. It is also considered incompressible perpendicular to the surface.

The results of the internal forces acting on a shell element are shown in Figure 2. The bending moments and shearing forces can be divided into two parts [1];

$$
\begin{aligned}
m_x &= m_{ox} + m_{\ell x} \\
m_y &= m_{oy} + m_{\ell y} \\
m_{xy} &= m_{oxy} + m_{\ell xy} \\
Q_x &= Q_{ox} + Q_{\ell x} \\
Q_y &= Q_{oy} + Q_{\ell y}
\end{aligned}
\tag{1}
$$

where the second terms are the forces and moments carried by the (local bending of the) upper (u) and lower faces (d):

$$m_{\ell x} = m_{\ell ux} + m_{\ell dx}$$
$$m_{\ell y} = m_{\ell uy} + m_{\ell dy}$$
$$m_{\ell xy} = m_{\ell uxy} + m_{\ell dxy} \qquad (2)$$
$$Q_{\ell x} = Q_{\ell ux} + Q_{\ell dx}$$
$$Q_{\ell y} = Q_{\ell uy} + Q_{\ell dy}$$

The membrane forces n_x, n_y, n_{xy} are carried by the faces only.

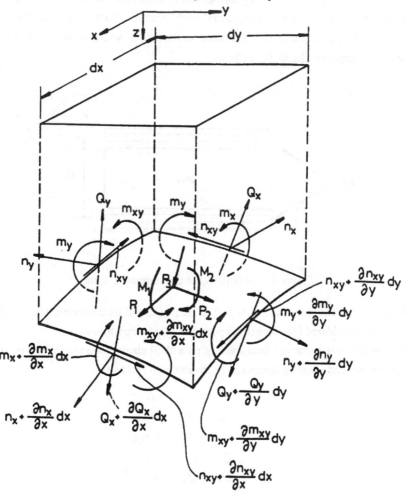

Figure 2. Internal forces on a shell element

The external loads acting on a shell element are the distributed forces p_1, p_2, p_3, and the distributed moments M_1, M_2 (Figure 2). It should be noted that the components of the load do not follow the co-ordinate axes x, y and z, but are tangents to the mid-surface in planes parallel to the coordinate planes (p_1, p_2, M_1, M_2) and normal to the surface (p_3).

The mid-surface of the shell is shallow. Consequently, we can use the assumptions of the shallow shell theory [3]. The function of the mid-surface is denoted by $z=z(x,y)$.

The displacements of an arbitrary point of the mid-surface are described by the components u, v, and w, which have the same directions as the load components p_1, p_2, p_3. We introduce the following functions:

$$\varphi_x = \frac{u_{ux} - u_{dx}}{d}, \qquad \varphi_y = \frac{u_{uy} - u_{dy}}{d} \tag{3}$$

where $u_{ux}, u_{dx}, v_{uy}, v_{dy}$ are the displacements of the upper and lower faces in the x and y directions. We can determine the displacements of an arbitrary point of the shell (which does not lie in the mid-surface) using these five functions (u, v, w, φ_x, φ_y) and their derivatives.

DIFFERENTIAL EQUATIONS OF SHALLOW SANDWICH SHELLS

We can write six equilibrium equations for each layer, which constitute in our case 18 equations. In fact the "sixth condition" (as Flügge [3] refers to the moment equilibrium about the axis perpendicular to the surface), is an immediate consequence of the reciprocity of the shear stresses. This relation, however, was used for the definitions of internal forces and so this equation becomes an identity. Six internal distributed forces arise, between the layers; eliminating them, we obtain nine equilibrium equations. Further, eliminating $Q_{\ell ux}$, $Q_{\ell dx}$, $Q_{\ell uy}$, and $Q_{\ell dy}$, from these nine equations we obtain five equations which can be written in the following matrix form:

$$p = \Theta \, n \tag{4}$$

where p and n are the vectors containing loads and internal forces. Their transposes are as follows

$$p^T = [p_1 \quad p_2 \quad p_3 \quad M_1 \quad M_2] \tag{5}$$
$$n^T = [n_x \quad n_y \quad n_{xy} \quad m_{ox} \quad m_{oy} \quad m_{oxy} \quad m_{\ell x} \quad m_{\ell y} \quad m_{\ell xy} \quad Q_{ox} \quad Q_{oy}]$$

and Θ is the following operator matrix

$$\Theta = \begin{bmatrix} -\dfrac{\partial}{\partial x} & 0 & -\dfrac{\partial}{\partial y} & 0 & 0 & 0 & 0 & 0 & 0 & 0 & 0 \\[2mm] 0 & -\dfrac{\partial}{\partial y} & -\dfrac{\partial}{\partial x} & 0 & 0 & 0 & 0 & 0 & 0 & 0 & 0 \\[2mm] -\dfrac{\partial^2 z}{\partial x^2} & -\dfrac{\partial^2 z}{\partial y^2} & -\dfrac{2\partial^2 z}{\partial x \partial y} & 0 & 0 & 0 & -\dfrac{\partial^2}{\partial x^2} & -\dfrac{\partial^2}{\partial y^2} & -\dfrac{2\partial^2}{\partial x \partial y} & -\dfrac{\partial}{\partial x} & -\dfrac{\partial}{\partial y} \\[2mm] 0 & 0 & 0 & \dfrac{\partial}{\partial x} & 0 & \dfrac{\partial}{\partial y} & 0 & 0 & 0 & -1 & 0 \\[2mm] 0 & 0 & 0 & 0 & \dfrac{\partial}{\partial y} & \dfrac{\partial}{\partial x} & 0 & 0 & 0 & 0 & -1 \end{bmatrix} \tag{6}$$

The deformations of the shell can be calculated from u, v, w, φ_x, φ_y as follows:

$$\varepsilon = \Theta^{*}\, u \tag{7}$$

where u and ε are the vectors of displacements and strains. Their transposes are as follows

$$u^{\mathrm{T}} = \begin{bmatrix} u & v & w & \varphi_x & \varphi_y \end{bmatrix} \tag{8}$$

$$\varepsilon^{\mathrm{T}} = \begin{bmatrix} \varepsilon_x & \varepsilon_y & \gamma_{xy} & \kappa_{ox} & \kappa_{oy} & \kappa_{oxy} & \kappa_{\ell x} & \kappa_{\ell y} & \kappa_{\ell xy} & \gamma_x & \gamma_y \end{bmatrix}$$

where ε_x, ε_y, γ_{xy} are the strains in the mid-surface, κ_x, κ_y, κ_{xy} are the changes in curvature of the mid-surface (which are equal to those of the faces), γ_x, γ_y are the shear strains, and the curvatures κ_{ox}, κ_{oy}, κ_{oxy} are defined by Eq.(7).

Θ^{*} is the adjoint of the operator matrix Θ. We obtain it from Θ in two steps, first we transpose Θ, then change the signs of the first order differential operators.

The relation between the internal forces and deformations can be expressed in the following form:

$$n = M\,\varepsilon \tag{9}$$

where M is the symmetrical stiffness matrix with constant elements:

$$M = \begin{bmatrix} T_{11} & T_{12} & T_{13} & C_{14} & C_{15} & C_{16} & C_{17} & C_{18} & C_{19} & 0 & 0 \\ T_{12} & T_{22} & T_{23} & C_{24} & C_{25} & C_{26} & C_{27} & C_{28} & C_{29} & 0 & 0 \\ T_{13} & T_{23} & T_{33} & C_{34} & C_{35} & C_{36} & C_{37} & C_{38} & C_{39} & 0 & 0 \\ C_{14} & C_{24} & C_{34} & B_{o11} & B_{o12} & B_{o13} & C_{47} & C_{48} & C_{49} & 0 & 0 \\ C_{15} & C_{25} & C_{35} & B_{o12} & B_{o22} & B_{o23} & C_{29} & C_{29} & C_{29} & 0 & 0 \\ C_{16} & C_{26} & C_{36} & B_{o13} & B_{o23} & B_{o33} & C_{29} & C_{29} & C_{29} & 0 & 0 \\ C_{17} & C_{27} & C_{37} & C_{47} & C_{57} & C_{67} & B_{\ell 11} & B_{\ell 12} & B_{\ell 13} & 0 & 0 \\ C_{18} & C_{28} & C_{38} & C_{48} & C_{58} & C_{68} & B_{\ell 12} & B_{\ell 22} & B_{\ell 23} & 0 & 0 \\ C_{19} & C_{29} & C_{39} & C_{49} & C_{59} & C_{69} & B_{\ell 13} & B_{\ell 23} & B_{\ell 33} & 0 & 0 \\ 0 & 0 & 0 & 0 & 0 & 0 & 0 & 0 & 0 & S_{11} & S_{12} \\ 0 & 0 & 0 & 0 & 0 & 0 & 0 & 0 & 0 & S_{12} & S_{22} \end{bmatrix} \tag{10}$$

and T, C, B, S are used to distinguish between tensile, coupling, bending, and shear stiffness components.

The vectors of internal forces n, deformations ε, and displacements u contain 27 unknown functions. The equations of equilibrium Eq.(4), strain displacement relations Eq.(7), and constitutive relations Eq.(10) constitute 27 equations. Eliminating the internal forces and deformations, we obtain the following equation:

$$v = \Theta \, M \, \Theta^* \, u \qquad (11)$$

GOVERNING EQUATIONS OF BUCKLING

For the investigation of stability, the change in geometry has to be taken into account.

The vector of internal forces before buckling is denoted by n_D, and the mid-surface of the shell is given by the function z_D which contains the deformations due to n_D as well. The displacements, deformations, and internal forces during buckling are denoted by u_B, ε_B and n_B respectively.

The shell is shallow, and consequently, the function of the mid-surface of the shell can be written in the following way:

$$z = z_D + w_B \qquad (12)$$

where w_B is the third element of u_B.

Operator matrices Θ and Θ^* depend on z. They can be split into two parts as follows:

$$\Theta = \Theta_1 + \Theta_2(z) \qquad \Theta^* = \Theta_1^* + \Theta_2^*(z) \qquad (13)$$

where

$$\Theta_2(z) = \begin{bmatrix} 0 & 0 & 0 & 0 & 0 & 0 & 0 & 0 & 0 & 0 & 0 \\ 0 & 0 & 0 & 0 & 0 & 0 & 0 & 0 & 0 & 0 & 0 \\ -\dfrac{\partial^2 z}{\partial x^2} & -\dfrac{\partial^2 z}{\partial y^2} & -\dfrac{2\partial^2 z}{\partial x \partial y} & 0 & 0 & 0 & 0 & 0 & 0 & 0 & 0 \\ 0 & 0 & 0 & 0 & 0 & 0 & 0 & 0 & 0 & 0 & 0 \\ 0 & 0 & 0 & 0 & 0 & 0 & 0 & 0 & 0 & 0 & 0 \end{bmatrix} \qquad (14)$$

and $\Theta_2^*(z)$ is the transpose of $\Theta_2(z)$.

The internal forces in the shell are

$$n = n_D + n_B = n_D + M\left[\Theta_1^* + \Theta_2^*(z_D + w_B)\right] u_B \qquad (15)$$

Introducing Eq.(12) and Eq.(15) into Eq.(4), and making use of the linearity of operator Θ_2, we obtain

$$p = [\Theta_1 + \Theta_2(z_D)]n_D + \Theta_2(w_B)n_D + \Theta M \Theta^* u_B \qquad (16)$$

The shell is in equilibrium before buckling, hence the left hand side of Eq.(16) equals to the first term on the right hand side. Cancelling these terms, and replacing the second on the right hand side using the equality: $\Theta_2(w_B)n_D = \Theta_3 u_B$, after algebraic manipulation, we obtain

$$\{\Theta_3 + [\Theta_1 + \Theta_2(z_D)]M[\Theta_1^* + \Theta_2^*(z_D)] + \Theta_2(w_B)M[\Theta_1^* + \Theta_2^*(z_D)] +$$
$$+[\Theta_1 + \Theta_2(z_D)]M \Theta_2^*(w_B) + \Theta_2(w_B)M \Theta_2^*(w_B)\} u_B = 0 \qquad (17)$$

where

$$\Theta_3 = \begin{bmatrix} 0 & 0 & 0 & 0 & 0 \\ 0 & 0 & 0 & 0 & 0 \\ 0 & 0 & -n_{xD}\dfrac{\partial^2}{\partial x^2} - n_{yD}\dfrac{\partial^2}{\partial y^2} - n_{xyD}\dfrac{2\partial^2}{\partial x \partial y} & 0 & 0 \\ 0 & 0 & 0 & 0 & 0 \\ 0 & 0 & 0 & 0 & 0 \end{bmatrix} \qquad (18)$$

and n_{xD}, n_{yD}, n_{xyD}, are the elements of n_D. Equation (17) is a Donnell-type [6] equation for the equilibrium of shallow sandwich shells.

The last three terms in Eq.(17) contain the second and third power terms of the displacements, which are negligible in the analysis of the bifurcation load. Omitting these, we obtain a homogeneous linear differential equation system from Eq.(17)

$$\{\Theta_3 + [\Theta_1 + \Theta_2(z_D)]M[\Theta_1^* + \Theta_2^*(z_D)]\} u_B = 0 \qquad (19)$$

From this equation the bifurcation load can be calculated, but generally only numerically.

BUCKLING LOAD FOR SHELLS WITH SECOND ORDER MID-SURFACE

We can obtain an analytical solution if (i) the mid-surface can be described by a second order function, i.e.

$$z_D = -\frac{1}{2 R_x} x^2 - \frac{1}{2 R_y} y^2 - \frac{1}{2 R_{xy}} xy \qquad (20)$$

and (ii) the membrane forces n_{xD}, n_{yD}, n_{xyD} are constant; because then the coefficients of the differential equation system Eq.(19) become constant. We assume the displacement functions iare of the following form

$$
\begin{aligned}
u_B &= u_1 \cos \alpha(x + c_1 y) \sin \beta(y + c_2 x) - u_2 \sin \alpha(x + c_1 y) \cos \beta(y + c_2 x) \\
v_B &= v_1 \sin \alpha(x + c_1 y) \cos \beta(y + c_2 x) - v_2 \cos \alpha(x + c_1 y) \sin \beta(y + c_2 x) \\
w_B &= w_1 \sin \alpha(x + c_1 y) \sin \beta(y + c_2 x) + w_2 \cos \alpha(x + c_1 y) \cos \beta(y + c_2 x) \quad (21) \\
\varphi_{Bx} &= \varphi_1 \cos \alpha(x + c_1 y) \sin \beta(y + c_2 x) - \varphi_2 \sin \alpha(x + c_1 y) \cos \beta(y + c_2 x) \\
\varphi_{By} &= \varphi_1 \sin \alpha(x + c_1 y) \cos \beta(y + c_2 x) - \varphi_2 \cos \alpha(x + c_1 y) \sin \beta(y + c_2 x)
\end{aligned}
$$

This buckling shape is a double trigonometric function in a skew coordinate system. α, β, c_1, c_2 are unknown constants. These functions may be introduced into Eq.(19). After algebraic manipulation we obtain:

$$
\left\{ \begin{bmatrix} A & C \\ C & A \end{bmatrix} R \begin{bmatrix} M_0 & M_n \\ M_n & M_0 \end{bmatrix} R \begin{bmatrix} A^T & C^T \\ C^T & A^T \end{bmatrix} - \begin{bmatrix} \lambda_1 B & \lambda_2 B \\ \lambda_2 B & \lambda_1 B \end{bmatrix} \right\} U = 0 \quad (22)
$$

where

$$
A = \begin{bmatrix}
-\alpha & 0 & \beta & 0 & 0 & 0 & 0 & 0 & 0 & 0 & 0 \\
0 & -\beta & \alpha & 0 & 0 & 0 & 0 & 0 & 0 & 0 & 0 \\
\dfrac{1}{R_x} & \dfrac{1}{R_y} & 0 & 0 & 0 & 0 & \alpha^2 + \beta^2 c_2^2 & \beta^2 + \alpha^2 c_1^2 & -2\alpha\beta(1 + c_1 c_2) & \alpha & \beta \\
0 & 0 & 0 & \alpha & 0 & -\beta & 0 & 0 & 0 & -1 & 0 \\
0 & 0 & 0 & 0 & \beta & -\alpha & 0 & 0 & 0 & 0 & -1
\end{bmatrix} \quad (23)
$$

$$
C = \begin{bmatrix}
c_2\beta & 0 & -c_1\alpha & 0 & 0 & 0 & 0 & 0 & 0 & 0 & 0 \\
0 & c_1\alpha & -c_2\beta & 0 & 0 & 0 & 0 & 0 & 0 & 0 & 0 \\
0 & 0 & \dfrac{1}{R_{xy}} & 0 & 0 & 0 & -2\alpha\beta c_2 & -2\alpha\beta c_1 & c_1\alpha^2 + c_2\beta^2 & -c_2\beta & -c_1\alpha \\
0 & 0 & 0 & -c_2\beta & 0 & c_1\alpha & 0 & 0 & 0 & 0 & 0 \\
0 & 0 & 0 & 0 & -c_1\alpha & c_2\beta & 0 & 0 & 0 & 0 & 0
\end{bmatrix}
$$

$$
B = \begin{bmatrix} 0 & & & & \\ & 0 & & & \\ & & 1 & & \\ & & & 0 & \\ & & & & 0 \end{bmatrix} \qquad R = \begin{bmatrix} 1 & & & & & \\ & 1 & & & & \\ & & \ddots & & & \\ & & & 1 & & \\ & & & & -1 & \\ & & & & & -1 \end{bmatrix} \quad (24)
$$

where M_0 and M_n contain elements of the stiffness matrix, according to the following scheme

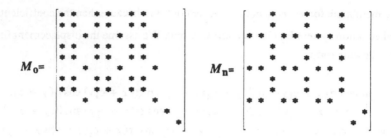

provided that $M = M_0 + M_n$, and where

$$\lambda_1 = -n_{xD}\left(\alpha^2 + c_2^2\beta^2\right) - n_{xyD}\left(c_1\alpha^2 + c_2\beta^2\right) - n_{yD}\left(c_1^2\alpha^2 + \beta^2\right)$$
$$\lambda_2 = 2\alpha\beta\left[n_{xD}c_2 + n_{xyD}\left(1 + c_1c_2\right) + n_{yD}c_1\right] \tag{25}$$

$$U^T = \begin{bmatrix} u_1 & v_1 & w_1 & \varphi_{x1} & \varphi_{y1} & u_2 & v_2 & w_2 & \varphi_{x2} & \varphi_{y2} \end{bmatrix} \tag{26}$$

We can obtain the buckling load from the condition of singularity of the coefficient matrix in Eq.(22):

$$\det\left\{ \begin{bmatrix} A & C \\ C & A \end{bmatrix} R \begin{bmatrix} M_0 & M_n \\ M_n & M_0 \end{bmatrix} R \begin{bmatrix} A^T & C^T \\ C^T & A^T \end{bmatrix} - \begin{bmatrix} \lambda_1 B & \lambda_2 B \\ \lambda_2 B & \lambda_1 B \end{bmatrix} \right\} = 0 \tag{27}$$

The buckling load calculated from Eq.(27) depends on parameters α, β, c_1, c_2. For the determination of the buckling load it has to be minimized with respect to these four parameters.

CONCLUSIONS

An algebraic equation (Eq.27) was derived, from which the buckling load of shallow sandwich shells can be calculated. This equation can be used for the investigation of local buckling of shallow shells [6]. In addition, it satisfies the boundary conditions in several cases. As an example, the stability analysis of (orthotropic or) generally anisotropic long sandwich cylinders and plates may be considered. The results can be applied to a wide range of cases, many more than are even considered in the literature [see 1, 2, 4, 5, 6, 7, 11 and their references]. These can be obtained as the special cases of our general result.

REFERENCES

1. Allen, H. G., Analysis and Design of Structural Sandwich Panels. Pergamon Press, Oxford, 1969.

2. Dulácska, E., Vibration and Stability of Anisotropic Shallow Shells. Acta Techn. Hung., 1969, **65,** 225-260.

3. Flügge, W., Stresses in Shells. 2nd edition. Springer, Berlin etc. 1973.

4. Hegedüs, I., Buckling of Axially compressed Cylindrical Sandwich Shells. Acta Techn. Hung., 1979, **89,** 377-387.

5. Jones, R. M. and Morgan, H. S., Buckling and Vibration of Cross-Ply Laminated Circular Cylindrical Shells. AIAA Journal, 1975, **13,** 664-671.

6. Kollár, L. and Dulácska, E., Buckling of Shells for Engineers. J. Wiley and Sons, Chichester, 1984.

7. Kollár, L. P., Buckling of Generally Anisotropic Shallow Shells with Transverse Shear Deformation. Int. Coll. Stability of Steel Structures, Hungary, (1990), Preliminary Report. Vol. III. 175-182.

8. Rao, K. M., Buckling Coefficients for Fiber-reinforced Plastic-Faced Sandwich Plates under Combined Loading. AIAA Journal, 1987. **25,** 733-739.

9. Stein, M., Nonlinear Theory for Plates and Shells Including the Effect of Transverse Shear. AIAA Journal, 1986. **24,** 1537-1544.

10. Tsai, W. S. and Hahn, H. T., Introduction to Composite Materials. Technomic, Lancaster, 1980.

11. Vinson, J. and Sierakowsky R. L., The Behaviour of Structures Composed of Composite Materials. Martinus Nijhoff Publishers, Dordrecht, 1986.

REMARKS ON THE EFFECTIVE-WIDTH CONCEPT FOR ORTHOTROPIC THIN SHELLS

MEYER-PIENING, H.-R.
Prof. Dr., Inst. f. Lightweight Struct. and Ropeways, ETH Zürich, Switzerland

GEIER, B., ROHWER, K.
Dr., DLR Braunschweig, Inst f. Strukturmechanik, Germany

ABSTRACT

Thin-walled stiffened panels tend to buckle under compressive and/or shear loads with resulting redistribution of stresses in the structure. Provided that the stiffeners are capable of sustaining the additional load, such conditions are acceptable in many structures. Since a complete postbuckling analysis for each state of loading is prohibitive, simple approximations are in use. With the appearance of orthotropic panels it is deemed desirable to extend such approximations to panels with orthotropic properties. A short review of available recommendations is given and a one term approach is used to demonstrate the effect of orthotropy. Further, recent panel test results from ETH Zürich are reviewed accordingly and two different FE-based postbuckling analyses (postbuckling at Zürich and initial postbuckling at Braunschwei) are performed to provide confirmation that the quality of a crude and simplifying formula is acceptable.

INTRODUCTION

Thin-walled technical structures such as aircraft consist mainly of thin plate- or shell-like sheet material and stiffeners. If these structures are subjected to in-plane compression and/or in-plane shear loads, they tend to assume an out-of-plane postbuckling pattern, following a bifurcation of equilibrium if the sheet material retains a geometric plane configuration, (no inititial imperfection) prior to such bifurcation load. This assumes that the stability of the stiffeners is not affected by the buckling of the panel.

Since actual structures are rarely manufactured to geometries which have sufficiently small imperfections to yield close agreement between theoretical buckling stress and the stress level at which the onset of buckling is detected, nonlinear analyses are performed to study the behaviour of such panels at stress levels around the theoretical buckling load. The smaller the radius of curvature of the panel relative to the overall dimensions in a plane perpendicular to the compression load the higher are the stress levels at buckling and the more violent is the snap-through into the post-buckled state.

For panels with shallow or zero curvature the onset of buckling is hardly detected and in many experiments obtained primarily by interpolating lateral deflection curves or measurements of bending strains if the strain gauges happen to be located at significant positions. It can be deduced that buckling of stiffened flat or shallow curved panels does not necessarily present a hazard to safe structural performance, and a large number of airplane panels do occasionally deformation and accounts for it then such structural behaviour may be considered permissible.

A consequence of the out-of-plane deflection is that the in-plane stress distribution in the sheet material is affected due to the fact that part of the overall shortening between the loaded boundaries is related to the deviation from the initially (almost) straight configuration while adjacent to the stable stiffeners the panel has to react primarily by inplane strain. As a result the assembly assumes a reduced stiffness, and the panel next to the stiffeners experiences higher than average inplane stress.

It is not highly desirable to analyze hundreds of thin-walled panels individually by nonlinear analyses, and simple design rules are, therefore, in use, - for pure shear the semi-tension fields and for uniaxial compression the concept of effective width. In the latter case the width of the sheet material adjacent to the stiffener is determined in such a way that at the design loads the resulting maximum axial *stress* in the panel at the attachments of the stiffeners assumes the correct value. Rough estimates are that this width is in the order of 20 to 30 times the wall thickness of the sheet for flat *isotropic* panels. There are expected to be different values for the effective width to yield either sufficiently correct values of maximum stress level ($w_b^{(\sigma)}$) or of structural stiffness ($w_b^{(\varepsilon)}$) or structural differential stiffness ($w_b^{(\Delta\varepsilon)}$).

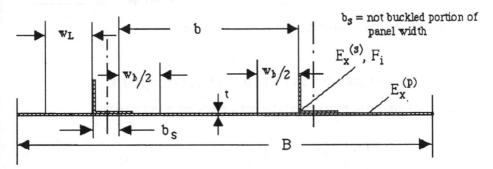

Figure 1. Nomenclature and Definitions

The stress-related effective width $w_b^{(\sigma)}$ has been frequently referred to in the literature and the most significant recommendations for isotropic panels are compiled from Ref. [1] to [23] in Fig. 2. Note that within this paper the effective width includes *both* edges of a supported panel as can be seen from Fig. 1, and is, consequently, twice the size of most commonly used values. It can be seen from Fig. 2 that even for isotropic panels the structure analyst has a wide choice to select from. Most authors declare that the results agree well with test results.

Thin structures implementing *fibre reinforced* material need also be designed to allow for some local buckling in order to avoid the need for sandwich type stiffening or too many layers or too close spacing of stiffeners. To avoid bond failures or delamination such panels might primarily remain in an early postbuckling condition at limit loads. It is, therefore, desirable to describe the behaviour around the buckling state in a sufficiently precise but simple way, accounting for non-isotropic panel properties to obtain design stress levels in the panel and in the stringers as well as overall stiffness reduction values. As the overall stiffness value will dictate the gross distribution of loads and the stiffness of the whole structure at that load level the reiteration of the process to obtain an improved information on the maximum stress level may remain desirable. For the effective width concept, two or three values of effective sheet width may have to be applied ($w_b^{(\sigma)}$), ($w_b^{(\varepsilon)}$) and ($w_b^{(\Delta\varepsilon)}$) for equivalence of stress, overall stiffness and differential stiffness at a given load level.

In the following, mathematical expressions are derived for the stiffness related effective width for orthotropic panels ($w_b^{(\varepsilon)}$) and ($w_b^{(\Delta\varepsilon)}$) and, later, the stress related effective width ($w_b^{(\sigma)}$) is treated for the simplifying case of uniaxial stress ($N_y = 0$).

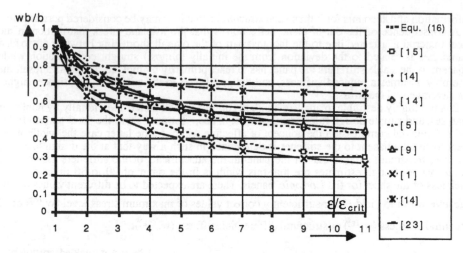

Figure 2. Variation of $w_b^{(\sigma)}$ with $\varepsilon/\varepsilon_{crit}$; Comparison of Theories

STIFFNESS RELATED EFFECTIVE WIDTH

The panel is assumed to be represented in the linear elastic range by the relation

$$\left\{ \begin{matrix} N \\ M \end{matrix} \right\} = \begin{bmatrix} A^{-1} & K \\ K^T & D \end{bmatrix} * \left\{ \begin{matrix} \varepsilon \\ \kappa \end{matrix} \right\} \tag{1}$$

In the following, the coupling terms K are assumed to be of negligeable quantity.
  In a recent thesis at the ETH Zürich, Müller [24] has used the equilibrium equation in the form:

$$D_{11}w_{xxxx} + 2(D_{12} + 2D_{33})w_{xxyy} + D_{22}w_{yyyy} = \Phi_{yy}w_{xx} + \Phi_{xx}w_{yy} - 2\Phi_{xy}w_{xy} \tag{2}$$

and the compatibility relation

$$A_{11}*\Phi_{yyyy} + (2*A_{12} + A_{33})*\Phi_{xxyy} + A_{22}*\Phi_{xxxx} = w^2_{xy} - w_{xx}*w_{yy} \tag{3}$$

  For simplicity reason, no complicated stress boundary conditions are presently implemented. With "a" being the panel length and "b" the width of the loaded edge, a one-term approximation for the deflection w is:

$$w = w_{mn} * \sin\frac{m*\pi*x}{a} * \sin\frac{n*\pi*y}{b} \tag{4}$$

the differential equation (3) is satisfied for the stress function

$$\Phi = \frac{N_x*y^2}{2} + \frac{w_{mn}^2*b^2*m^2}{32a^2A_{11}*n^2}*\cos\frac{2n\pi y}{b} + \frac{w_{mn}^2*a^2*n^2}{32b^2A_{22}*m^2}*\cos\frac{2m\pi x}{a} \tag{5}$$

and the amplitude for the deflection pattern is obtained from Equ. (2) by Galerkin's method

$$w_{mn}^2 = - \frac{16\, a^4 b^4 A_{11} A_{22}}{A_{11} a^4 * n^4 + A_{22} b^4 * m^4} * \left(\frac{D_{11} * m^4}{a^4} + \frac{2(D_{12} + 2 * D_{33}) * m^2 * n^2}{a^2 b^2} + \frac{D_{22} * n^4}{b^4} + \frac{N_x * m^2}{\pi^2 a^2} \right) \quad (6)$$

It can be realized that there is a real value for w_{mn} only if the compression load is greater than a critical value. Again it is stated that this is so far only a one term approach and that special stress boundaries in x and y are neglected; they would require the superposition of homogeneous solutions of the compatibility condition. Of course, the axial and circumferential half wave numbers m and n are selected as to minimize the corresponding average buckling line load $N_{x,crit.}$.

The axial strain follows from

$$\varepsilon_x = A_{11} * N_x + A_{12} * N_y \quad (7)$$

or after integration over y for Ny = 0

$$\frac{1}{b} \int_0^b \frac{\partial u}{\partial x}\, dy = A_{11} * N_x - \frac{1}{2b} * \int_0^b \left(\frac{\partial w}{\partial x} \right)^2 dy \quad (8)$$

The average shortening (δ) in load direction prior to buckling ($w_{mn} = 0$) is obtained from integration of Equ. (7) over x:

$$\delta = -u = - N_x * a * A_{11} \quad (9)$$

and beyond buckling from Equ. (8)

$$\delta = -u = -N_x * a * A_{11} + \frac{w_{mn}^2 * m^2 * \pi^2}{8a} \quad (10)$$

In order to determine a value for the effective width on the basis of equivalent *strain*, the relation for the unbuckled state is used

$$\varepsilon_{x, y=0, b} = A_{11} * N_x + A_{12} * N_y = u/a \quad (11)$$

Assuming that $N_y = 0$ and $N_x < 0$ for compression it follows from the definion of Φ

$$\phi_{yy\,(y=0,b)} = \frac{1}{A_{11}} * \frac{u}{a} = \frac{N_x}{w_b} * b \quad (12)$$

or, finally,

$$w_b^{(\varepsilon)} = - N_x * \frac{ab A_{11}}{\delta} \; ; \; w_b^{(\Delta\varepsilon)} = -\frac{\partial N_x}{\partial \delta} * a * b * A_{11} \quad (13)$$

Prior to buckling, the effective width $w_b^{(\varepsilon)}$ is equal to b. After buckling the above value is used.

It is understood that before implementing the panel stiffness into an overall analysis, the stringer stiffness is to be added to include its participation to the net load.

The assumption of $N_y = 0$ may be replaced by other criteria, e.g. $\varepsilon_y = 0$ or by a more correct evaluation of the local stress field on the basis of refined stress boundary conditions.

It can be realized that the effective width according to the above equations is generally load dependent, i.e., the panel loses effective stiffness as the amplitude of the buckles increase. However, if dynamic motion of the structure around a state of stress is considered - like flutter in a highly stressed flight condition - the tangent of the load deflection curve becomes of interest. After substitution of the expression for w_{mn} according to equ. (6) as derived for only one term in the deflection pattern one finds that w_{mn}^2 is proportional to N_x and the slope of the postbuckling curve becomes constant as does the effective width $w_b^{(\Delta\varepsilon)}$. This has also been pointed out by Stein [22] and is primarily a consequence of the crude approach as demonstrated by Levy [8]. Denoting this value as a differential stiffness the following relation is found

$$-\frac{\partial N_x}{\partial \delta} = \frac{A_{11}*n^4*a^4 + A_{22}*m^4*b^4}{aA_{11}*\left(A_{11}n^4a^4 + 3A_{22}m^4b^4\right)} \quad (14)$$

or

$$w_b^{(\Delta\varepsilon)}/b = \frac{A_{11}*n^4*a^4 + A_{22}*m^4*b^4}{\left(A_{11}n^4a^4 + 3A_{22}m^4b^4\right)} \quad (15)$$

More terms in the initial approximation yield somewhat smaller values for $w_b^{(\Delta\varepsilon)}$, where the superscript (Δ) indicates the differential or dynamic nature of the stiffness relation. This value is independent of $\varepsilon/\varepsilon_{cr}$ and shown in Fig. 3 for a variation of A_{22}/A_{11} and some aspect ratios a/b, which still are assumed to correspond to m = n = 1.

It may be interesting to note that the differential postbuckling stiffness can be evaluated by W. T. Koiter's theory of the initial postbuckling behaviour. The computer program based on that theory was set up by the second author [25]. Application to the test example of Fig. 5 showed good agreement with FE-based and experimental results as provided in this paper. The advantage of the indicated method has to be seen in its great flexibility with respect to panel geometry (e.g. curvature), boundary conditions and load cases.

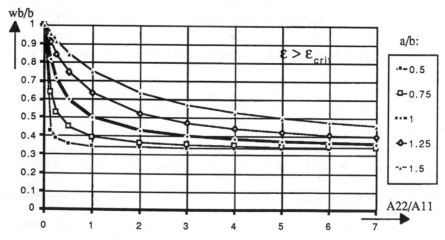

Figure 3. Variation of *differential stiffness* related $w_b^{(\Delta\varepsilon)}$ with E_x/E_y resp. A_{22}/A_{11}, Equ. (15)

STRESS RELATED EFFECTIVE WIDTH

Considering only one rectangular panel with hinged edges, the effective width of this panel is the remaining width of an unbuckled portion of the panel which has a constant in-plane stress field and the same maximum axial *stress* component within the effective width as the buckled panel underneath the stringer or at the edge support.

The result is identical to $w_b^{(\varepsilon)}$ since the correct overall stiffness will yield a correct strain in the unbuckled zone next to the stringers and, consequently, the correct axial stress if no other stress components, i.e. N_y, are present.

The derivation on the basis of the stress field suffers from the fact that the stress field is not as realistic as the deflection pattern due to the before mentioned mismatch of stress boundary conditions, i.e., just in the zones where the stress value is to be evaluated. Taking the stress value from the stress function at the boundaries $y = 0$, b and relating it to the buckling line load $N_{x,\ crit}$ as obtained from Equ. (6), one finds

$$w_b^{(\sigma)}/b = \frac{\alpha*\left(A_{11}a^4n^4 + A_{22}b^4m^4 \right)}{\alpha*\left(A_{11}a^4n^4 + 3A_{22}b^4m^4 \right) - 2A_{22}b^4m^4} \tag{16}$$

with $\alpha = \varepsilon_x / \varepsilon_{xkrit.}$ or $N_x/N_{x,\ crit.}$

Fig. 4 shows the variation of the effective width $w_b^{(\sigma)}$ with the ratio $\varepsilon_x / \varepsilon_{x,crit}$ for some ratios E_x/E_y, or, A_{22}/A_{11} and aspect ratio $a/b = 1$, implementing the relation $A_{11} = 1/(E^{(p)}_x*t)$

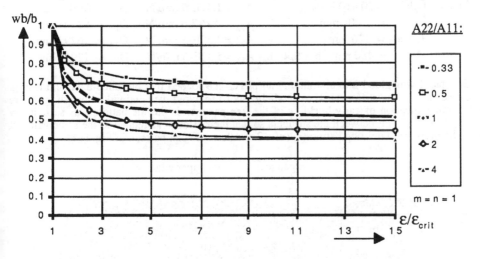

Figure 4. Variation of *stress related* $w_b^{(\sigma)}$ with E_x/E_y resp. A_{22}/A_{11}, Equ. (16)

EVALUATION OF CFR THERMOPLAST PANEL TEST RESULTS

The test panel as tested at the ETH Zürich consisted of a flat panel with four angle stiffeners arranged in loading direction, as shown in Fig. 5.

The basic properties are

Stringer: $E_x^{(s)}$ = 44 278 MPa $0 < b_s < \approx 30$ mm (see Fig. 1),

          F_i = 86.5 mm²; (30+30)*1.44   subsequently selected as 30 mm

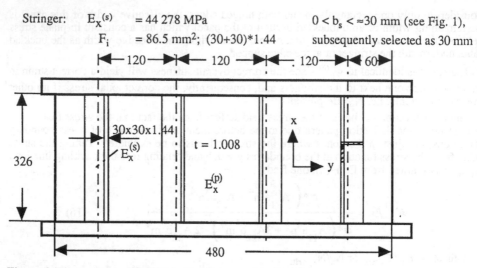

Figure 5. Dimensions of CFR Test Panel as tested in Zürich

Panel: $E_x^{(p)}$ = 70 410 MPa A_{11} = 1.41E-5 mm/N D_{11} = 7207 Nmm

        t   = 1.008 mm A_{12} = -826E-6 mm/N D_{12} = 885 Nmm

        L = a = 326 mm A_{22} = 4.59E-5 mm/N D_{22} = 1679 Nmm

        B   = 480 mm; b = B/4 - b_s A_{33} = 5.32E-5 mm/N D_{33} = 1292 Nmm

The panel was glued into aluminum end pieces and loaded by controlled axial shortening. The free ends of the stiffeners were parallely guided to avoid their individual buckling. There was some indication of early buckling of the free panel ends but the pattern was sufficiently equal to the buckling between the stringers so that their effect was related to the same value of w_b. The measured buckling pattern(s) (m = 2 and 3, n = 1) can be seen in Fig. 6.

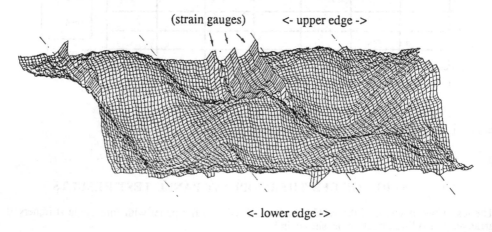

Figure 6. Post-Buckling Pattern for CFR Panel Test (3D Optical Measurement)

The individual effective widths are determined with the following assumptions:

* The skin section of width b_s being directly adjacent to the stiffener is considered to be part of the stiffener since this portion remains unbuckled.
* Adjacent to each side of the stiffener/skin area the effective width portion is denoted by $w_b/2$, while b_s has still to be selected
* At the two outer free (unsupported) skin sections the effective width is denoted by w_L, for simplicity assumed to be $\approx w_b/2$ subsequently.

Consequently, the total effective width of the test panel is $w_{b\ total} \approx 6 * w_b/2 + 2 * w_L \approx 4*w_b$

Stiffness Related Effective Widths $w_b^{(\varepsilon)}$, $w_b^{(\Delta\varepsilon)}$

Total stiffness of the *unbuckled* stiffened set-up is calculated to be

$$E*F = 4 * F_i * E_x^{(s)} + B * t * E_x^{(p)}$$
$$= 4 * 86.5 * 44\ 278 + 480 * 1.008 * 70\ 410 = 49\ 387\ 362\ N$$

The test results yield approximtaely (see Fig. 7):

$$E*F = P*a/\delta = 2\ 003 * 326 / .0141 = 46\ 313\ 496\ N$$

for small loads, corresponding to an initial (steepest) slope of the test curve after alignment.

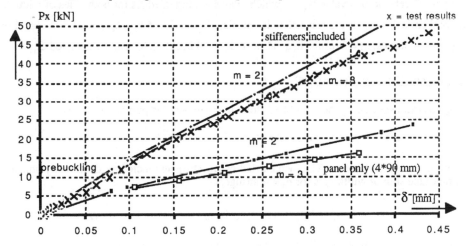

Figure 7. Load-Shortening Curve for Stiffened CFR Panel, Equ. (10) and Test Result

The experimental overall buckling load is between 11 (partially) and 14 kN, see Fig. 7.

At loads beyond buckling the load shortening curve displays a reduced slope which is accounted for by reducing the effective width accordingly: (Here $b_s = 30$mm and $w_L = .5*w_b$.) From the recorded slopes of the test curve the effective width values are calculated:

$$(E*F)^{(\varepsilon)} = (E_x^{(s)} *F_i + E_x^{(p)}*b_s*t)*4 + (6 * w_b^{(\varepsilon)}/2 + 2 * w_L^{(\varepsilon)})* t * E_x^{(p)} = (P*\delta/a)_{(\varepsilon)}$$

or for differential stiffness

$$(E*F)^{(\Delta\varepsilon)} = (E_x^{(s)} *F_i + E_x^{(p)}*b_s*t)*4 + (6 * w_b^{(\Delta\varepsilon)}/2 + 2 * w_L^{(\Delta\varepsilon)})* t * E_x^{(p)} = (\partial P/\partial\varepsilon)_{(\varepsilon)}$$

$$w_b^{(\varepsilon)}/b = \frac{1}{4*t*b*E_x^{(p)}} * \left\{ (P*a/\delta)_{(\varepsilon)} - (E_x^{(s)} * F_i + E_x^{(p)} *b_s*t)*4 \right\} \quad (17)$$

or for differential stiffness, respectively,

$$w_b^{(\Delta\varepsilon)}/b = \frac{1}{4*t*b*E_x^{(p)}} * \left\{ (\partial P/\partial\varepsilon)_{(\varepsilon)} - (E_x^{(s)} * F_i + E_x^{(p)} *b_s*t)*4 \right\} \quad (18)$$

The evaluation of Equ. (18) is contained in Fig. 8.

The average buckling line load $N_{x, \text{crit.}}$ for an individual panel is obtained from Equ. (6)

$$- N_{x, \text{crit.}} = \frac{\pi^2 a^2}{m^2} * \left(\frac{D_{11} m^4}{a^4} + \frac{2(D_{12}+2D_{33})m^2 n^2}{a^2 b^2} + \frac{D_{22} n^4}{b^4} \right) \quad (19)$$

$$= \underline{17.46 \text{ N/mm}} \text{ for width b } (b_s = 30 \text{ mm, no load-carrying stiffners)}.$$

The lowest buckling load was found for m = 3 and n = 1. The loading ratio or strain ratio in the diagrams refer to this value $N_{x, \text{crit}}$ which relates to hinged edges for both sides of width b. The total critical load for the test article is calculated to be 12.15 kN, $\delta_{\text{crit}} = 0.0802$ mm. For the given properties, m = 2 results in an almost identical buckling load and associated δ_{crit}.

Stress Related Effective Width $w_b^{(\sigma)}$

At a given axial shortening the associated stress in the unbuckled part of the skin is

$$\sigma^{(p)} = \varepsilon_x * E_x^{(p)} = -\delta * E_x^{(p)}/a$$

(assuming zero lateral stress). The load carried by the stiffeners including the directly attached portion of the skin of width b_s is for $n_s = 4$ stringers

$$P^{(sp)} = \delta *n_s*(E_x^{(s)}*F_i+E_x^{(p)}*b_s*t)/a \quad (20)$$

Hence, the load carried by an effective total panel width (for $w_L = w_b/2$)

$$w_b^{(\sigma)}/b = (P-P^{(sp)})*a/(\delta *n_s* E_x^{(p)} *b* t) \quad (21)$$

thus being related to the portion b of the individual panel width which is assumed to buckle.

The variation of the values of the effective widths as obtained from the test is presented in Fig. 8. The corresponding half wave numbers were m = 2 or 3 and n = 1. It can be seen that for the calculation of the differential stiffness at small values $\delta/\delta_{\text{crit}}$ a considerably smaller effective width is to be considered than for the evaluation of the maximum stress in the structure. The theoretical load-shortening curve for this test panel as shown in Fig. 7 is calculated with the above value of $w_b^{(\sigma)}$ / b and reasonable fit with FE and test results was found for $b_s = 30$ mm ($\approx$ all of the bonded stiffener leg). w_L appears to be smaller than $w_b/2$, see also [23]. As m = 2 and m = 3 appear simul-taneously in the test, both related analytical results are indicated in Fig. 8, see also Fig.9. Further indicated are results as given by Shanmugam [26] if applied for given properties, L/b = 1 and initial imperfection A_o = b/1000.

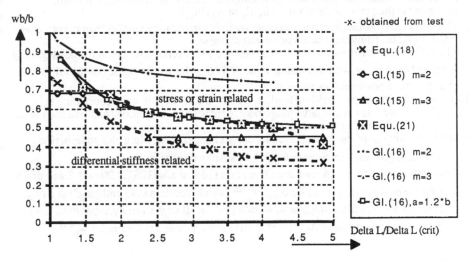

Figure 8. Effective Widths as Obtained from the CFR Test Panel, and Equs. (15) and (16).

FINITE ELEMENT ANALYSIS

A MARC K3 finite element analysis has been performed, implementing 420 plate elements Type 75 with composite, oriented layers. The configuration was loaded in increments using a special subroutine controling the stepwidth to keep the stability of the calculation during snap-through. The FE buckling load was obtained from this calculation to be in the vicinity of -15kN, compared with the analytically estimated value of -17.76kN. The postbuckling pattern had reasonable similarity with the observed pattern, see Figs. 6 and 9.

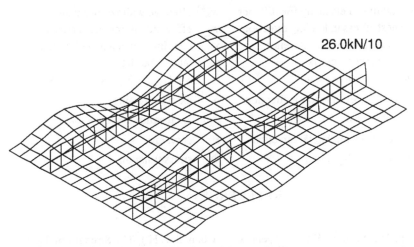

Figure 9. Finite Element Model (MARC) and Buckling Pattern at -26 kN (left edge = C_L)

SUMMARY

1. The effective width decreases with increasing panel stiffness in loading direction, compared to the perpendicular value (E_x/E_y or A_{22}/A_{11}).
2. The differential stiffness related effective width is significantly smaller than the effective width based on equivalent stress or overall stiffness.
3. The differential stiffness and its related effective width is not constant in the postbuckling range but decreasing with increasing value of δ / δ_{crit} due to geometric nonlinearities and changes in the postbuckling pattern.
4. A fair prediction of the buckling load, its pattern and of the crit. axial shortening is required, as well as a reasonable judgement for panel edge conditions and free span width b if use is made of effective width charts.
5. When the tangent of the load shortening curve changes slowly (e.g. due to imperfections), Equ. (18) underestimates $w_b^{(\Delta\varepsilon)}$ for small values of δ / δ_{crit}.
6. Good test agreement is also obtained from Equ.(16) for $a/b=326/(3*90) = 1.2$ and $m = n = 1$ for the evaluated range of δ / δ_{crit}. $w_b^{(\Delta\varepsilon)}$ would result in 0.45 and match at $\delta / \delta_{crit}= 2.5$. Application of Shanmugams approach [26] for all edges simply supported, $L = b$ and initial imperfections of $A_0 = b/1000$ yield similar close agreement for $w_b^{(\sigma)}$.

LIST OF SYMBOLS

a	free length of panel	P_x	total (compression) load
A_{ij}	coefficients of elasticity	$P^{(sp)}$	load portion carried by stiffener
b	considered width of panel		and unbuckled skin portion
b_s	unbuckled width of panel	t	nominal wallthickness of panel
B	total width of test article	u	axial displacement, + for elongation
D_{ij}	bending stiffness coefficients	w	out-of-plane deflection
$E_x^{(p)}$	axial modulus of elasticity of panel	w_{mn}	amplitude of w for pattern m,n
$E_x^{(s)}$	axial modulus of elasticity for stiffener	$w_b^{()}$	effective width of one panel
EF	resulting stiffness in loading direction	w_L	eff. width at free panel edges
F_i	crosssection area of one stiffener	x	coordinate in loading direction
K	coupling stiffness matrix	y	lateral coordinate
L	length of test article (= a)	α	$= \varepsilon / \varepsilon_{crit} = N_x / N_{x\ crit}$ for $N_y = 0$
m	halfwave number in loading direction	δ	axial shortening
M	line moments $\{M_x, M_y, M_{xy}\}$	ε	axial strain ($= u/a = -\delta/a$)
n	halfwave number in lateral direction	κ	curvature
N	line loads $\{N_x, N_y, N_{xy}\}$	Φ	stress function

REFERENCES

1. v. Kármán, Th., Sechler, E.E., (Appendix by Donnell, L.H.), The Strength of Thin Plates in Compression, Trans ASME, vol. 54, no. 2, 1932; cit. [5], [14], [16], [17]
2. Sechler, E.E., The Ultimate Strength of Thin Flat Sheets in Compression, Publ. 27 of the Cal. Inst. of Tech.,1933

3. Lahde, R., Wagner, W., Luftfahrt-Forschg,. vol.13, 1936; cit.[5], [14]
4. Sechler, E.E., Stress Distribution in Stiffened Panels under Compression, J. Aeron. Sci., June 1937
5. Sechler, E.E., Dunn, L.G., Airplane Structural Analysis and Design, The Galcit Aeronautical Series, J. Wiley & Sons, Inc., 1942
6. Niles, S.A.B., Newell, J.S. Airplane Structures, Vol.I, J. Wiley & Sons, Inc.,1943
7. Schuman, L., Black, G., Strength of Rectangular Flat Plate under Edge Compression, NACA Rep. 356, 1930; cit. [6], [16]
8. Levy, S., Bending of Rectangular Plates with Large Deflections, NACA Rep. 737, 1946
9. Marguerre, K., The Apparent Width of the Plate in Compression, NACA Tech. Memo. 833; cit.[14]
10. Peery, D.J., Aircraft Structures, McGraw-Hill, 1950
11. Sechler, E.E., Elasticity in Engineering, Galcit Aeron. Engg., J. Wiley & Sons, 1952
12. Kuhn, P. Stresses in Aircraft and Shell Structures, McGraw-Hill, 1956
13. v.d.Neut, A., Postbuckling Behaviour of Structures, NATO AGARD Rep. 60, 1956; cit.[18]
14. Timoshenko, S.P., Gere, J.M., Theory of Elastic Stability, McGraw-Hill, 1961
15. Bruhn, E.F., Analysis and Design of Flight Vehicle Structures, Tri State Co. 1965
16. Roark, R.J., Young, W.C., Formulas for Stress and Strain, 5th ed., McGraw-Hill, 1975
17. Brush, D.O., Almroth, B.O., Buckling of Bars, Plates and Shells, McGraw-Hill, 1975
18. Donnell, L.H., Beams, Plates and Shells, McGraw-Hill, 1976
19. Peery, D.J., Azar, J.J., Aircraft Structures, McGraw-Hill, 1982
20. Timoshenko, S.P., Woinowsky-Krieger, S., Theory of Plates and Shells, McGraw-Hill, 1981
21. Librescu, L., Stein, M., A Geometrically Nonlinear Theory of Shear Deformable Laminated Composite Plates and its Use in the Postbuckling Analysis, ICAS-88-5.2.4
22. Stein, M., Sydow, P.D., Librescu, L., Postbuckling Response of Long Thick Plates Loaded in Compression Incl. Higher Order Transverse Shearing Effects, ICAS-90-4.9.4
23. Wiedemann, J. Leichtbau, Band 1: Elemente; Springer-Verlag 1986
24. Müller, J., Mittragende Breite bei druckbelasteten, ebenen, dünnen und orthotrop laminierten "Blechfeldern"; Diplomarbeit ETH Zürich, Inst. f. Leichtbau u. Seilbahntechnik, 1990 (unpublished)
25. Geier, B., On the Initial Postbuckling Behaviour of Anisotropic Panels, Manuscript to be published by Elsevier 1991
26. Shanmugam, N.E., Effective Widths of Orthotropic Plates Loaded Uniaxially, Computers & Structures Vol. 29, No. 4, pp.705-713, 1988

APPENDIX

Formulas for Fig. 2

Bruhn [14]:	$= 1/\sqrt{\varepsilon/\varepsilon_{cr}}$
Timoshenko, Gere:	$= 0.661 + 0.339/(\varepsilon/\varepsilon_{cr})$
Timoshenko, Gere [14]:	$= 0.81 / \sqrt{(\varepsilon/\varepsilon_{cr})} + 0.19$
Sechler [5]:	$= 0.5*[1 + 1 / (\varepsilon/\varepsilon_{cr})]*1 / (\varepsilon/\varepsilon_{cr})^n$, $n = 1$
Marguerre [8]:	$= 1 / (\varepsilon/\varepsilon_{cr})^{1/3}$
Karman, Sechler [1]:	$= 0.894 / \sqrt{(\varepsilon/\varepsilon_{cr})}$
Timoshenko, Gere [14]:	$= 0.623 + 0.377/(\varepsilon/\varepsilon_{cr})$
Wiedemann [23], p. 77:	$=0.75+0.25/(\varepsilon/\varepsilon_{cr})$; laterally restrained
Wiedemann [23], p. 77:	$= 0.45 + 0.55/(\varepsilon/\varepsilon_{cr})$; free panel edge
Stein, M [22]:	$=1 / [3 - 2*(\varepsilon/\varepsilon_{cr})]$

FIBRE ORIENTATION EFFECTS ON THE BUCKLING BEHAVIOUR OF IMPERFECT COMPOSITE CYLINDERS

by

Carlo POGGI
Alberto TALIERCIO
Antonio CAPSONI

Dipartimento di Ingegneria Strutturale

Politecnico di Milano - Italy

ABSTRACT

A theoretical and numerical investigation on the influence of the fibre orientation on the buckling behaviour of composite cylinders is presented. Composite cylinders with different laminations and subject to axial compression are examined. The results show the great influence of fibre orientation on the buckling load and highlight the presence of a high number of almost coincident buckling modes in the cases of quasi-isotropic lay-up or of a particular value of the fibre angle. In the latter case the initial slope of the post-buckling branch has a very high negative value and the cylinders consequently result to be very imperfection sensitive. This confirms the fact that any optimisation process should include an imperfection sensitivity parameter.

INTRODUCTION

The use of composite shell structures has been increasing in the last two decades mainly in the aircraft and spacecraft industry. In this field cylindrical shells are structural members frequently subjected to destabilising loads such as axial load and torsion in the presence of unavoidable initial geometric imperfections [1].

The analysis of the buckling behaviour of composite cylindrical shells and panels may be performed by means of theoretical models or by finite element codes. This packages must include the particular constitutive equations of the laminate that can exhibit various couplings between membrane forces and bending or twisting. Furthermore, because of the great flexibility of the laminae, the ratio of the Young modulus over the shear modulus can be large and the assumption, according to which the deflection due to transverse shear force is neglected, may be not acceptable and a Mindlin or a Reissner-type theory becomes compulsary.

It is known that the dominant factor in significantly reducing the classical buckling load of circular cylinders is due to the presence of shape imperfections in the wall of the shell [2]. Obviously the nature of these imperfections is random in nature and in the absence of a random model it is impossible to formulate reliable design criteria. As a consequence it is evident the importance in determining, theoretically or numerically, the effects of the imperfection modes on the buckling load of composite cylinders. Usually the amplitude of the considered imperfection modes is determined using existing tolerances specifications but it needs a detailed study of the effects of manufacturing on imperfection characteristics. This study, undertaken for a series of nominally identical cylinders [3] is a part of a general project on the stability of composite shells [4].

This paper presents a study on the effects of the lamination geometries on the buckling and post-buckling behaviour of imperfect composite cylindrical shells. The calculations have been performed both using a finite element programme able to handle anisotropic shell elements and applying Koiter's imperfection sensitivity theory extended to composite cylinders [5]. The use of a program for symbolic algebra [6] allowed to solve the governing differential equations exactly.

THEORETICAL MODEL

The theoretical model, based on Koiter's stability theory extended to anisotropic shells, has been developed in [7] and [8] where further details of the formulation can be found. Here the main characteristics of the approach will be summarised.

Governing equations

Consider a linear elastic laminates cylindrical shell with the geometric dimensions reported in Fig. 1. Let the mid-surface be the reference surface and u,v,w define the displacements of the mid-surface.

Adopting the Love hypothesis and the nonlinear Donnell kinematic relations the following equilibrium and compatibility equations can be derived in terms of the transversal displacement w and of the stress function F. These assumptions are satisfactory in the analysis of buckling phenomena where the effects of the shear flexibility are negligible [9].

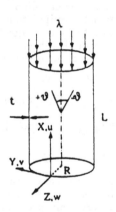

Figure 1. Cylinder geometry

Adopting a stress function F defined as

$$F_{,yy} = N_x \quad ; \quad F_{,xx} = N_y \quad ; \quad F_{,xy} = -N_{,xy} \tag{1a,c}$$

the following governing equations for composite cylindrical shells expressed in terms of the transversal displacement w and of the stress function F may be derived

$$L_1(F) - L_2(w) = -1/2 \, N_1(w,w) \tag{2a,b}$$

$$L_3(w) + L_2(F) = N_1(w,F)$$

where L_1, L_2, L_3 are linear differential operators and N_1 is a nonlinear differential operator defined as follows

$$L_1(.) = A_{22}^* (.)_{,xxxx} + 2 \left(A_{12}^* + \tfrac{1}{2} A_{66}^* \right) (.)_{,xxyy} + A_{11}^*(.)_{,yyyy} \tag{3a}$$

$$L_2(.) = B_{21}^* (.)_{,xxxx} + \left(B_{11}^* + B_{22}^* \right) (.)_{,xxyy} + B_{12}^* (.)_{,yyyy} + \tfrac{1}{R} (.)_{,xx} \tag{3b}$$

$$L_3(.) = D_{11}^* (.)_{,xxxx} + 2 \left(D_{12}^* + 2 D_{66}^* \right)(.)_{,xxyy} + D_{22}^*(.)_{,yyyy} \tag{3c}$$

$$N_1(H,K) = H_{,xx}K_{,yy} - 2 H_{,xy} K_{,xy} + H_{,yy}K_{,xx} \tag{3d}$$

The simply supported boundary conditions (SS3) are assumed, that, if expressed in terms of F and w, assume the form

$$w = 0 , \quad F_{,xx} = \theta_1 \lambda , \quad w_{,xx} = \theta_2 \lambda , \quad F_{,yy} = -\lambda \quad \text{for} \quad x = 0, L$$

where

$$\theta_1 = \frac{(B_{21}^* B_{11}^* + A_{21}^* D_{11}^*)}{(A_{22}^* D_{11}^* + B_{21}^{*\,2})} \qquad\qquad \theta_2 = \frac{(A_{12}^* B_{21}^* - A_{22}^* B_{11}^*)}{(A_{22}^* D_{11}^* + B_{21}^{*\,2})}$$

Initial post-buckling analysis

The Koiter's general theory of elastic stability applied to anisotropic shells allows to produce good indications of the nonlinear behaviour of imperfect composite shells [5]. On the other hand, the application of asymptotic procedures for the buckling and postbuckling analysis of shells involves manipulations of long and complicated expressions that become almost unhandy in the case of composite cylinders. The use of modern symbolic manipulation programmes facilitates this task and allows to derive error-free expressions in a quick and easy way [7]. A package, recently adapted for personal computer, has been used in this work to derive and solve the various sets of differential equations involved in the problem.

The following asymptotic expressions are adopted

$$w = w^{(0)} + \zeta \, w^{(1)} + \zeta^2 \, w^{(2)} + \ldots$$

$$F = F^{(0)} + \zeta F^{(1)} + \zeta^2 F^{(2)} + \ldots \tag{4a-c}$$

$$\lambda = \lambda_c + \zeta \lambda_1 + \zeta^2 \lambda_2 + \ldots$$

where ζ is the perturbation parameter and $w^{(1)}, F^{(1)}$ and $w^{(2)}$, $F^{(2)}$ the first and second order contribution. Introducing (4a-c) into (2a,b) and into the boundary conditions, a series of linear problems is obtained.

Let us assume a linear prebuckling solution of the type

$$F^{(0)} = -\tfrac{1}{2} \lambda \, y^2 \tag{5a}$$

$$w^{(0)} = -\lambda \, A_{12}^* \, R \tag{5b}$$

where the boundary conditions are neglected under the hypothesis of pure membrane behaviour of the shell. A doubly periodic function makes the bifurcation problem symmetric ($\lambda_1 = 0$) and the first order perturbation equations assume the form.

$$L_1[F^{(1)}] - L_2[w^{(1)}] = 0 \tag{6a,b}$$

$$L_3[w^{(1)}] + L_2[F^{(1)}] + \lambda_c \, w_{,xx}^{(1)} = 0$$

While the second-order perturbation equations result to be

$$L_1(F^{(2)}) - L_2(w^{(2)}) = -\tfrac{1}{2} N_1(w^{(1)}, w^{(1)}) \tag{7a,b}$$

$$L_3(w^{(2)}) + L_2(F^{(2)}) + \lambda_c \, w_{,xx}^{(2)} = N_1(w^{(1)}, F^{(1)})$$

with the relevant boundary conditions (not reported here).

Assuming that the buckling mode is represented by the function

$$w^{(1)} = \sin \frac{m\pi x}{L} \sin \frac{ny}{R} \tag{8}$$

the expression of the first order term of the stress function can be derived from the compatibility equation (6a).

$$F^{(1)} = \bar{C}_{mn} \sin \frac{m\pi x}{L} \sin \frac{ny}{R} \tag{9}$$

being $\bar{C}_{mn} = \dfrac{(\frac{m\pi}{RL})^2 + \bar{B}_{mn}}{\bar{A}_{mn}}$

Note that $\bar{A}_{mn}$ e $\bar{B}_{mn}$ are algebraic operators containing only geometric parameters and the circumferential and axial wave numbers

$$\bar{A}_{mn} = A_{22}^*(\frac{m\pi}{L})^4 + 2(A_{12}^* + \frac{1}{2}A_{66}^*)(\frac{m}{L}\frac{n\pi}{R})^2 + A_{11}^*(\frac{n}{R})^4 \tag{10a}$$

$$\bar{B}_{mn} = B_{21}^*(\frac{m\pi}{L})^4 + (B_{11}^* + B_{22}^*)(\frac{m}{L}\frac{n\pi}{R})^2 + B_{12}^*(\frac{n}{R})^4 \tag{10b}$$

The classical buckling load is the lowest of the eigenvalues λ_c given by

$$\lambda_c = -\frac{\bar{C}_{mn}}{R} + (\frac{L}{m\pi})^2[\bar{D}_{mn} + \bar{C}_{mn}\bar{B}_{mn}] \tag{11}$$

where

$$\bar{D}_{mn} = D_{11}^*(\frac{m\pi}{L})^4 + 2(D_{12}^* + 2D_{66}^*)(\frac{mn\pi}{LR})^2 + D_{22}^*(\frac{n}{R})^4 \tag{12}$$

The solution of the second-order problem is sought assuming a series of functions orthogonal to the first-order series.

$$w^{(2)} = \sum_{i=1}^{\infty} \left\{ c_{0i} + c_{2i} \cos \frac{2ny}{R} \right\} \sin \frac{i\pi x}{L}$$

$$F^{(2)} = \sum_{i=1}^{\infty} \left\{ b_{0i} + b_{2i} \cos \frac{2ny}{R} \right\} \sin \frac{i\pi x}{L} \tag{13a,b}$$

Applying the Galerkin procedure the value of the second order coefficient λ_2 may be worked out in an exact form. This was made possible by the use of the aforementioned symbolic manipulation programmes.

$$\lambda_2 = \sum_{i=1}^{\infty} \frac{4n^2}{\pi R^2} \left[\frac{2i(b_{0i} + 2c_{0i}C_{mn})}{(i^2 - 4m^2)} + \frac{b_{2i} + 2c_{2i}C_{mn}}{i} \right] \tag{14}$$

The expressions of the coefficients are not reported here but may be found in [8]. The advantage of producing an exact form of these coefficient is evident. Eq. (14) contains only geometric parameters and it can be implemented even into pocket calculators.
The obtained coefficients allow to analyse the imperfection sensitivity of the composite cylinders within the limits of the proposed initial post-buckling theory.

THE FINITE ELEMENT MODEL

The problem was also tackled using a finite element shell model based on a Mindlin-type theory that can be used for laminated plates and shells provided a proper value of the transverse shear stiffness is defined. This theory was preferred even if, as it is shown in [9], the shear flexibility on buckling phenomena is small and the Kirchhoff theory is still precise enough. As it will be discussed in the next section the finite element model provides results very close to the theory but a great number of elements and a quite high number of iterations are needed even in the estimate of the eigenvalues.

The same model was adopted to investigate the effects due to the presence of a geometric imperfection on the buckling load. The Riks' method was used to analyse the effect of a single or a combination of imperfection modes. In this case the computational effort becomes very heavy and almost precludes the possibility to derive imperfection sensitivity curves by means of a parametric study.

The comparison between the theory and the finite element method is a good validation for the first tool that, within the limits of the accepted approximation, can easily provide information useful for the buckling design of composite cylinders.

APPLICATIONS AND EXAMPLES

The geometric parameters of the cylinders considered in the present work are those of the series of specimens described in [4]. Each cylinder has the following dimensions.

TABLE 1

Cylinder geometry and elastic properties of each lamina

t	1.04 mm
R	350 mm
L	700 mm
E_{11}	23450 Nmm^{-2}
E_{22}	23450 Nmm^{-2}
G_{12}	1520 Nmm^{-2}
ν_{12}	0.20

The layers are made with a orthogonal Kevlar fabric embedded in an epoxy resin matrix. It must be noted that the cylinders present two thicker parts at the top and bottom to facilitate the fixing of the specimens into the testing machine. As a consequence the actual length is reduced to 550 mm. The indices of the material properties refer to axis x_1 and x_2 directed as the fibre axis of the lamina so that x_3 is normal to the lamina midplane.

The attention has been focused mainly on the following stacking sequences that correspond to the available cylindrical specimens.
a) cross-ply cylinders made with 4 laminae at 0°/90° to the cylinder axis
b) angle-ply cylinders made with 4 laminae at ±45° to the cylinder axis
c) quasi-isotropic cylinders made with 8 laminae with a (45/0/90/-45°)$_s$

Note that because of the particular elastic properties of each lamina the orientation +/-45° and 0°/90° become coincident and the shell surface behaves as a single orthotropic layer. This is not the case for other inclinations.

The influence of the fibre orientation on the linear buckling load of an axially compressed cylinder with stacking sequences in the range between 0° and 45° (i.e. from a cross-ply lamination 0°/90° to an angle-ply lamination ±45°) has been analysed using the derived expression of the critical parameter, eq. (11). It was assumed that the coupling stiffnesses are very small and thus negligible. It is known that this is actually true only for $\theta=0$° and 45°or for symmetric angle-ply laminates made with many layers. The results of the analysis are reported in Fig. 2.

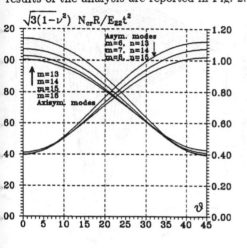

Fig. 2   Nondimensional buckling loads.

The eigenvalue problem yields a single buckling load that is associated to axisymmetric modes with short axial wavelength for fibre orientation angles between 22.5° and 45°. The buckling mode becomes asymmetric for θ angles in the range between 0° and 22.5°. From the analysis reported in fig. 2 it appears that the optimal orientation of the fibres is around 22° where the buckling load is almost the double of the cross-ply lamination but it must be noted that in that area several almost coincident buckling modes are present. This suggests to analyse the imperfection sensitivity of each lamination. This is possible on the basis of the previously discussed formulation that furnishes the following expression of the limit load

$$(1 - \tfrac{\lambda_s}{\lambda_c})^{3/2} = \tfrac{3}{2} \sqrt{3} \, (-\tfrac{\lambda_2}{\lambda_c})^{1/2} \tfrac{\lambda_s}{\lambda_c} \bar{\zeta} \tag{15}$$

where $\bar{\zeta}$ is the amplitude of an imperfection of the form

$$\bar{w} = \zeta \sin(\tfrac{m\pi x}{L}) \sin(\tfrac{ny}{R}) \tag{16}$$

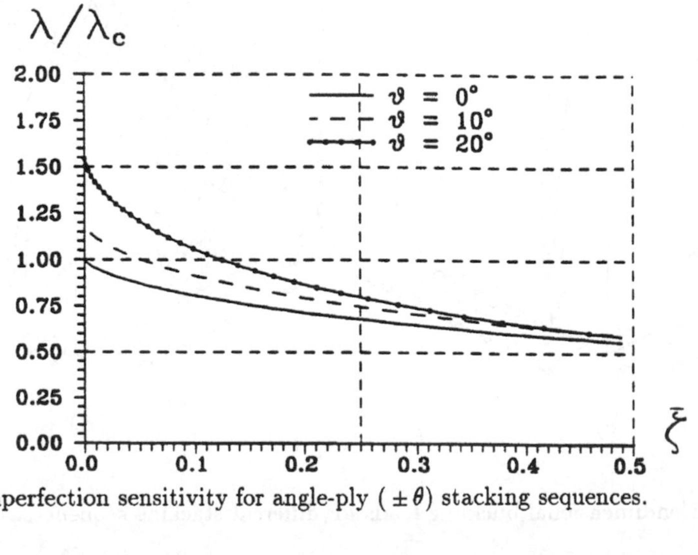

Figure 3 - Imperfection sensitivity for angle-ply ($\pm\theta$) stacking sequences.

In fig. 3 it is evident how the effects of an imperfection on the buckling load is much more detrimental in the case of a lamination equal to 22° and how the limit loads assume very similar values when the imperfection amplitude is only 50% of the shell thickness.

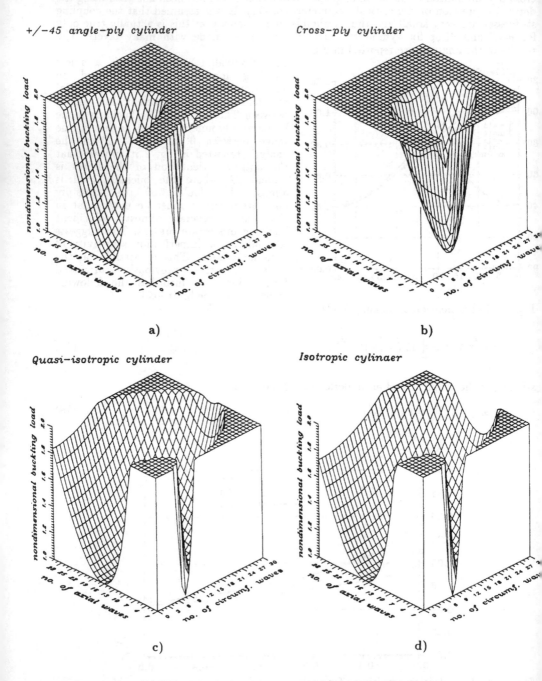

Figure 4. Nondimensional buckling loads for different stacking sequences.

A three dimensional representation of the eigenvalues for an angle-ply (± 45)$_s$ and a cross-ply stacking sequence is reported in Fig. 4a,b . It is evident how in both these cases a single critical mode is well localised. On the contrary, the quasi-isotropic cylinder (fig. 4c) shows various simultaneous buckling mode in a fashion very similar to the isotropic cylinder reported in fig. 4d. In both these figures the typical valley corresponding to the Koiter circle is clearly shown.

This characteristic shows the importance in determining the imperfection sensitivity of the quasi isotropic lamination that is expected to be as imperfection sensitive as the isotropic one [10]. It must be underlined that such an analysis should be carried out by means of numerical methods as the multimode imperfection sensitivity analysis becomes too complicated in the case of more than two interactive modes.

It is known that the actual buckling mode can be something different from that obtained analytically with a single mode analysis mainly because of the coupling of several modes. For this reason and for comparison purpose, the same problem has been studied also using a finite element programme.

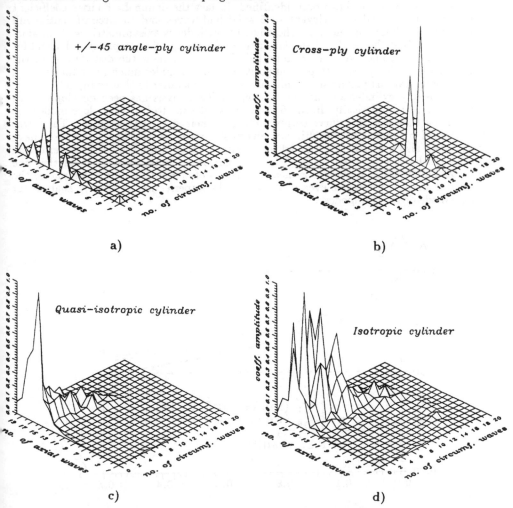

Figure 5. Fourier decomposition of the numerical buckling modes.

TABLE 2

Linear buckling loads for axially compressed composite cylinders

Axial load (KN)			
Lay-up	F.E.M.	theory	Mode
$(0°/90°)_s$	38.33	37.07	ASYM
$(45°/-45°)_s$	37.02	37.03	AXISYM
$(45°/0°/90°/-45°)_s$	70.35	69.93	AXISYM

It is evident how the results are very similar in terms of buckling loads. The comparison of the shape of the eigenmodes with the theory was possible performing a Fourier analysis on the finite element eigenmodes. The Fourier coefficients of each mode are reported in Fig. 5a-d. In the case of angle-ply (±45°) (fig. 5a) and cross-ply laminations the same buckling modes have been identified. In fact the dominant Fourier coefficient for the cross-ply cylinder is relevant to 6 axial half-waves and 13 circumferential full waves while the dominant mode for the angle-ply cylinder is axisymmetric with 13 axial waves. In Fig. 5c,d the results of similar analyses for the quasi-isotropic cylinder and for an equivalent isotropic cylinder are reported. It is evident how the distribution of the active modes for the quasi-isotropic composite cylinder includes much more modes than the two configurations previously examined and is very similar to that of fig. 5d.
For the cross-ply cylinder an imperfection sensitivity analysis has been developed both theoretically and numerically. In fig. 6 the curves relevant to two very close modes are reported together with the curve corresponding to their simultaneous effect. The latter was obtained by means of a multimode analysis reported in [7]. The same cases have been studied also using the F.E. model for a defined value of the geometric imperfection and the results are reported in fig. 6 for comparison. It must be underlined that the nonlinear numerical analysis is very heavy in terms of computing time. This makes the theoretical analysis preferable even if limited to the imperfection sensitivity analysis for a single mode or for two modes.

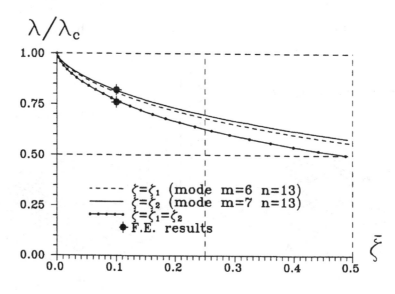

Figure 6. Imperfection sensitivity of a cross-ply cylinder.

123

CONCLUSIONS

The effects of the orientation of the laminae and of the stacking sequence on the buckling load of axially compressed cylindrical shells have been analysed comparing theoretical and numerical results. The former were obtained applying the Koiter's theory for the initial post-buckling analysis to composite cylinders. The results have highlighted the fact that the advantages (in terms of buckling load) obtained assuming a particular stacking sequence can be only fictitious if the post-buckling range is explored. This information become very important in the case of an optimisation process.

REFERENCES

[1] Simitses G.J., Shaw D., Sheinman I. "Imperfection Sensitivity of Laminated Cylindrical Shells in Torsion and Axial Compression", Composite Struct. No. 4, 1985, pp.335-360.

[2] Budiansky B. "Theory of buckling and post-buckling behaviour of elastic structures" Advances in Applied Mech., Vol 14, 1974, pp.1-65.

[3] Chryssanthopoulos M., Giavotto V., Poggi C. " Statistical imperfection models for buckling analysis of composite shells"- Proc. of Int. Colloquium "Buckling of Shell Structures, on Land, in the Sea and in the Air", 1991, Lyon, France.

[4] Giavotto V., Poggi C., Chryssanthopoulos M. "Buckling of imperfect composite shells under compression and torsion" Proc. of American Helicopter Society Int. Meeting on Rotorcraft Basic Research, Atlanta, Georgia, March 25-27, 1991, pp.43-1,43-13.

[5] Arbocz J., Hol J.M.A.M. "Koiter's Stability theory in a computer aided engineering environment", Int. J. Solids Structures, Vol. 26, No. 9/10, 1990, pp.945-973.

[6] Rizzi N., Tatone A. "Symbolic manipulation in buckling and postbuckling analysis" Comp. Struct., Vol. 21, No. 4, 1985, pp. 691-700.

[7] Capsoni A., Poggi C.,"The Role of Symbolic Algebra in the Initial Post-buckling and Imperfection Sensitivity Analysis of Axially Compressed Composite Cylindrical Shells", Tech. Rep., Dip. di Ingegneria Strutturale, Politecnico di Milano, 1990.

[8] Capsoni A. " Analisi del comportamento post-critico di gusci cilindrici in materiale composito mediante metodi perturbativi", Tesi di Laurea, Dipartimento di Ingegneria Strutturale, Politecnico di Milano, 1991.

[9] Geier B., Rohwer K. "On the analysis of the buckling behaviour of laminated composite plates and shells" Int. J. Num. Meth. Eng., Vol. 27, 1989, pp. 403-427.

[10] Byskov E., Hutchinson J.W. " Modal interaction in axially stiffened cylindrical shells" AIAA J., Vol. 15, No. 7, (1977) pp. 941-948.

BUCKLING OF CIRCULAR CONICAL SHELLS OF COMPOSITE MATERIALS UNDER AXIAL COMPRESSION AND EXTERNAL LOAD

LIYONG TONG* and TSUN. KUEI. WANG**

*Department of Mechanical Engineering, University of Victoria, Victoria,
B.C.,Canada V8W 2Y2
**Department of Applied Mechanics, Beijing University of Aeronautics and
Astronautics, Beijing, P.R. China

ABSTRACT

Using Donnell type shell theory a simple and exact procedure is presented for linear buckling analysis of composite laminated conical shells, with orthotropic stretching-bending coupling, under axial compressive loads and external pressure. The solution is in the form of a power series in terms of a particularly convenient coordinate system. By analyzing the buckling of a series of conical shell, the validity of the presented procedure is confirmed. A simple formula of critical axially compressive load for composite laminated conical shells with orthotropic stretching-bending coupling is suggested.

INTRODUCTION

Due to their extensive use, particularly in aeronautical industry, the buckling of conical shells has been studied by many researchers. Much literature exists on the buckling of isotropic conical shells under compressive axial loads, and under external pressure as well as combined loading. However, there have been few studies for orthotropic and laminated conical shells. In the following we develop a simple and exact procedure for buckling analysis of composites laminated conical shells under axial compression and external pressure. The procedure consists of the following steps: 1) the buckling equations are developed and expressed in terms of displacements; 2) using a new technique, exact solutions are constructed in series form for the governing equations; 3) convergence properties of the series solution are determined. Finally several examples are presented.

DONNELL-TYPE GOVERNING EQUATIONS

Consider a conical shell, R_1 and R_2 indicate the radii of the cone at its small and large ends, respectively, α denotes semivertex angle of the cone and L is the cone length along its generator. We now introduce the $x - \phi$ coordinate system; x is measured along the cone's generator starting at middle length and ϕ is the circumferential coordinate. The displacements of the shell's middle surface are denoted by U and V along x and ϕ directions, respectively, and by W along the normal of the surface (inward positive). In terms of these variables the cone's radius at any point along its length may be expressed as

$$R(x) = R_o + x\sin\alpha; \qquad R_o = (R_1 + R_2)/2 \tag{1}$$

Now let the cone be subjected to an axially compressive load P and an external normal pressure q. Under this loading the membrane stress resultants, at the critical state, may be expressed as

$$N_{xo} = -\frac{P + q\pi(2R_o + x\sin\alpha)x\sin\alpha}{2\pi R(x)\cos\alpha}; \qquad N_{\phi o} = -\frac{qR(x)}{\cos\alpha} \tag{2}$$

Supposing no twist coupling created in each layer of the laminated cone, and utilizing thin-walled shallow shell theory of Donnell-type with moderately large deformation for linear buckling analysis, we have the following equilibrium equations

$$
\begin{aligned}
L_{11}U + L_{12}V + L_{13}W &= 0 \\
L_{21}U + L_{22}V + L_{23}W &= 0 \\
L_{31}U + L_{32}V + L_{33}W + L_N W &= 0
\end{aligned}
\tag{3}
$$

where

$$L_{11} = A_{11}\frac{\partial^2}{\partial x^2} + \frac{A_{11}\sin\alpha}{R(x)}\frac{\partial}{\partial x} - \frac{A_{22}\sin^2\alpha}{R^2(x)} + \frac{A_{66}}{R^2(x)}\frac{\partial^2}{\partial\phi^2}$$

$$L_{12} = \frac{(A_{12} + A_{66})}{R(x)}\frac{\partial^2}{\partial x\partial\phi} - \frac{(A_{22} + A_{66})\sin\alpha}{R^2(x)}\frac{\partial}{\partial\phi}$$

$$L_{21} = \frac{(A_{12} + A_{66})}{R(x)}\frac{\partial^2}{\partial x\partial\phi} + \frac{(A_{22} + A_{66})\sin\alpha}{R^2(x)}\frac{\partial}{\partial\phi}$$

$$L_{22} = A_{66}\left[\frac{\partial^2}{\partial x^2} + \frac{\sin\alpha}{R(x)}\frac{\partial}{\partial x} - \frac{\sin^2\alpha}{R^2(x)}\right] + \frac{A_{22}}{R^2(x)}\frac{\partial^2}{\partial\phi^2}$$

$$L_{13} = -\frac{A_{12}\cos\alpha}{R(x)}\frac{\partial}{\partial x} + \frac{A_{22}\sin\alpha\cos\alpha}{R^2(x)}$$

$$-B_{11}\frac{\partial^3}{\partial x^3} - \frac{(B_{12} + 2B_{66})}{R^2(x)}\frac{\partial^3}{\partial x\partial\phi^2} - \frac{B_{11}\sin\alpha}{R(x)}\frac{\partial^2}{\partial x^2} +$$

$$\frac{(B_{12} + B_{22} + 2B_{66})\sin\alpha}{R^3(x)}\frac{\partial^2}{\partial\phi^2} + \frac{B_{22}\sin^2\alpha}{R^2(x)}\frac{\partial}{\partial x}$$

$$L_{23} = -\frac{A_{22}\cos\alpha}{R^2(x)}\frac{\partial}{\partial\phi} - \frac{(B_{12} + 2B_{66})}{R(x)}\frac{\partial^3}{\partial x^2\partial\phi} - \frac{B_{22}}{R^3(x)}\frac{\partial^3}{\partial\phi^3} -$$

$$\frac{B_{22}\sin\alpha}{R^2(x)}\frac{\partial^2}{\partial x\partial\phi} -$$

$$L_{31} = -\frac{A_{12}cos\alpha}{R(x)}\frac{\partial}{\partial x} - \frac{A_{22}sin\alpha cos\alpha}{R^2(x)}$$

$$-B_{11}\frac{\partial^3}{\partial x^3} - \frac{(B_{12}+2B_{66})}{R^2(x)}\frac{\partial^3}{\partial x\partial\phi^2} - \frac{2B_{11}sin\alpha}{R(x)}\frac{\partial^2}{\partial x^2} +$$

$$\frac{B_{22}sin^2\alpha}{R^2(x)}\frac{\partial}{\partial x} - \frac{B_{22}sin\alpha}{R^3(x)}\frac{\partial^2}{\partial\phi^2} - \frac{B_{22}sin^3\alpha}{R^3(x)}$$

$$L_{32} = -\frac{A_{22}cos\alpha}{R^2(x)}\frac{\partial}{\partial\phi} - \frac{(B_{12}+2B_{66})}{R(x)}\frac{\partial^3}{\partial x^2\partial\phi} - \frac{B_{22}}{R^3(x)}\frac{\partial^3}{\partial\phi^3} +$$

$$\frac{B_{22}sin\alpha}{R^2(x)}\frac{\partial^2}{\partial x\partial\phi} - \frac{B_{22}sin^2\alpha}{R^3(x)}\frac{\partial}{\partial\phi}$$

$$L_{33} = \frac{A_{22}cos^2\alpha}{R^2(x)} + \frac{2B_{12}cos\alpha}{R(x)}\frac{\partial^2}{\partial x^2} + \frac{2B_{22}cos\alpha}{R^3(x)}\frac{\partial^2}{\partial\phi^2} + \frac{B_{22}cos\alpha sin^2\alpha}{R^3(x)} +$$

$$D_{11}\frac{\partial^4}{\partial x^4} + \frac{2(D_{12}+2D_{66})}{R^2(x)}\frac{\partial^4}{\partial x^2\partial\phi^2} + \frac{D_{22}}{R^4(x)}\frac{\partial^4}{\partial\phi^4} +$$

$$\frac{2D_{11}sin\alpha}{R(x)}\frac{\partial^3}{\partial x^3} - \frac{2(D_{12}+2D_{66})sin\alpha}{R^3(x)}\frac{\partial^3}{\partial x\partial\phi^2} - \frac{D_{22}sin^2\alpha}{R^2(x)}\frac{\partial^2}{\partial x^2} +$$

$$\frac{2(D_{12}+D_{22}+2D_{66})sin^2\alpha}{R^4(x)}\frac{\partial^2}{\partial\phi^2} + \frac{D_{22}sin^3\alpha}{R^3(x)}\frac{\partial}{\partial x}$$

$$L_N = \frac{1}{R(x)}\frac{\partial}{\partial x}[R(x)N_{xo}\frac{\partial}{\partial x}] + \frac{1}{R^2(x)}\frac{\partial}{\partial\phi}(N_{\phi o}\frac{\partial}{\partial\phi}) \tag{4}$$

The force and moment stress resultants can be expressed in terms of displacements U, V and W. The related boundary conditions may be expressed as

$$
\begin{aligned}
N_x &= 0 \quad or \quad U = 0 \\
N_{x\phi} &= 0 \quad or \quad V = 0 \\
M_x &= 0 \quad or \quad \frac{\partial W}{\partial x} = 0 \quad when \quad x = \pm\frac{L}{2} \\
Q_x &= 0 \quad or \quad W = 0
\end{aligned}
\tag{5}
$$

and the four subclass simply supported conditions are denoted as follows:

$$
\begin{aligned}
SS1: &\quad N_{x\phi} = N_x = M_x = W = 0 \\
SS2: &\quad N_{x\phi} = U = M_x = W = 0 \\
SS3: &\quad V = N_x = M_x = W = 0 \\
SS4: &\quad V = U = M_x = W = 0
\end{aligned}
\tag{6}
$$

The above set of governing equations degenerate to those of composites cylindrical shells when α is set equal to zero, and also to those of orthotropic conical shells without stretching- bending coupling. It is also worth noting that if the starting point of the x axis is changed to the cone's vertex, where the radius is equal to zero, the $x-\phi$ coordinate system will coincide with the $s-\phi$ coordinate system used by many previous researchers.

Evidently the system of governing equations presented in the foregoing is complex and to our knowledge exact solutions have not been given for these equations. In the following section we outline a strategy for constructing general solutions for these equations and for obtaining critical buckling loads for orthotropic conical shells.

EXACT SOLUTIONS

An inspection of the differential operators $L_{i,j}(i,j = 1,2,3)$ and L_N in Eqs (4) reveals the following properties: The coefficients of all these operators are functions of x only, i.e., they are independent of ϕ, and they include terms of the following form: $1/R^k(x)$, $k = 0,1,2,3,4$. For the operators $L_{i,j}$ $(i = 1,2; j = 1,2,3)$, k takes values from zero to two. For L_{3j} $(j = 1,2,3)$ the value of k takes values from zero to four. These useful properties allow us to change the equations into a more convenient form. Multiplying the first two equations of (3) by $R^3(x)$ and the third by $R^4(x)$ we may obtain the modified equations. Then for the modified equations, let us assume their solutions of following form:

$$U = \sum_{m=0}^{\infty} a_m x^m \cos n\phi \qquad V = \sum_{m=0}^{\infty} b_m x^m \sin n\phi \qquad W = \sum_{m=0}^{\infty} c_m x^m \cos n\phi \qquad (7)$$

where n is an integer representing the circumferential wave number of the buckled shell, a_m, b_m and c_m are constants to be determined later. On substituting the above solutions into the modified equations, we develop three linear algebraic equations, by matching the terms of same order in x, and further we obtain the following recurrence relations

$$b_{m+2} = G_{2,1}a_{m+1} + G_{2,2}a_m + G_{2,3}a_{m-1} + G_{2,4}b_{m+1} + G_{2,5}b_m$$
$$+G_{2,6}b_{m-1} + G_{2,7}c_{m+2} + G_{2,8}c_{m+1} + G_{2,9}c_m + G_{2,10}c_{m-1}$$
$$a_{m+2} = G_{1,1}a_{m+1} + G_{1,2}a_m + G_{1,3}a_{m-1} + G_{1,4}b_{m+1} + G_{1,5}b_m + G_{1,6}b_{m-1}$$
$$+G_{1,7}c_{m+3} + G_{1,8}c_{m+2} + G_{1,9}c_{m+1} + G_{1,10}c_m + G_{1,11}c_{m-1}$$
$$c_{m+4} = G_{3,1}a_{m+3} + G_{3,2}a_{m+2} + G_{3,3}a_{m+1} + G_{3,4}a_m + G_{3,5}a_{m-1}$$
$$+G_{3,6}a_{m-2} + G_{3,7}b_{m+2} + G_{3,8}b_{m+1} + G_{3,9}b_m + G_{3,10}b_{m-1}$$
$$+G_{3,11}b_{m-2} + G_{3,12}c_{m+3} + G_{3,13}c_{m+2} + G_{3,14}c_{m+1} + G_{3,15}c_m$$
$$+G_{3,16}c_{m-1} + G_{3,17}c_{m-2} + G_{3,18}c_{m-3}$$
$$(m = 0,1,2,......) \qquad (8)$$

where the coefficients $G_{i,j}$ $((i,j) = (1,10),(2,11)$ and $(3,18))$ are given in the Appendix. From the above recurrence relations, one can not directly obtain c_{m+4} because the term a_{m+3} is involved in the last equation. However, combining the first two equations with $m + 1$ with the last equation with m and rearranging them, we may have explicit recurrence relations to express a_m, b_m $(m \geq 2)$ and c_m $(m \geq 4)$ and further more $U(x)$, $V(x)$ and $W(x)$ in terms of a_0, a_1, b_0, b_1, c_0, c_1, c_2 and c_3, which can be determined by boundary conditions at both ends of the cone.

Before going into details of the solution procedure, let us consider the convergence properties of the series solutions $U(x)$, $V(x)$ and $W(x)$. Careful analysis of the recurrence equations and the coefficients $G_{i,j}$ given in the Appendix shows that: 1) The power series defined in (7) are altering series, i.e., the terms of the series change sign consecutively This property can obviously be verified through numerical calculations; 2) When m becomes large enough, the recurrence equations (8) can be approximately written as follows

$$b_{m+2} \doteq -\frac{3\sin\alpha}{R_o}b_{m+1} - \frac{3\sin^2\alpha}{R_o^2}b_m - \frac{\sin^3\alpha}{R_o^3}b_{m-1}$$

$$c_{m+4} \doteq -\frac{4sin\alpha}{R_o}c_{m+3} - \frac{6sin^2\alpha}{R_o^2}c_{m+2} - \frac{4sin^3\alpha}{R_o^3}c_{m+1} - \frac{sin^4\alpha}{R_o^4}c_m \tag{9}$$

$$a_{m+2} \doteq -\frac{3sin\alpha}{R_o}a_{m+1} - \frac{3sin^2\alpha}{R_o^2}a_m - \frac{sin^3\alpha}{R_o^3}a_{m-1}$$

$$+m\frac{B_{11}}{A_{11}}(c_{m+3} + \frac{3sin\alpha}{R_o}c_{m+2} + \frac{3sin^2\alpha}{R_o^2}c_{m+1} + \frac{sin^3\alpha}{R_o^3}c_m) \tag{10}$$

Through these approximate recurrence equations, we can verify that the convergence radius for them is

$$r_c = \frac{R_1 + R_2}{2sin\alpha} \tag{11}$$

That is as long as x is within the circle of radius r_c, convergence will be assured. For the shells considered here, the maximum value of $|x|$ is $L/2$. Thus for our purposes, the condition of convergence is $L/2 \leq r_c$, which may finally be written as $R_1 \geq 0$. Hence the three constructed series will converge to their corresponding solutions if the small radius is not zero, i.e., if the conical shell is a truncated one. A complete cone is treated as a truncated cone with a very small radius at its apex. Thus for all of practical purposes, there are no limitations on the geometric parameters of the shell considered. Accordingly, the solution obtained provides exact solutions for the three displacements U, V and W for the buckling of cones under axial compressive loads and external pressure. The three displacements U, V and W can be be used to calculate the stress resultants N_x, N_ϕ and $N_x\phi$ and the bending moments M_x, M_ϕ and $M_x\phi$, and further more the transverse shear forces Q_x and Q_ϕ can be obtained. This solution is exact because it satisfies the governing equations rigorously and it also can match the eight boundary conditions accurately by enforcing the eight arbitrary constants. The critical buckling loads and the corresponding buckling patterns can finally be obtained by equating the determinants of the coefficients matrix obtained after impositions of the eight boundary conditions to zero.

NUMERICAL RESULTS AND DISCUSSIONS

Numerical Results for Isotropic Cones:

In this section numerical results are presented for buckling of isotropic conical shells under axial compression with different parameters under various kinds of boundary conditions. Before presenting the results, let us introduce the following notation $\rho_{cr} = P_{cr}/P_{cl}$, where P_{cr} is the critical buckling load obtained from the present method, and P_{cl} is the classical value of the critical buckling load suggested by Seide in Ref.[1]. The present values ρ_{cr} and their comparison with those in Ref.[2] are shown for isotropic cones with different values L/R_1, semivertex angles α and different boundary conditions, i.e., SS_1 and SS_3 in Table 1, $SS4$ in Table 2. Good agreement for ρ_{cr} can be observed between the present results and those from Ref.[2]. There is however a difference in the circumferential wavenumber.

All calculations show that only 20 terms of the series (7), are sufficient for accurate value of ρ_{cr}. Further calculations using 15 terms show little difference in results obtained.

Numerical Results for Laminated Cones:

In this section a particular set of laminated cones, namely, regularly antisymmetric cross-plied cones., are numerically studied. In this case, the coupling terms B_{ij} becomes:

$B_{11} = -B_{22} = \pm(Q_{11} - Q_{22})h^2/4\bar{N}$, $B_{12} = B_{66} = 0$, where Q_{ij} $(i,j = 1,2)$ are material parameters[4], $\bar{N}$ is the total number of layers, h denotes the total wall-thickness. The elementary material parameters $E_x/E_\phi = 40.0$, $\mu_{x\phi} = 0.25$ and $G_{x\phi}/E_\phi = 0.5$ of each carbon/epoxy layer will be used in the following analysis. For this particular set of laminated cones, we compute the same ρ_{cr} as before except in Seide's formula μ replaced by $\mu_{x\phi}$ and E replaced by E_x.

Critical load ratios ρ_{cr} and the associated circumferential wavenumbers n are shown for these cones with $R/h = 100$ and with different value L/R_o, semivertex angles α and various kinds of boundary conditions, i.e., SS_4 in Table 3, SS_3 in Table 4, SS_1 in Table 5 and SS_2 in Table 6. It can be seen that ρ_{cr} increases and n tends to decrease when the total number of layer $\bar{N}$ becomes larger for cones with ratio L/R_o and α fixed. physically speaking, the larger the stretching-bending coupling becomes, the less the critical load ratio tends to be. This phenomena is same as that for antisymmetric cross-plied flat plate. The ratio L/R_o has a strong influence on the critical load ratio ρ_{cr}, i.e., ρ_{cr} tends to be a constant when L/R_o increases and other parameters remain unchanged. Another important fact worth noting is that the effect of α on ρ_{cr} and n is similar to that of $\bar{N}$ on ρ_{cr} and n. However, for short cones this effect is quite obvious, while for long cones this effect becomes quite weak. For example, when $L/R_o = 1.0$, ρ_{cr} varies smoothly as α increases. This significant fact informs us that for long cones ρ_{cr} is almost independent of α, namely, in term of $cos^2\alpha$ the variation of ρ_{cr} can be described properly. Thus a simple formula for critical buckling load of laminated cone is recommended as $P_{co} = P_{cl}cos^2\alpha$, where P_{co} denotes the critical buckling load of laminated cone, P_{cl} represents the critical buckling load of laminated cylindrical shell with radius equal to the average radius of assorted cones, α is the semivertex angle of the cone. Numerical results show that good agreement between the exact solution and the recommended formula can be obtained.

CONCLUSIONS

The salient points in this study include: 1) Derivation of a systematic exact solutions for buckling analysis of laminated conical shells under axial compression and external pressure, using power series method; 2) The solutions are applicable to all types of boundary conditions and to various kinds of laminated conical shells; 3) The effects of stretching-bending coupling, semivertex angle and geometrical constants on the buckling loads have been noted; 4) A simple formula for critical buckling load of laminated cones is suggested and verified..

REFERENCES

1. P. Seide,'Axisymmetrical buckling of circular cones under axial compression',J. Appl. Mech.,(1956)p626-628

2. M. Baruch, O. Harari and J. Singer,'Low buckling loads of axially compressed conical shells', J. Appl. Mech.,(1970) p384-392

3. Liyong Tong,'Buckling and Vibration of Conical Shells Composed of Composite Materials',Ph. D. Thesis, Beijing University of Aeronautics and Astronautics, April 1988

4. R. M. Jones,'Mechanics of Composite Materials',Script Book Company, 1975

APPENDIX

$$G_{1,1} = -\frac{(3m+1)sin\alpha}{(m+2)R_o}; \quad G_{1,2} = -\frac{m(3m-1)sin^2\alpha}{(m+2)(m+1)R_o^2} + \frac{A_{22}sin^2\alpha + A_{66}n^2}{A_{11}R_o^2(m+2)(m+1)}$$

$$G_{1,3} = \frac{(A_{22}sin^2\alpha + A_{66}n^2)sin\alpha - A_{11}(m-1)^2sin^3\alpha}{A_{11}R_o^2(m+2)(m+1)}$$

$$G_{1,4} = -\frac{(A_{12}+A_{66})n}{A_{11}R_o^2(m+2)}$$

$$G_{1,5} = -\frac{[2(A_{12}+A_{66})(m-1)-(A_{22}+A_{66})]nsin^2\alpha}{A_{11}R_o^2(m+2)(m+1)}$$

$$G_{1,6} = -\frac{[(A_{12}+A_{66})(m-1)-(A_{22}+A_{66})]nsin^2\alpha}{A_{11}R_o^3(m+2)(m+1)}$$

$$G_{1,7} = \frac{B_{11}(m+3)}{A_{11}}; \quad G_{1,8} = \frac{B_{11}(3m+1)sin\alpha}{A_{11}R_o}$$

$$G_{1,9} = [B_{11}m(3m-1)sin^2\alpha - (B_{12}+2B_{66})n^2 - B_{22}sin^2\alpha + A_{12}R_ocos\alpha]/A_{11}R_o^2(m+2)$$

$$G_{1,10} = \{B_{11}m(m-1)^2sin^3\alpha - [(B_{12}+2B_{66})n^2 + B_{22}sin^2\alpha] msin\alpha + R_o(2A_{12}m - A_{22})sin\alpha cos\alpha + (B_{12}+B_{22}+2B_{66})n^2sin\alpha\}/A_{11}R_o^3(m+2)(m+1)$$

$$G_{1,11} = \frac{[A_{12}(m-1)-A_{22}]sin^2\alpha cos\alpha}{A_{11}R_o^3(m+2)(m+1)}; \quad G_{2,1} = \frac{(A_{12}+A_{66})n}{A_{66}R_o(m+2)}$$

$$G_{2,2} = \frac{[2(A_{12}+A_{66})m + A_{22}+A_{66}]nsin\alpha}{A_{66}R_o^2(m+2)(m+1)}$$

$$G_{2,3} = \frac{[(A_{12}+A_{66})(m-1)-(A_{22}+A_{66})]nsin^2\alpha}{A_{66}R_o^3(m+2)(m+1)}; \quad G_{2,4} = G_{1,1}$$

$$G_{2,5} = \frac{A_{22}n^2 + A_{66}[1 - m(3m-1)]sin^2\alpha}{A_{66}R_o^2(m+2)(m+1)}$$

$$G_{2,6} = \frac{[A_{22}n^2 - A_{66}m(m-2)sin^2\alpha]sin\alpha}{A_{66}R_o^3(m+2)(m+1)}$$

$$G_{2,7} = -\frac{(B_{12}+2B_{66})n}{A_{66}R_o}; \quad G_{2,8} = -\frac{[2(B_{12}+B_{66})m + B_{22}]nsin\alpha}{A_{66}R_o^2(m+2)}$$

$$G_{2,9} = [B_{22}(n^2 - msin^2\alpha) - A_{22}R_ocos\alpha - (B_{12}+2B_{66}) m(m-1)sin^2\alpha]n/A_{66}R_o^3(m+2)(m+1)$$

$$G_{2,10} = -\frac{A_{22}nsin\alpha cos\alpha}{A_{66}R_o^3(m+2)(m+1)}$$

$$G_{3,1} = \frac{B_{11}}{D_{11}(m+4)}; \quad G_{3,2} = \frac{2(2m+1)B_{11}sin\alpha}{D_{11}R_o(m+4)(m+3)}$$

$$G_{3,3} = \frac{(6B_{11}m^2 - B_{22})sin^2\alpha - (B_{12}+2B_{66})n^2 + A_{12}R_ocos\alpha}{D_{11}R_o^2(m+4)(m+3)(m+2)}$$

$$G_{3,4} = \{2B_{11}m(m-1)(2m_1)sin^3\alpha + 3A_{12}R_o m sin\alpha cos\alpha -$$
$$2(B_{12} + 2B_{66})mn^2 sin\alpha - B_{22}(n^2 + (2m-1)sin^2\alpha]sin\alpha +$$
$$A_{22}R_o sin\alpha cos\alpha\}/(D_{11}R_o^4 m_{41})$$

$$G_{3,5} = \{B_{11}(m-1)^2(m-2)sin^4\alpha + 3A_{12}R_o(m-1)sin^2\alpha cos\alpha -$$
$$(B_{12} + 2B_{66})n^2(m-1)sin^2\alpha - B_{22}[n^2 + (m-2)sin^2\alpha]sin^2\alpha$$
$$+2A_{22}R_o sin^2\alpha cos\alpha\}/(D_{11}R_o^4 m_{41})$$

$$G_{3,6} = \frac{[A_{22} + A_{12}(m-2)]sin^3\alpha cos\alpha}{D_{11}R_o^4 m_{41}}; \qquad G_{3,7} = \frac{(B_{12} + 2B_{66})n}{D_{11}R_o(m+4)(m+3)}$$

$$G_{3,8} = -\frac{[3(B_{12} + 2B_{66})m - B_{22}]n sin\alpha}{R_o(m+4)(m+3)(m+2)}$$

$$G_{3,9} = \{3(B_{12} + 2B_{66})nm(m-1)sin^2\alpha + A_{22}R_o n cos\alpha -$$
$$B_{22}n[n^2 + (2m-1)sin^2\alpha]\}/D_{11}R_o^3 m_{41}$$

$$G_{3,10} = \{(B_{12} + 2B_{66})n(m-1)(m-2)sin^3\alpha + 2A_{22}R_o n sin\alpha cos\alpha -$$
$$B_{22}n[n^2 + (m-2)sin^2\alpha]sin\alpha\}/D_{11}R_o^4 m_{41}$$

$$G_{3,11} = \frac{A_{22}n sin^2\alpha cos\alpha}{D_{11}R_o^4 m_{41}}; \qquad G_{3,12} = -\frac{2(2m+1)sin\alpha}{R_o(m+4)}$$

$$G_{3,13} = [\frac{2(D_{12} + 2D_{66})n^2 + D_{22}sin^2\alpha}{D_{11}R_o^2} - \frac{P}{2\pi D_{11}R_o cos\alpha} +$$
$$\frac{6m^2 sin^2\alpha}{R_o^2} - \frac{2B_{12}cos\alpha}{D_{11}R_o}]/(m+4)(m+3)$$

$$G_{3,14} = \{-\frac{q(m+1)tan\alpha}{D_{11}} - \frac{3Pm sin\alpha}{2D_{11}R_o^2 \pi cos\alpha} -$$
$$\frac{2m(m-1)(2m-1)sin^3\alpha}{R_o^3} - \frac{6B_{12}m sin\alpha cos\alpha}{D_{11}R_o^2} +$$
$$\frac{[2(D_{12} + 2D_{66})n^2 + D_{22}sin^2\alpha](2m-1)sin\alpha}{D_{11}R_o^3}\}/(m+4)(m+3)(m+2)$$

$$G_{3,15} = \{[2(D_{12} + 2D_{66})n^2 + D_{22}sin^2\alpha]sin^2\alpha m(m-2)$$
$$-A_{22}R_o^2 cos^2\alpha - D_{11}m(m-1)^2(m-2)sin^4\alpha - D_{22}n^4$$
$$+\frac{qn^2 R_o^3}{cos\alpha} + 2(D_{12} + D_{22} + 2D_{66})n^2 sin^2\alpha - 4qR_o^3 m tan\alpha sin\alpha$$
$$-\frac{3PR_o m(m-1)sin^2\alpha}{2\pi cos\alpha} - 3.5qR_o^3 m(m-1)tan\alpha sin^2\alpha$$
$$-6B_{12}R_o m(m-1)cos\alpha sin^2\alpha - B_{22}R_o cos\alpha(sin^2\alpha - 2n^2)\}/(D_{11}R_o^4 m_{41})$$

$$G_{3,16} = [-(\frac{P}{2\pi cos\alpha} + 2B_{12}cos\alpha + \frac{4.5qR_o^2}{cos\alpha})(m-1)(m-2)sin^3\alpha$$
$$-6qR_o^2(m-1)tan\alpha sin^2\alpha - 2A_{22}R_o sin\alpha cos^2\alpha$$
$$-B_{22}(sin^2\alpha - 2n^2)sin\alpha cos\alpha + \frac{3qR_o^2 n^2 sin\alpha}{cos\alpha}]/(D_{11}R_o^4 m_{41})$$

$$G_{3,17} = [-0.5q tan\alpha R_o(m-2)(5m-7)sin^3\alpha + 3qn^2 R_o tan\alpha sin\alpha$$
$$-A_{22}sin^2\alpha cos^2\alpha]/(D_{11}R_o^4 m_{41})$$

$$G_{3,18} = [-0.5q(m-2)(m-3)tan\alpha sin\alpha - \frac{qn^2}{cos\alpha}]sin^3\alpha/(D_{11}R_o^4 m_{41})$$

where $m_{41} = (m+4)(m+3)(m+2)(m+1)$

Table 1. ρ_{cr} for SS_3 and SS_1, $(R_1/h = 100.0 \quad \mu = 0.3)$

L/R_1	0.2	0.2	0.5	0.5
α	present	Ref.[2]	present	Ref.[2]
$1°$	0.5028	0.4991	0.5131	0.5131
$5°$	0.5051	0.5021	0.5139	0.5139
$10°$	0.5073	0.5075	0.5146	0.5147
$30°$	0.5567	0.5567	0.5138	0.5139
$60°$	0.8701	0.8701	0.4486	0.4486
$80°$	2.3832	2.3830	0.5405	0.5407

Table 2. ρ_{cr} and n for SS_4 $(R_1/h = 100.0 \quad \mu = 0.3)$

L/R_1	0.2	0.2	0.5	0.5
α	present	Ref.[2]	present	Ref.[2]
$1°$	1.0000(7)	1.005(7)	1.0019(8)	1.002(8)
$5°$	1.0000(7)	1.006(7)	1.0017(8)	1.002(8)
$10°$	1.0000(7)	1.007(7)	1.0013(8)	1.002(8)
$30°$	1.0000(5)	1.017(5)	1.0007(7)	1.001(7)
$60°$	1.1442(0)	1.144(0)	1.0011(5)	1.044(7)
$80°$	2.4500(0)	2.477(0)	1.0106(3)	1.015(5)

Table 3. ρ_{cr} and n for SS_4 boundary condition

L/R_o	N	$10°$	$30°$	$45°$	$60°$
0.2	2	0.1665(9)	0.1827(9)	0.2131(8)	0.2831(7)
	4	0.2218(8)	0.2442(8)	0.2858(7)	0.3820(6)
	6	0.2258(8)	0.2489(8)	0.2912(7)	0.3895(6)
	∞	0.2270(8)	0.2504(7)	0.2933(7)	0.3931(6)
0.5	2	0.07926(9)	0.08389(9)	0.08703(6)	0.09380(6)
	4	0.1101(6)	0.1106(6)	0.1150(6)	0.1226(5)
	6	0.1119(6)	0.1128(6)	0.1173(5)	0.1248(5)
	∞	0.1115(6)	0.1127(6)	0.1166(5)	0.1247(5)
1.0	2	0.06990(7)	0.06984(7)	0.06949(6)	0.07034(6)
	4	0.1021(6)	0.1029(6)	0.1067(5)	0.1119(5)
	6	0.1070(6)	0.1086(6)	0.1121(5)	0.1193(4)
	∞	0.1107(6)	0.1129(6)	0.1165(5)	0.1246(4)
1.2	2	0.06877(8)	0.06999(7)	0.06962(6)	0.06837(5)
	4	0.1065(7)	0.1057(5)	0.1042(5)	0.1082(4)
	6	0.1087(6)	0.1100(5)	0.1097(5)	0.1139(4)
	∞	0.1173(6)	0.1134(5)	0.1139(5)	0.1187(4)

Table 4. ρ_{cr} and n for SS_3 boundary condition

L/R_o	N	$10°$	$30°$	$45°$	$60°$
	2	0.09407(9)	0.1011(9)	0.1146(8)	0.1462(7)
	4	0.1943(8)	0.2133(8)	0.2488(7)	0.3290(0)
0.2	6	0.2124(8)	0.2339(8)	0.2732(7)	0.3618(0)
	∞	0.2268(8)	0.2500(7)	0.2927(7)	0.3876(0)
	2	0.06742(7)	0.06726(7)	0.06751(6)	0.06849(6)
	4	0.1016(6)	0.1018(6)	0.1054(6)	0.1111(5)
0.5	6	0.1066(6)	0.1075(6)	0.1117(5)	0.1185(5)
	∞	0.1106(6)	0.1120(6)	0.1158(5)	0.1242(5)
	2	0.06986(9)	0.06984(6)	0.06743(6)	0.06949(6)
	4	0.1017(6)	0.1028(6)	0.1063(5)	0.1117(5)
1.0	6	0.1067(6)	0.1086(6)	0.1122(5)	0.1188(4)
	∞	0.1107(6)	0.1125(6)	0.1165(5)	0.1240(4)

Table 5. ρ_{cr} and n for SS_1 boundary condition

L/R_o	N	$10°$	$30°$	$45°$	$60°$
	2	0.08559(8)	0.09250(8)	0.1054(7)	0.1360(6)
	4	0.1820(7)	0.2009(6)	0.2364(6)	0.3185(5)
0.2	6	0.1996(7)	0.2206(6)	0.2605(6)	0.3523(5)
	∞	0.2137(7)	0.2363(6)	0.2797(6)	0.3790(4)
	2	0.05964(7)	0.05975(7)	0.05844(6)	0.05950(5)
	4	0.08999(6)	0.09145(6)	0.09212(5)	0.1007(4)
0.5	6	0.09508(6)	0.09653(5)	0.09749(5)	0.1069(4)
	∞	0.09915(6)	0.1000(5)	0.1017(5)	0.1118(4)
	2	0.06619(6)	0.06396(6)	0.06207(6)	0.06277(5)
	4	0.09505(6)	0.09643(6)	0.09842(5)	0.1036(5)
1.0	6	0.1002(6)	0.1023(6)	0.1039(5)	0.1075(4)
	∞	0.1043(6)	0.1070(6)	0.1084(5)	0.1098(3)

Table 6. ρ_{cr} and n for SS_2 boundary condition

L/R_o	N	$10°$	$30°$	$45°$	$60°$
	2	0.1594(8)	0.1752(8)	0.2050(7)	0.2738(6)
	4	0.2084(7)	0.2303(6)	0.2718(6)	0.3672(5)
0.2	6	0.2121(7)	0.2344(6)	0.2770(6)	0.3746(4)
	∞	0.2137(7)	0.2363(6)	0.2798(6)	0.3791(4)
	2	0.07371(8)	0.07732(8)	0.08535(8)	0.08840(5)
	4	0.09824(6)	0.1002(6)	0.1017(5)	0.1121(4)
0.5	6	0.09983(6)	0.1015(5)	0.1029(5)	0.1131(4)
	∞	0.09973(6)	0.1007(5)	0.1025(5)	0.1126(4)
	2	0.06380(7)	0.06433(7)	0.06319(6)	0.06366(5)
	4	0.09506(6)	0.09659(6)	0.09849(5)	0.1043(5)
1.0	6	0.1002(6)	0.1023(6)	0.1040(5)	0.1117(4)
	∞	0.1044(6)	0.1071(6)	0.1086(5)	0.1148(4)

THE SYMMETRICAL BUCKLING OF SPHERICAL SHELLS

CHU QUOC THANG AND E. DULACSKA
INSTITUT OF ARCHITECTURAL DEVELOPMENT
Budapest, Asboth u. 9-11
HUNGARY-1075

INTRODUCTION

The latest decades' bibliography concerning the buckling of shells dedicated much to the buckling of spherical shells both in the aeronautical and building domain.

It is well known that the linear approach offers a theoretical resolution for the issues of shell stability by means of eigenvalue-problem. However, the theoretical results thus obtained are usually much greater than the experimental results. The surmounting of the difference between theoretical and experimental results , called for the working out of shell theory according to the non-linear (namely large deformation) approach. Due to the mathematical difficulties which arise in the resolution of a non-linear differential equation-system, researchers have hitherto resorted to different approximations in the study of shell-buckling.

It is also known that geometrically perfect dome shells have only a theoretical existence. In reality the shells possess some initial imperfections(eg. initial wave, remanent tensions) owing to productional or technical reasons. The impact of these imperfections affects the behaviour of the shell. As a result of this the actual critical load of the shell is usually less than the critical value calculated for a theoretically perfect shell.

The bibliography seemed to have resolved the problem of non-linear shell-buckling for the most part. First it was Kármán and Tsien [1] who studied the buckling of perfect and complete shell (without initial imperfection) in 1939. They were followed by Mushtari and

Surkin [2] ,Gabriljanc and Feodosiev [3], Wolmir [4] ,Thompson [5] ,Van Koten and Haas [6] ,Poso Frutos and Pozo Vindel [7].

While the buckling of imperfect and complete spherical shells with initial geometrical imperfections was examined by Hutchinson [8] ,Buschnell [9] ,Koga and Hoff [10] by means of different methods.The analysis of research methods and procedures hitherto applied illustrates that they were not always correct in everything.For example some of them failed to guarantee the boundary conditions along the buckling periphery.Others neglected the differences between the extension of the initial wave and the buckling wave or didn't calculate the lower critical load for the imperfect shell.

This paper treats on the problem of complete perfect spherical shells parallel with buckling of imperfect shells ,with initial geometrical imperfections (initial wave) ,pressuposing the linearly elastic feature of the material.By taking note of the boundary conditions along the buckling periphery the research aims at a more correct resolution than the ones mentioned.At the same time the research has as its goal the study of the degree in which the initial waviness affects shell buckling behaviour.

We pointed out the different extensions of initial waviness (i.e. those with the smallest critical load for the shell), consequently we made suggestions as far as the critical load for shell project is concerned when determining buckling-safety.

ASSUMPTIONS

The subject suggested for the energy-method analysis is a perfect or imperfect spherical shell subjected to uniform load caused by external pressure.The calculations were made considering the following assumptions :

1.The material of the shell remains linearly elastic all along the buckling, in accordance with the hookian law of stress-strain.

2.The geometrical imperfection (initial waviness) possible has a circular symmetry and occupies a relatively restricted area, in other words the top view size of initial waviness is much less than the R radius of the shell ($c_0 \ll R$).

3. The deformation caused by buckling has a circular symmetry and a relatively small extent (i.e. c<<R).

We regard the shell as consisting of two parts: the buckling, flat spherical calotte and the remaining scanty, spherical diaphragm-like part. The paper assumes that buckling deformation is symmetrical and can be expressed in a polynomial way, which meets the requirements of statical and strain junction conditions along the buckling circumference.

NOTATIONS

R =the radius of curvature of the spherical shell's mean surface

h =the wall thickness of the shell

E =the elastic modulus

ν =the cross-contractional factor

D =the bending stiffness of the shell

σ_r, σ_θ =meridian-oriented and ring-oriented specific stress

$\varepsilon_r, \varepsilon_\theta$ =meridian-oriented and ring-oriented specific strain

κ_r, κ_θ =meridian-oriented and ring-oriented variation of curvature

F =stress function

W, U_r =the normal and tangential displacement component of the shell's mean surface

W_m =the diaphragm-like normal displacement of the shell's mean surface

W_h =buckling displacement

W_0 =initial waviness

A, δ =the amplitude of the buckling wave and of the initial wave

a, d =the dimensionless amplitude(amplitude parameter) of the buckling wave and of the initial wave

c, c_0 =the extension of the buckling wave and of the initial wave

K, K_0 =the dimensionless extension (extension parameter) of the buckling wave and of the initial wave

$\alpha = K/K_0$ =the ratio of extension of the buckling wave and of the initial wave p   =the uniform load, which is normal to shell's surface

p_{cr}^{lin} =linear critical load

p_{cr}^{u}, p_{cr}^{l} =upper and lower critical load

$P, P_{cr}^{u}, P_{cr}^{l}$ =dimensionless load, dimensionless upper and lower critical load

U, u   =complete potential energy and substitutional dimensionless complete potential energy

U_m, U_h, U_p, = membrane deformation energy, bend deformation energy and the work of external load

$g_i (i=1,2...,10)$ = the constants in the expression of the 'u' substitutional dimensionless complete potential energy

$B_i(\alpha)$ $(i=1,...,4)$='α' variable functions in the 'u' expression

THE FUNDAMENTAL EQUATIONS OF THE PROBLEM

The general differential equations based on non linear (large deformation) theory for flat spherical shells are analoguous with Marguer's equations for strips suffering big deformations [11].

Owing to the circularly symmetric nature of the flat spherical shell equations for mean surface strain-displacement are set up in a polar coordinate system, as follows :

$$\varepsilon_r = \frac{dU_r}{dr} - \frac{W}{R} + \frac{1}{2}\left(\frac{dW}{dr}\right)^2, \quad \varepsilon_\theta = \frac{U_r}{r} - \frac{W}{R}, \quad \kappa_r = \frac{d^2W}{dr^2}, \quad \kappa_\theta = \frac{1}{r}\cdot\frac{dW}{dr},$$

Note that R is the radius of the shell's mean surface, W is the radial oriented component of the shell's buckling strain, while U_r is the tangential component.

The normal stresses of the shell derive from the F stress function:

$$\sigma_r = \frac{1}{r}\cdot\frac{dF}{dr}, \quad \sigma_\theta = \frac{d^2F}{dr^2}.$$

The stress-strain equations are the following :

$$\varepsilon_r = \frac{1}{E}\left(\sigma_r - \nu\cdot\sigma_\theta\right), \quad \varepsilon_\theta = \frac{1}{E}\left(\sigma_\theta - \nu\cdot\sigma_r\right).$$

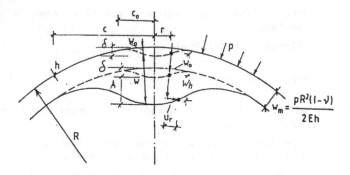

Fig 1. Notations

The compatibility equation for the spherical shell possessing W_0 initial waviness is the following :

$$\frac{d}{dr}\left[\nabla^2 F\right] = -E\cdot\left[\frac{1}{R}\cdot\frac{dW}{dr} + \frac{1}{2r}\cdot\left[\frac{dW}{dr}\right]^2 + \frac{1}{r}\cdot\frac{dW}{dr}\cdot\frac{dW_0}{dr}\right],$$

where $\nabla^2() = \frac{1}{r}\cdot\frac{d}{dr}\left[r\cdot\frac{d}{dr}()\right]$ is the Laplacian operator circular symmetric cases in the polar coordinate system.

Thus 'U' complete potential energy can be defined as: $U = U_h + U_m - U_p$, where the energy of bend deformationis :

$$U_h = \frac{D}{2}\cdot\int_0^c\left[\left[\nabla^2 W\right]^2 - 2\left(1-\nu\right)\cdot\frac{1}{r}\cdot\frac{dW}{dr}\cdot\frac{d^2W}{dr^2}\right]\cdot 2\pi r dr,$$

and

$$D = \frac{E\cdot h^3}{12(1-\nu^2)}$$

is the bending stiffness of the shell, while 'c' is the buckling half wavelength and 'h' is the wall-thickness of the shell.

The energy of the membrane strain is the following :

$$U_m = \frac{h}{2E}\cdot\int_0^c\left[\left[\nabla^2 F\right]^2 - 2\left(1+\nu\right)\cdot\frac{1}{r}\cdot\frac{dF}{dr}\cdot\frac{d^2F}{dr^2}\right]\cdot 2\pi r dr.$$

The work of external load is :

$$U_p = \int_0^c pW 2\pi r dr.$$

STRAIN FUNCTION AND STRESS FUNCTION

Let's test the complete sphere ,which possesses a W_0 initial waviness and which is loaded by a 'p' uniform external load. We assume that the shape of the initial wave is axially symmetrical and can be set up in the following way (fig.1) :

$$W_0 = \delta \cdot \left[1 - \left(\frac{r}{c_0} \right)^2 \right]^2 ,$$

where 'δ' is the amplitude of the initial wave and c_0 is the top view radius of the initial wave.

Along the circumference of the buckling surface the peripheral junction conditions are the following : at $r = 0$ $\left(\frac{dW_h}{dr} \right)_{r=0} = 0,$

at $r = c$ $\left[W_h \right]_{r=c} = 0$, $\left(\frac{dW_h}{dr} \right)_{r=c} = 0$, $\left(\frac{d^2 W_h}{dr^2} \right)_{r=c} = 0$, $\left(\frac{d^3 W_h}{dr^3} \right)_{r=c} = 0$

If one considers the buckling function of an eight-degree polynomial while meeting the requirements of the above mentioned peripheral conditions and reducting and simplifying he will arrive at the following mathematical expression :

$$W = W_m + W_h = \frac{pR^2}{2Eh} \cdot \left[1 - \nu \right] + A \cdot \left[1 - \left(\frac{r}{c} \right)^2 \right]^4 .$$

The first member of the buckling function is the displacement according to the membrane state. In the function 'A' is the amplitude of the buckling wave and 'c' is the buckling half wavelength.

The stress function can be calculated if one substitutes the chosen initial waviness and buckling waviness functions into the compatibility equation. Thus the complete potencial energy becomes known.

In resolving the F stress function two cases may occur :
- In the first case the shell's buckling valley entirely surrounds the initial wave ,which means that the 'c' buckling wave extension is greater than the c_0 initial wave extension ($c > c_0$).
- In the second case the situation is reversed ,thus $c < c_0$.

In the first case the shell consists of three parts :

a) one part ($0 \leq r \leq c_0$) possesses a W_0 initial wave and changes its shape together with the W buckling wave.

b) another part ($c_0 \leq r \leq c$) doesn't possess an initial buckle; it only bends with the W buckling wave.

c) finally in the region $r > c$ the shell is in membrane state.

The compatibility equation for the first part of the shell ($0 \leq r \leq c_0$):

$$\frac{d}{dr}\left[\nabla^2 F_I\right] = -E\left[\frac{1}{R}\cdot\frac{dW}{dr} + \frac{1}{2r}\left(\frac{dW}{dr}\right)^2 + \frac{1}{r}\cdot\frac{dW}{dr}\cdot\frac{dW_0}{dr}\right],$$

For the second part of the shell ($c_0 \leq r \leq c$), regarding that $W_0 = 0$:

$$\frac{d}{dr}\left[\nabla^2 F_{II}\right] = -E\left[\frac{1}{R}\cdot\frac{dW}{dr} + \frac{1}{2r}\left(\frac{dW}{dr}\right)^2\right].$$

In determining F_I and F_{II} stress functions one has to take note of the following junctional conditions :

—stresses can nowhere be infinite,

— on the junctional dead line of the two parts (I and II) the stresses are equal:

$$\left(\sigma_r^{(I)}\right)_{r=c_0} = \left(\sigma_r^{(II)}\right)_{r=c_0}, \qquad \left(\sigma_\theta^{(I)}\right)_{r=c_0} = \left(\sigma_\theta^{(II)}\right)_{r=c_0}.$$

Due to the diaphragm state junction the border line of the buckling surface cannot move tangentially:

$$\left(U_r\right)_{r=c} = 0.$$

In the second type of strain ,as the buckling wave is entirely placed in the initial wave,the F function can be found by integrating the compatibility equations and regarding the first and the third condition of the above mentioned ones.

The next stage is the introduction of dimensionless parameters:

$$\eta = \frac{r}{c}, \qquad a = \frac{A}{h}, \qquad d = \frac{\delta}{h}, \qquad \alpha = \frac{c}{c_0} = \frac{K}{K_0},$$

$$K = \sqrt[4]{12\left[1-\nu^2\right]}\cdot\frac{c}{\sqrt{R\cdot h}}, \qquad K_0 = \sqrt[4]{12\left[1-\nu^2\right]}\cdot\frac{c_0}{\sqrt{R\cdot h}},$$

$$\bar{P} = \frac{p}{p_{cr}^{lin}} = \frac{\sqrt{3\cdot\left[1-\nu^2\right]}}{2}\cdot p\cdot\frac{R^2}{E\cdot h^2}\text{(load parameter)},$$

$$u = \frac{U}{\pi \cdot Eh^4} \cdot$$

Thus the 'u' substitutional dimensionless complete potential energy can be formulated :

$$u = K^2 \left[\bar{P}^2 \cdot g_1 + a^2 \cdot g_2 \right] + \frac{1}{K^2} \left[a^4 \cdot g_3 + a^3 d \cdot g_4 \cdot B_3(\alpha) + \right.$$

$$\left. + a^2 d \cdot g_5 \cdot B_1(\alpha) + a^2 \cdot g_6 \right] + a^3 \cdot g_4 + a^2 \cdot \bar{P} \cdot g_8 + a^2 d \cdot g_9 B_4(\alpha) +$$

$$+ \bar{P} \cdot ad \cdot B_2(\alpha) \cdot g_{10} \quad ,$$

where g_i are constants and $B_i(\alpha)$ are polynomials with 'α' variable.

THE STUDY OF PERFECT SHELLS

In case of perfect shells d=0, fact which enables a more simple way of expressing complete potencial energy.

When determining the critical load the following conditions have to be fulfilled:

the equilibrium condition $\quad \dfrac{\partial u}{\partial a} = 0$,

and the minimum conditions $\quad \dfrac{\partial \bar{P}}{\partial K} = 0 \quad , \quad \dfrac{\partial \bar{P}}{\partial a} = 0.$

In case of perfect spheres the relationship between the $\bar{P}$ load parameter and the 'a' buckling strain can be calculated from the $\dfrac{\partial u}{\partial a} = 0$ equilibrium condition:

$$\bar{P} = \quad - \frac{1}{g_8} \cdot \left[K^2 \cdot g_2 + \frac{1}{K^2} \left(2a^2 \cdot g_3 + g_6 \right) + \frac{3}{2} a \cdot g_7 \right].$$

while the 'K' parameter can be defined by the $\dfrac{\partial \bar{P}}{\partial K} = 0$ minimum condition:

$$K^2 = \sqrt{ \frac{1}{g_8} \cdot \left(2a^2 \cdot g_3 + g_6 \right) }.$$

If one substitutes the K^2 value obtained into the previous $\bar{P}$ equation, he will come to an equation which expresses the relationship between the $\bar{P}$ load and the 'a' buckling wave amplitude.

To avoid the derivation of the above mentioned complex, square root containing formula one can resort to a graphical method. Thus he will obtain the load-strain curve together with its minimum point and bifurcation point which corresponds to the condition $\frac{\partial \bar{P}}{\partial a} = 0$.

In the case studied the value of the minimum (i.e. the lower) critical load for $\nu = 0$ is the following:

$$p^1_{cr} = 0.255 \cdot p^{lin}_{cr} = 0.294 \cdot E \cdot \frac{h^2}{R^2}.$$

The buckling wave amplitude corresponding to this is $A = 5.56h$ and the buckling half wavelength parameter is $K = 7.09$.

It is clear that both the buckling wave amplitude and the half wavelength grow in time of buckling, the latter illustrating buckling extension.

THE STUDY OF IMPERFECT SPHERES

In case of spheres with initial waviness, the equilibrium condition $\frac{\partial u}{\partial a} = 0$ leads to the $\bar{P}$ load parameter:

$$\bar{P} = \frac{-1}{2a \cdot g_8 + d \cdot B_2(\alpha) \cdot g_9} \left\{ 2K^2 \cdot ag_2 + \frac{1}{K^2} \left[4a^3 g_3 + 3a^2 d \cdot g_4 B_3(\alpha) + \right. \right.$$
$$\left. \left. + 2ad^2 \cdot g_5 B_1(\alpha) + 2ag_6 \right] + 3a^2 g_7 + 2ad \cdot g_3 B_4(\alpha) \right\}.$$

Starting from the minimum condition one can draw a set of load-strain curves for different K (i.e. α) values for the previously given K_0 and for the initial d waviness. The second figure shows the set of curves with $K_0 = 3$ and d = 0.2. The envelope of the set has a precise physical content. If the point with given $\bar{P}$ and 'a' coordinates is above the envelope the sphere's state of equilibrium is excluded. In the A and B extreme points of the envelope the $\bar{P}$ load is at its extreme values; these are the lower and upper critical load of the shell. Figure 3 shows the effect of 'd' variation when $K_0 = 3$.

The numerical results for varying parameters of d (initial wave) in range 0.1—3.5 and of K_0(initial wave extension) in range 1.0—8.0 can be seen on table 1.

TABLE 1.

The $\bar{P}^u_{cr}$ and $\bar{P}^l_{cr}$ load of a shell possessing different "d" and K_0

k_0	d	$\bar{P}^u_{cr}$	a	α	$\bar{P}^l_{cr}$	a	α
1,0	0,1	0,7716	0,3	3,3	0,255	5,6	7,2
	0,3	0,6230	0,4	3,4	0,2581	5,6	7,1
	0,7	0,5210	0,6	3,7	0,2660	5,7	7,2
2,0	0,1	0,6423	0,4	1,7	0,255	5,5	3,5
	0,3	0,4428	0,6	1,8	0,2586	5,3	3,5
	0,7	0,3358	0,7	2,2	0,2771	4,1	3,2
3,0	0,8	0,2305	0,8	1,6	0,2215	1,4	1,6
	1,0	0,2128	0,9	1,5	0,2071	1,5	1,7
	1,5	0,2087	1,1	1,9	0,2084	1,5	1,9
	2	0,2511	1,3	2,1	0,2503	1,65	2,1
4,0	1,0	0,1630	1	1,2	0,1565	2,2	1,4
	1,5	0,1182	1,1	1,4	0,1182	1,2	1,4
	2,0	0,1320	1,1	1,6	0,1295	1,65	1,1
	2,5	0,1622	1,4	1,75	0,1615	1,8	1,75
5,0	1,5	0,1133	0,9	1,05	0,0966	2,5	1,25
	2,5	0,0731	1,0	1,65	0,0693	1,65	1,35
	3,5	0,1190	1,55	1,6	0,1180	2,1	1,6
6,0	2,5	0,0590	0,85	1	0,0576	3,0	1
	3,5	0,0349	1,05	1,1	0,0349	1,25	1,1
7,0	2,5	0,1326	0,86	1	0,0324	3,34	1,045
	3,0	0,0756	0,8	1	0,0158	3,8	1,15
	3,5	0,0424	0,75	1	0,0116	3,25	1,2
8,0	2,5	0,2907	1,15	1	0,0230	5,6	1,075
	3,0	0,1895	1,1	1	0,0035	5,35	1,1
	3,5	0,1197	1	1	-0,0084	4,85	1,1

144

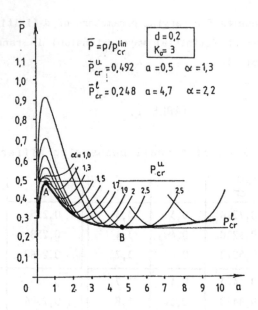

Fig 2. The upper and lower critical load when $K_0=3$ and $d=0.2$

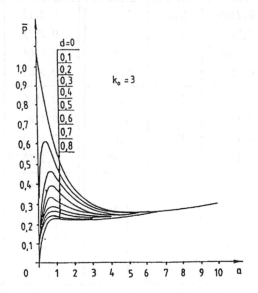

Fig 3. The effect of "d" variation When $K_0=3$

The set of curves for $K_0=8$ and $d=3.5$ and the envelope are illustrated on figure 4.

The variation of the upper critical load is shown by fig.5 while the lower critical load depending on K_0 and 'd' can be seen on fig.6.

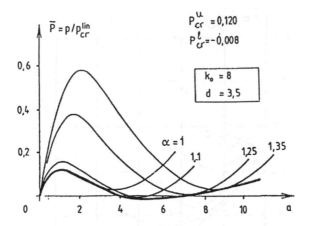

Fig 4. The upper and lower critical loads when K_o=8 and d=3.5

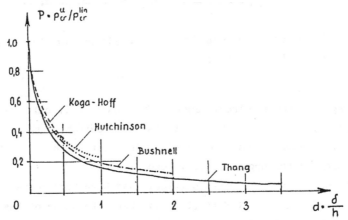

Fig 5. The decrease of the upper critical load as resulting from the growth of the waviness amplitude

The calculations indicate that the P_{cr}^l lower critical load tends to be negative for d ≥ 3.5 and for K_0 = 8. According to this, if the initial waviness is greather , the shell remains buckled even when there is no more load.

The above mentioned results show that the value of the lower critical load will be zero or negative in some instances and therefore ceases to be the lower limit of the upper critical load. When the lower critical load is zero or negative the shell will remain buckled even there is no load.

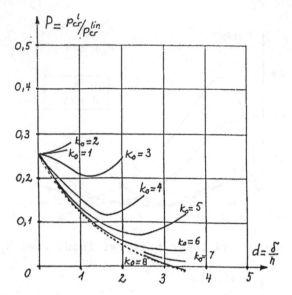

Fig 6. The effect of the initial waviness on the lower critical load of the shell

Owing to these circumstances the lower critical load is not definitely positive. Hence it is not suitable to control the stability-security of the shell. The latter has to be judged according to the upper critical load.

The results for the upper critical load have been compared with the results of Koga and of Hoff [10] together with those of Buschnell [9] and Hutchinson [8] (see fig.5). The more precise calculations that we made led to smaller values of the upper critical load.

Practical projecting often needs some simple formulae offering a good approximation for exact results. Such approximating formulae for

$$d = \frac{\delta}{h} \leq 3 \quad \text{are :}$$

$$p_{cr}^{u} = p_{cr}^{lin} \cdot \frac{1}{1 + 5.5d - 1.75d^2 + 0.85d^3} \quad ,$$

$$p_{cr}^{l} = \frac{p_{cr}^{lin}}{4} \cdot \left(1 - 0.6d + 0.1d^2\right)$$

The difference of these approximations is less than $0.01 p_{cr}^{lin}$.

CONCLUSIONS

The studies on the non-linear buckling of the shell were not always correct in every respect. For example some would fail in assuring boundary conditions along the buckling circumference. Others were not satifactory in applying the energy criteria.

According to the calculations based on the refining the equations and the resolution, the results are the following:

—The upper critical load (the result of a more accurate examination) is smaller than previously known critical loads.

—In some circumstances the lower critical load can be negative, therefore it cannot be the lower limit of the upper critical load. As a result of this it is not the lower critical load which is suitable for the study of stability security, but the upper critical load.

The following approximations have been roughly made. However they can still be well applied:

$$p_{cr}^{u} = p_{cr}^{lin} \cdot \left[\frac{1}{1 + 6d} \right],$$

$$p_{cr}^{l} = 0.23 \cdot p_{cr}^{lin} \cdot \left[1 - 0.33d \right]$$

The fault of the formulae is less than $0.05 \cdot p_{cr}^{lin}$.

REFERENCES

(1) Kármán, Th. von and Tsien, H. S.: The Buckling of Spherical Shell by External Pressure. Journ. Aeronaut. Sci 7, 43 (1939).

(2) Mushtari, H.M. and Surkin, R.G.: O nelineinoi teorii ustoichivosti uprugogo ravno vesiia sfericheskoi obolochki pri deistvii ravnomerno raspredelennogo vneshnego davleniia. Prikl. Mat. i Mech. 14 (1950), p. 573.

(3) Gabriliants, A. G. and Feodosiev, F. I.: Ob osesimmetrichnyk formakh ravnovesiia uprugoi sferichesk obolochki, nakhodia--shcheisia pod deistviem ravnomerno raszpredelennogo davleniia. Prkl. Mat. i Mech. 25. No. 6 (1961).

(4) Wolmir, A. S.: Biegsame Platten und Schalen. Verlag für Bauwesen, Berlin, 1962.

(5) Thompson, J. M. T.: The Post Buckling of Spherical. Shell by Computer Analysis. Proc. World Conference on Shell Structures, San Francisco, 1961. National Academy of Sciences Washington, D. C. (1964).

(6) Koten, H. van and Haas, A. M.: The Stability of Doubly Curved Surfaces Having a Positive Gaussian Curvature Index, Heron, 17. No. 4. =1970-70).

(7) Pozo Frutos, F. del and Pozo Vin-Del, P., del.: Elastic buckling of nonperfect spherical shells of constant thickness Proc. IASS World Congress, Madrid, 1 (1979), p. 1173-1179.

(8) Hutchinson, J. W.: Imperfection Sensitivity of externally pressurized spherical shells. Journ. Appl. Mech. 34, 49-55 (1967).

(9) Buschnell, D.: Nonlinear Axisymmetric Bechavior of Shells of Revolution. AIAA Journ. 5, 432-439 (1967).

(10) Koga, T. and Hoff, N. J.: The Axisymmetric Buckling of Initially Imperfect Complete Spherical Shell. Int. Journ. Solids Struct. 5 (1969).

(11) Marguerre, K.: Zur Theorie der gekrümmte Platte grosser Formänderung. Proc. 5 th Int. Congr. Appl. Mech. Cambridge, Massachusetts (1938), p. 93.

A CONCEPT OF DISTURBANCE ENERGY FOR THE BUCKLING ANALYSIS OF IMPERFECT SHELLS

HEINZ DUDDECK AND WALTER WAGENHUBER
Institut für Statik, Technical University Braunschweig
Beethovenstr. 51, 3300 Braunschweig, Germany

ABSTRACT

The buckling of thin shells is in general very sensitive to imperfections. A consistent numerical analysis for the determination of the buckling load has to consider the "correct" imperfections. Instead of assuming load or deformation imperfections, a new approach is presented which applies the minimum of perturbation energy with respect to all possible disturbance pattern as criteria for the stability degree of the shell. Thus, by starting from the perfect shell the engineer may compute the least energy necessary to reach the critical state of equilibrium and may introduce this approach for determining safety against buckling. A nonlinear eigenvalue problem is solved to find the adjacent buckling mode of smallest energy. The concept has been developed so far only for elastically buckling shells of arbitrary shape. As examples to demonstrate the method cylindrical shells under axial loads are investigated.

1. GENERAL APPROACH

In design codes, as in the ECCS-Recommendations "Buckling of steel shells", the sensitivity against imperfections is taken into account by special reduction factors entirely derived from statistically evaluated experimental results. In the last years sophisticated numerical methods have been developed which cover the geometrically and physically non-linear load-deformation path of buckling shells far into the post-buckling region, see e.g. /1/ and for the available approaches /2/. Numerical nonlinear analyses have to consider imperfect shells. So far it is not known which kind (displacements, additional loads, or inner eigen-stresses) and mode (distribution along the shell-surface) and value (maximum amount of disturbance) of imperfections should be applied in the buckling analysis, especially for those shells for which experimental results are not available.

The paper proposes to define safety against buckling by a degree of stability, expressed by the energy required to reach the instable post-buckling state of equilibrium at the same load level, see Fig. 1. The rate of change of this energy can be taken as a criterion for the sensitivity of the shell against imperfections. This approach takes the following main steps (for more details see /3/):

1. For the perfect structure (without any imperfections at all) and a given load level (P_0 in Fig. 2), the reference state of stresses and deformations is determined by a linear (or also a non-linear) theory of buckling. The corresponding point in Fig. 2 lies on the stable pre-buckling branch of the load deformation curve.
2. Then a nonlinear eigen-value algorithm is applied to determine that post-buckling state belonging to P_0 which will need a minimum of elastic energy to cross over the peak of the load-displacement function. Hereby all the possible disturbances and imperfections are considered without restrictions (of course, staying within the scope of the theory and the numerical method chosen).
3. The maximum of the potential energy of the perfect shell up to the first bifurcation point (taken as the basic reference value) and the perturbation energy, required additionally from P_0 to reach the critical state, are evaluated.

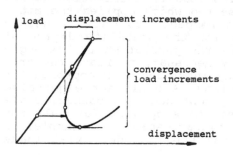

Fig. 1. Idealized pre- and post-buckling curves

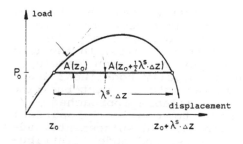

Fig. 2. Secant of the eigen-value-increment towards the corresponding critical state

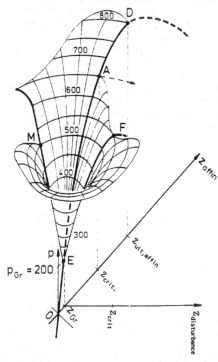

Fig. 3. Critical paths for different imperfection vectors $\mathbf{z}$

4. The degree of stability at point P_0 is then defined by the
   quotient of
   - perturbation energy, divided by
   - the basic maximum energy of the perfect structure.

   In Fig. 3 the complex loading-deformations functional of a
buckling structure is graphically visualized in a three-
dimensional figure. The vertical axis represents the load vector (here with arbitrary values). All possible deformation vectors are represented within the horizontal plane. The nonlinear
analysis of the perfect structure with given disturbances
(theory of second order) follows the curve O - E - A - D, where
D = 800 would be the ultimate load for a deformation pattern   z
in affinity to the given disturbances. The basic state P_{Gr} =
200 may be investigated with regard to its stability degree. A
linear eigen-value algorithm would yield the bifurcation point
A = 650 because the analysis takes into account only deformations in affinity to the basic state. However, when in E are
also considered perturbation vectors of arbitrary form, then
the curves O - E - F or O - E - M may be evaluated. In Fig. 3
the point M = 500, would be the ultimate load with a minimum of
energy. The real structure would experience and the analysis
would yield M = 500, if in both cases the general perturbations
would at least partic1ly include those deformations directed
towards M. Fig. 3 visualizes also the possible large scattering
of experimental results. As can be shown, for axially loaded
cylindrical shells a disturbance energy of only 0.2 % of the
maximum basic energy may result in an ultimate load of even 20%
of that corresponding to the theoretical bifurcation load, and
that a disturbance energy of 0.3 % yields the design values derived from experimental results (as in the German Code DIN
18.800 T.4).

2. THEORETICAL FUNDAMENTALS /3/

The geometrically non-linear theory of thin shells for arbitrary shape is expressed for numerical solutions by the finite-
element-method of the mixed stress-deformation formulation. The
elastic potential can be written in

$$\Pi = z^t \left[\frac{1}{2} A^L + \frac{1}{6} A^{NL}(z) \right] z - z^t \cdot p \qquad (2.1)$$

   The symbols are representing:

$z = (u, n, m)$: vector of displacements and stress resultants,

A^L = system matrix of linear elements,

A^{NL} = system matrix of geometrically non-linear parts,

p = vector of outer actions and imposed deformations.

The second variation of the potential (2.1), which corresponds
to the condition that the matrix of the generalized stiffness
is zero, can be expressed as an eigen-value problem:

152

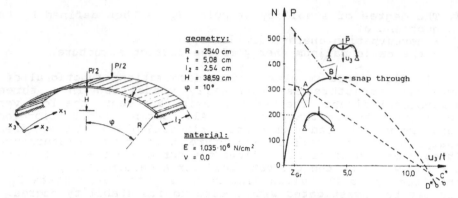

Fig. 4. Thin shell arch as validation example, symmetrically loaded

Fig. 5. Buckling of the shell arch with asymmetric A and symmetric B modes

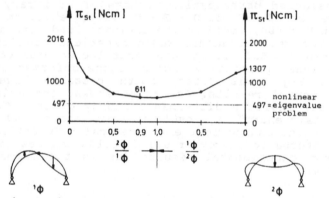

Fig. 6. Required disturbance energy starting at load level P = 200 N for different combinations of the first and second eigenvalue modes

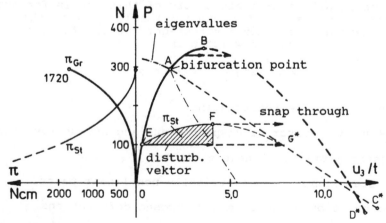

Fig. 7. Disturbance vector at load level P = 100 N and corresponding disturbance energy π_{St}

$$\left[A(z_0) + \lambda^S \cdot A^{NL}\left(\frac{1}{2}\Delta z\right)\right] \Delta z = \Delta p = 0. \qquad (2.2)$$

The iteration procedure is searching for the smallest eigen-value by using the approximation of the secant stiffness matrix (index S), evaluated at the middle value of the incremental displacement increase to the closest state of equilibrium on the instable branch, see Fig. 2.

As visualized in Fig. 3, it may be necessary that the disturbance vector should have components being not in affinity to the first eigen-mode of the bifurcation solution. This is the more important the farther the distance is from the basic reference state to the bifurcation point. By assuming that the closest critical state can be found by analysing the closest "snap through" point (by the tangential matrix T at the ultimate load point), the non-linear eigen-value problem

$$\left[A\ (z_0) + \lambda^T \cdot A^{NL}(\Delta z)\right] \Delta z = 0. \qquad (2.3)$$

has to be solved where the disturbance vector Δz which is also minimized with respect to the elastic energy and the eigen-value λ are the only unknowns. For more details of the theoretical background see /3, 4/. The close connection of an eigen-value problem to the determination of the closest state of equilibrium confirms the principle of Leipholz /5/ that the degree of stability is proportional to the energy necessary to reach the adjacent state of equilibrium. Thus, the necessary disturbance energy can be taken directly for evaluating the degree of stability at the investigated state of stresses and deformations of a structure.

3. EXAMPLE FOR VALIDATION OF THE PROPOSED METHOD

The buckling of the shell arch of Fig. 4 is investigated, demonstrating especially the results of the iterative procedure by using the deformation modes of the first and second eigenvalues, shown in Fig. 5 with their bifurcation points A and B and load-displacement functions. By an eigen-value analysis the first bifurcation point is found at P = 296 N (point A). Applying displacement increments the instable curve A - C* can be computed. For a perfect structure, load increments from A to B up to the second bifurcation point at P = 345 N may be applied. From here on the structure snaps through along the post-buckling branch (path B - D*).

In Fig. 6 the differences in the computed disturbance energy are demonstrated when instead of the minimum disturbance vector the eigen-modes only are considered. The shell arch (Fig. 4) is loaded up to P = 200 N. Then the first eigen-value mode Φ is chosen for the disturbance vector Δz in (2.3) and incrementally increased until a critical state is reached. The corresponding disturbance energy is 2016 Ncm. If only the second eigen-mode is considered as disturbance, then an energy of 1307 Ncm is required to reach the snap through point. If combinations of both eigen-modes are investigated, then the energy is much smaller, having its minimum of 611 Ncm at $^2\phi$ = 0.9.$^1\phi$ of the relative maximum value of both modes. If - as proposed in section 2 - the vector to the closest state of equilibrium is chosen, then the required energy is only 497 Ncm, s. Fig. 6. This is 20 % smaller than the lowest value of all combinations of the first and second eigen-modes.

By applying the non-linear eigen-value analysis, the disturbance vector and the corresponding disturbance energy can be determined for any point on the stable path from O to A, as shown in Fig. 5 and 7 for the load level P = 100 N. Numerical experiences showed that for P smaller than 100 N the convergence is becoming slow because of the larger distance to the adjacent state of equilibrium. For P larger than 100 N about 50 to 10 iteration steps had to be computed for convergence of the results. In Fig. 7 the energies are also plotted: for the basic state on the stable curve from O to A (with the maximum of 1720 Ncm at the bifurcation point), and for the required disturbance engergy π_{st}, which corresponds to half the area of E - F - G*. It declines with greater load level where a snap-through buckling is possible with less efforts. The diagram for the disturbance energy π_{st} provides the engineer with the information, that any occuring disturbance energy larger than plotted will cause the shell to snap through to the closest instable equilibrium state and hence towards collapse. The algorithm yields the critical disturbance vector in its modul form as well as in its numerical value. For the mixed FE-method the disturbance vector is composed of strains as well as of stress-resultants. The computed disturbance energy is not the absolutely smallest of all the possible ones. The assumption of taking the closest snap-through state as start for an iteration process is an approximation. Experiences by this method showed, however, that the errors are negligible small, especially for more sensitive structures.

For generalizations sake, it is more sensible to introduce the dimensionless ratio

$$\pi_{St} / \pi_{Gr, krit.} = \text{relative disturbance energy} \qquad (3.1)$$

as a criterion of the degree of stability and the imperfection sensitivity of a structure with regard to loss of stability.

4. EXAMPLES: AXIALLY LOADED CYLINDRICAL SHELLS

4.1. Cylindrical shell r/t = 100

The numerically investigated shell is given by:

length	l = 100 mm,	E = 206.000 N/mm^2,
radius	r = 100 mm,	ν = 0,30.
thickness	t = 1 mm,	

lower boundary: no displacements, no warping, hinged,
upper boundary: hinged, free for warping, axially loaded.
FE-discretization: 8 x 24 elements for half the shell.

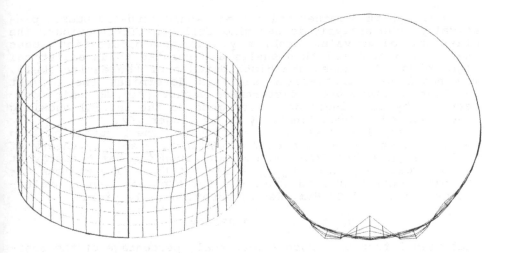

Fig. 8. Critical disturbance vector (40 times enlarged) for a
cylindrical shell of r/t = 100

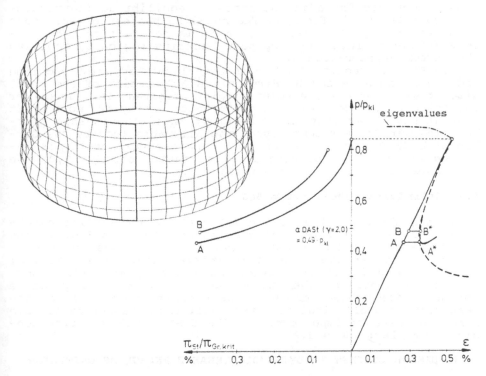

Fig. 9. Buckling pattern at
p = 900 N/mm (20 times enlarged) for the shell
r/t = 100

Fig. 10. Load-strain-curve and
disturbance energy of shell
r/t = 100

Although this rather stiff shell would actually buckle plastically, the analysis is assuming full elastic behaviour. The
classical eigen-value problem yields the bifurcation modes
which for almost all thin shells are related to unrealistical
high loads. The modes are distributed over the entire shell.
The non-linear eigen-value analysis determines the closest
equilibrium state and a disturbance vector (Fig. 8) which corresponds by its single dent very good to the initial buckling
mode observed in experiments by M. Esslinger /6/.

The basic load level is p = 900 N/mm. The bifurcation load
for the first eigen-value of the nonlinear stability theory of
shells is p_{Kr} = 970 N/mm and the corresponding basic energy at
the bifurcation point $\pi_{Gr,krit.}$ = 148.000 Nmm. The minimum disturbance energy at load level p = 900 N/mm has been   evaluated as π_{St} = 45,22 Nmm. Hence, equation (3.1) yields

$$\pi_{St} / \pi_{Gr,krit.} \ = 0.0306 \ \%.$$

That means, that only such a very small percentage of the energy, theoretically imposable up to the first bifurcation point,
is required to let the shell collapse.

Having found the disturbance vector of least energy, it is
easy to compute the adjacent state of equilibrium (corresponding to point G* in Fig. 7). The buckling pattern of this case
is shown in Fig. 9. From here on it is not difficult to procede furtheron into the deep post-buckling region by applying
the Newton-Raphson-method.

In Fig. 10 results for two of the closest adjacent buckling
modes are shown: A and B. Here only a 30° section of the shell
discretisized by 4 x 8 elements is investigated. The curves for
the relative loads - relative disturbance energy are more apart
than it may be expected by the almost indistinguishable postbuckling curves. For comparison, the allowable load by the German DASt-Ri 013 which is based on experimental results is included in Fig. 10. Applying a safety factor of γ = 2,0, the
allowable load factor is α = 0.49.

4.2. Cylindrical shell r/t = 500

Analysing the same shell as in 4.1, yet with a smaller thickness of only t = 0.2 mm, the results are given in Fig. 11 and
12. The shell buckles elastically. The FE-net has to be more
refined to cover the more local initial dent. The numerical
analysis, especially to provide convergence, asks for a more
refined FE-approach. The steeper decline of the energy curve
for small disturbance energies (as compared to Fig. 10 of the
less slender shell) indicates very well that thinner shells are
more sensitive to imperfections. The DASt-factor is correspon-
dingly smaller: α = 0.28.

5. DESIGN CONCEPT APPLYING DISTURBANCE ENERGY AS CRITERION

The non-linear eigen-value analysis offers the chance to determine directly the degree of stability for any structure by computing the disturbance energy required for a snap through towards a non-stable state of equilibrium. This approach covers
automatically the search for the most unfavourable disturbance

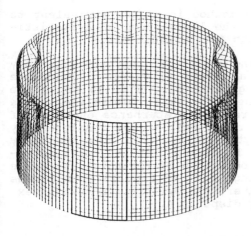

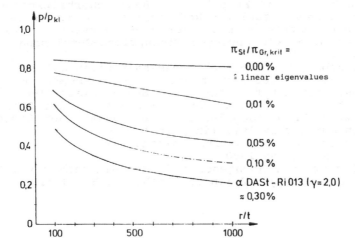

Fig. 11. Critical disturbance
vector for a shell r/t = 500
(100 times enlarged)

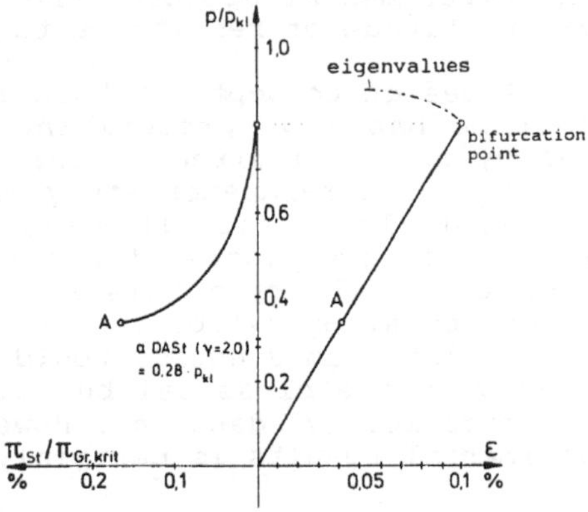

Fig. 12. Load-strain-curve and
disturbance energy of shell
r/t = 500

Fig. 13. Buckling loads (related to the classical bifurcation
value) for different relative disturbance energy and different
slenderness of cylindric shells under axial loading

vector. Therefore, the analytic method is almost equivalent to the experimental approach where the lower bound of a statistical multitude of results is taken for design basis.

A design concept for buckling shells of arbitrary shape may well be based on prescribing a lowest allowable disturbance energy computed directly through analysis. Fig. 13 shows the results of a parameter study where for axial loaded shells the allowable loads for different related disturbance energies of equ. (3.1) are plotted for all ratio r/t. A code committee may decide to calibrate this analytic approach to experimentally based codes. By calibrating to the German DASt the required minimum disturbance energy would be 0,30 % of the total imposable energy at the classical buckling load. In Fig. 13 both curves are practically identical. However, the calibration towards experimental results is certainly disputable.

REFERENCES

/1/ Kröplin, B. and Dinkler, D., Quasi-viskose Nachbeulberechnung bei axial gestauchten Platten und Zylinderschalen. Ing.Arch. **51**, 1982, 415-420.

/2/ Schmidt, H. and Krysik, R., Beulsicherheitsnachweis für baupraktische stählerne Rotationsschalen mit beliebiger Meridiangeometrie - mit oder ohne Versuche? Festschrift "Heinz Duddeck", Springer Produktions-Gesellschaft, Berlin 1988, 271-288.

/3/ Wagenhuber, W., Imperfektionssensitivität und rechnerischer Nachweis der Beulsicherheit dünner Schalen, Bericht Nr. 89-59, Institut für Statik, TU Braunschweig (1989).

/4/ Wagenhuber, W. and Duddeck, H., Numerischer Stabilitätsnachweis dünner Schalen mit dem Konzept der Störenergie. Ingenieur-Archiv 1991 (in print).

/5/ Leipholz, H., Stabilität elastischer Systeme. Verlag G. Braun, Karlsruhe 1980.

/6/ Esslinger, M., Hochgeschwindigkeitsaufnahmen vom Beulvorgang dünnwandiger Zylinder. Stahlbau, **H. 39** (1970), 73-76.

VERIFICATION OF COMPUTER PROGRAMS
BY DISCUSSION OF THE RESULTS

Maria Esslinger[1], Bodo Geier[2], and Gert Poblotzki[3]

[1] Bussardweg 2, 3300 Braunschweig, Germany
[2] Deutsche Forschungs- und Versuchsanstalt Braunschweig,
   Flughafen, 3300 Braunschweig, Germany
[3] Altewiekring 37a, 3300 Braunschweig, Germany

ABSTRACT

With the aid of calculated examples it is demonstrated, how computer programs
can be verified by checking wether the prebuckling deflections and deformations
as well as the buckling modes fit together convincingly. The examples discussed
are a long cylinder, loaded by pure axial load and pure bending and a short cylinder with a shrunk ring.

INTRODUCTION

There are two methods for verifying a computer program. One of them is to
compare the results with known ones or with new results produced by other
programs running in a different computer environment. This is what we call an
external check.

The other method is to output prebuckling deflections and stresses as well
as buckling modes, and to check wether they fit together convincingly. This is
what we call an internal check. Applying it frequently develops one's capacity for
judging computed results. That internal check will be dealt with in the following.

The program under consideration is for shells of rotation. It applies transfer
matrices in the direction of the meridian and expansion into Fourier series in the
circumferential direction. With regard to the boundary conditions the Fourier
series is divided into three parts, viz. the terms with $\cos 0\vartheta$, which include axi -
symmetric loads, the terms with $\cos 1\vartheta$, which include bending moments and
shear forces, and higher order terms with $\cos n\vartheta$ or $\sin n\vartheta$, which include self
equilibrating systems of stress resultants.

LONG CYLINDER

Our first example is a long cylinder. We are going to discuss two load cases: A centric axial load and a bending moment.

The cylinder is represented in Fig. 1. On the top and on the bottom it is closed by rigid plates. The symmetry about the middle plane is exploited in the analysis. Young's modulus is 5500 N/mm^2. The computation is performed for elastic response.

We start with the centrally loaded cylinder by computing the Euler load
$$P_E = 1160 \text{ N},$$
which corresponds to the axial membrane force
$$S_E = 3.89 \text{ N/mm}.$$
Then we compute with the program to be verified the buckling load of the cylinder under the condition that the cross section is not allowed to deform i.e. for a buckling mode with the circumferential wave number $n = 1$. The result is
$$S = 3.85 \text{ N/mm}.$$

It is known that a long cylinder the wall of which is not connected with the end plates, can buckle under internal pressure. Under the condition, that the cross section is not allowed to deform, this calculation yields the internal pressure
$$p = 0.1632 \text{ N/mm}^2,$$
which corresponds to $\quad S = 3.87 \text{ N/mm}.$
The three values agree with each other.

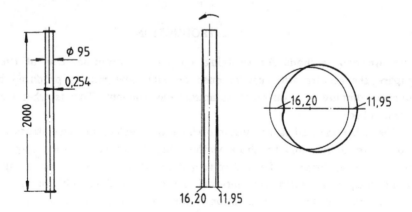

a. Along the meridian   b. In the symmetry plane

Fig. 1   Cylinder   Fig. 2   Prebuckling deformation with bending, mm

Now we consider the prebuckling deformations for bending, Fig. 2. The cylinder and the displacements were drawn in the same scale. In Fig. 2a you see the deformations along the meridian; the extreme fibre on the compression side and the extreme fibre on the tension side are deflected differently. Fig. 2b shows that the cross section in the plane of symmetry is deformed in a distorted ellipse.

We turn to the membrane forces in the prebuckling region, Table 1, Here you see for each of the two load cases, at the upper end (top) and in the symmetry plane (middle), three numbers, viz.

The Fourier term S_0, i.e. the axial force

the Fourier term S_1, which equilibrates the bending moment and

the maximal compressive force resulting from the load distribution ΣS_n

TABLE 1
Axial Membrane Forces N/mm

	axial load top	middle	bending top	middle
S_0	3.656	3.656	0	0
S_1	0	0	1.772	1.832
ΣS_n	3.65	3.656	1.215	1.927

For the pure axial load the bending moment $S_1 = 0$. The axial membrane force S_0 is constant over the entire length of the meridian and everywhere equal to ΣS_n.

With bending the axial force $S_0 = 0$. The Fourier term S_1 is somewhat higher in the symmetry plane than at the upper edge. This does not mean that the bending moment has increased. It is merely a consequence of the cross section having become narrower. At the upper edge the maximum compressive force ΣS_n is much smaller than the Fourier term S_1 which equilibrates the bending moment. The distribution of the membrane forces ΣS_n along the circumference is shown in Fig. 3a. Its shape is a consequence of preventing the warping of the cross section by the rigid end plate. It is difficult to understand the consequence of preventing the warping without knowing how the cross section would deform with free warping. Therefore we calculated the warping for classical boundary conditions.

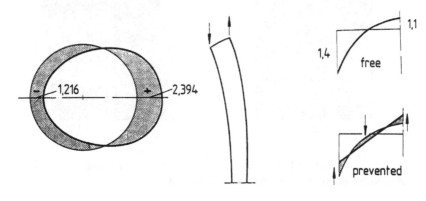

a. Force distribution, N/mm   b. Cylinder   c. Warping, mm

Fig. 3 Prevention of warping

In Fig. 3b you see a cylinder deflected by a bending moment. For classical boundary conditions the compressive and tensile forces at the load introduction are antimetric. It is known that a rod with an initial deflection is weaker under compression than under tension. Consequently at the edge of the cylinder the downward displacement at the compressed fibre is greater than the upward displacement at the tension fibre. In this way the warping shown in the upper part of Fig. 3c is caused; this figure is a schematic sketch with exaggerated scale.

If the warping is to be cancelled by a rigid end plate, that plate must be inserted in such a way that the produced forces are in equilibrium. This is shown in the lower part of Fig. 3c. On the left, i.e. on the compressive side, the cylinder wall has to be pulled upward; in this way the prevailing compressive force is reduced. On the right, i.e. on the tension side, the cylinder wall must also be pulled upward, and so the prevailing tensile force is increased. In the middle the cylinder wall has to be compressed.

In summary: By the prevention of warping the compressive force on the left is reduced, the tensile force on the right is increased and in the middle a small compressive force is induced. This yields the force distribution shown in Fig 3a.

We return to Table 1 and consider the axial membrane forces acting under bending in the symmetry plane. There the maximum compressive force ΣS_n differs only little from the Fourier term S_1, since the edge disturbance has largely decayed. ΣS_n is somewhat greater than S_1, in contrast to the relation at the upper edge. This reversal is due to the fact, that the decay of the edge disturbance is wavy rather than asymptotic.

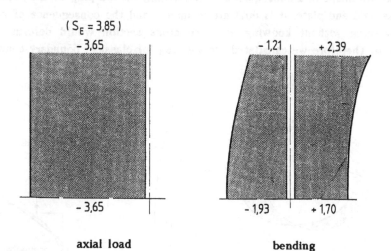

$(S_E = 3,85)$

$-3,65$ $-1,21$ $+2,39$

$-3,65$ $-1,93$ $+1,70$

axial load   bending

Fig. 4   Maximum compressive forces ΣS_n, N/mm

The distribution of the maximum compressive forces ΣS_n along the meridian is shown in Fig. 4 for both load cases. With pure axial force it is constant over the entire length. For bending the compressive forces are greatest in the symmetry plane . Knowing this we are going now to look at the buckling modes.

Fig. 5a presents the buckling mode along the meridian. For axial load it extends over the entire length. For bending it is concentrated near the symmetry plane. The buckling deformation of the cross section in the symmetry plane is shown in Fig. 5b. For axial load it extends over the entire circumference with the wave number n = 3. This is the buckling mode found by the program; it contrasts to the mode n = 1 corresponding to the Euler load. For bending the buckling deformation is concentrated in the region of the greatest compression forces. Its circumferential wave number corresponds to n = 5.

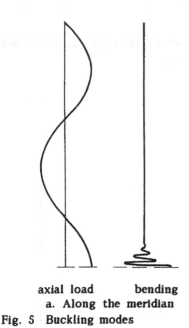
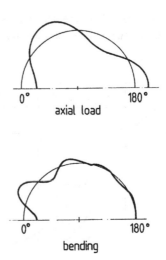

axial load

bending

axial load   bending
a. Along the meridian   b. In the symmetry plane
Fig. 5 Buckling modes

We return to Fig. 4 and consider now the magnitude of the buckling forces. For the centric axial load the buckling force S = 3.65 N/mm found with free choice of the buckling mode is smaller than the Euler load S_E = 3.85 N/mm calculated under the constraint that the cross section is not allowed to deform.

Comparing the local buckling forces in the two load cases reveals that for bending the local axial membrane force is about half as great as for pure axial load. The low buckling force at bending is a consequence of the deformation of the cross section in the prebuckling range. Once more we look at Fig. 2b, where it is shown that the cross section in the middle plane has been flattened and moreover there is an indentation which acts as an initial imperfection.

To formulate a computer job without any input error is a matter of luck. Safety can be increased by performing a series of computations. We computed the load case bending for cylinders with decreasing length. The results are summarized on Fig. 6, showing membrane forces S as function of the cylinder length. There are two curves, the forces S_1 and ΣS_n at the upper edge of the cylinder. S_1 corresponds to the applied bending moment and thus represents the buckling load. (let it be understood that this is not the local axial membrane force in the area where the cylinder buckles) The difference between S_1 and ΣS_n is a measure for the effect of the warping. It can be noticed that with decreasing length the buckling force increases and the effect of warping gets smaller.

The upper limit corresponding to the well known approximation formula for the buckling load of axially loaded cylinders

$$S = 0.605 \ E \ t^2 \ r$$

is not attained, because the edges are constrained by the rigid end plate. With Fig. 6 the connection to known results is etablished.

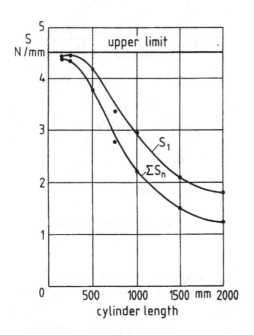

Fig. 6 Series computation

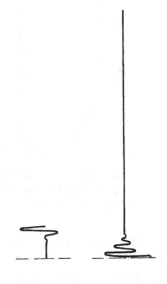

l = 250 mm   l = 2000 mm

Fig. 7 Buckling mode of a short and a long cylinder

The short cylinders buckle near the edges rather than in the region about the symmetry plane. In Fig 7 the buckling modes of a short and a long cylinder are compared.

SHRUNK RING

Our next example is a cylinder with a shrunk ring, Fig. 8. Young's modulus is $E = 0.2 \cdot 10^6$ N/mm^2 and the yield limit is $\sigma_y = 240$ N/mm^2. The computation is performed with an "elastic perfectly-plastic" material law. The state of stress and deformation is axisymmetric. The ring is situated in a vertical symmetry plane of the structure. If the stresses are symmetric with respect to that symmetry plane, it would be sufficient to consider the hatched area.

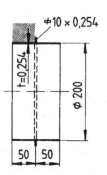

Fig. 8 Cylinder with shrunk ring

The symmetry of the stresses cannot be taken for granted. In preparing this paper I started with a shrinkage of 0.5 mm and thereby I did not obtain symmetry. The shell had become plastic to such an extent that the stress striving in a comprehensive mud could nowhere find firm support. Probably in a real structure the same would have happened. After a small series of computations I chose a shrinkage of 0.2 mm; now plasticity is moderate and the stresses are exactly symmetrical. The first check is to verify that the jump of the transverse shear force in the shell equals the radial force on the ring, and that the correlative displacements of the cylinder and the ring correspond to the prescribed shrinkage. Both these conditions are satisfied. Fig. 9 shows the displacements. The shell is shrunk by 0.115 mm, while the ring is extended by 0.085 mm.

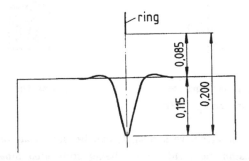

Fig 9   Radial displacements, mm

Next we consider the stresses at the external and internal surface of the shell and the variation of the stresses across the shell wall just beneath the ring.

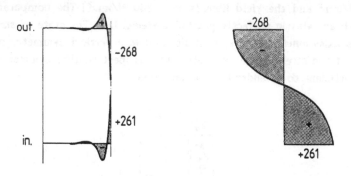

outer and inner surface   across the wall

Fig. 10   Longitudinal stresses σ_φ , N/mm^2

In Fig. 10 you see the longitudinal stresses σ_φ. These are bending stresses, compressive stresses at the outer and tensile stresses at the inner surface. They are greatest below the ring and decay rapidly from there. At the ring they are somewhat above the yield stress. That is no pitfall of the calculation , since we have a two dimensional stress state, and it is the effective stress

$$\sigma_e = \sqrt{\sigma_\varphi^2 + \sigma_\vartheta^2 - \sigma_\varphi\,\sigma_\vartheta}$$

which must not exceed the yield stress.

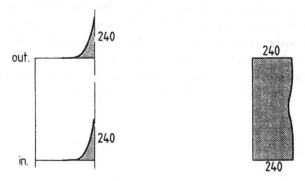

outer and inner surface   across the wall

Fig. 11   Effective stress σ_e , N/mm^2

Fig. 11 shows the effective stress. It can be seen that on the outer and inner surface it just attains the yield stress, being somewhat smaller in the middle of the wall.

Finally we turn to the circumferential stresses σ_ϑ, Fig. 12. You probably expect that the shell is compressed below the ring and that the compressive stresses are constant over the shell wall. That expectation does not correspond to the computed result. There are tensile stresses on the inner surface!

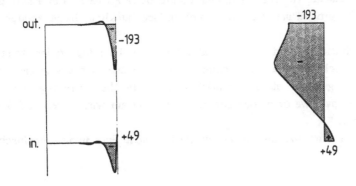

outer und inner surface   across the wall

Fig. 12 Circumferential stresses σ_ϑ, N/mm^2

The check of the computation is simple. If the effective stress and the long-itudinal stress are known, the circumferential stress can be calculated. On the outer surface where there are high longitudinal compressive stresses, the limitation of the effective stress requires compressive stresses in the circumferential direction . On the inner surface, where there are high longitudinal tensile stresses, the limitation of the effective stress requires tensile stresses in the circumferential direction.

And how does this fit to the deformations? Very well! The program yields equal strain ε_ϑ = -0.00115 on the outer and inner surface. This corresponds to a radial displacement of $\varepsilon_\vartheta \cdot r$ = -0.115 mm, which is exactly the value shown in Fig. 2b. At the inner surface the negative strain is a consequence of Poisson's ratio related to the longitudinal tensile stresses.

The single slender ring was an example for the check of the program. In practical structures several flat rings will be used.

Finally I will give you a hint by which a Prüfingenieur can be appeased. A wide ring should be replaced in the calculation by two or more narrow rings, Fig. 13. This will reduce the calculated bending stresses.

Fig. 13 Hint

FINAL REMARK

A conference on shell buckling supplies an occasion to find partners. We are looking for a partner who is willing to study experimentally the buckling due to foundation settlement. The test cylinders should be fabricated from a soft plastic foil to make the ground settlements, which induce buckling, large enough to be measurable.

Our contribution would be the calculation of buckling under foundation settlement viz. buckling under axial compression or shear stresses in the cylinder wall, and buckling under circumferential stresses in the stiffening ring at the upper edge. Moreover we offer our computation program with a detailed description of the input.

The research work was supported by the Deutsche Forschungsgemeinschaft.

REFERENCES

[1] Esslinger, M., Geier B., Wendt U., Berechnung der Spannungen und Deformationen von Rotationsschalen im elastoplastischen Bereich. Stahlbau 53 (1984), H.1 S. 17-24

[2] Esslinger, M., Wendt, U., Hey,H., Stabilitätsnachweis für eine 20 m hohe wassergefüllte Rotationsschale in Wespentaillenform, Chemie-Ingenieur-Technik, 56 (1984) 11, S. 872-873

[3] Esslinger, M., Weiß, H.P., Berechnung einer ringversteiften Rotationsschale mit Vorbeulen, Chemie-Ingenieur-Technik, 58 (1986), H.10, S. 826-827

[4] Esslinger, M., Weiß. H.P., Postbuckling Calculation of Extremely Thinwalled Vessels under External Pressure, Buckling of Structures, Elsevier Science, Publishers B.V. Amsterdam 1988

[5] Esslinger, M., Poblotzki, G., Spannungs- und Stabilitätsrechnung von Rotationsschalen unter nicht-axialsymmetrischer großflächiger Belastung im elastoplastischen Bereich, Chemie-Ingenieur-Technik 62 (1990), H.9, S.776-777

[6] Esslinger,M., Weiß, H.P., Eingabebeschreibung für die Programme F04B10 und F04B12, Nichtlinerare, nichtaxialsymmetrische Spannungs- und Beulrechnung für Rotationsschalen mit und ohne Vorbeulen, DLR Braunschweig, IB 131-89/36

[7] Esslinger M., Geier B., Poblotzki G., Beispiele zur Berechnung von Rotationsschalen im elastoplastischen Bereich, Stahlbau 60 (1991), H. 6

[8] Esslinger M., Poblotzki G., Beulen unter Winddruck, Stahlbau, accepted for publication

ON THE NUMERICAL SIMULATION OF THE BUCKLING OF IMPERFECT SHELLS

by

J. Arbocz and J.M.A.M. Hol

Delft University of Technology, The Netherlands

Abstract

The development of "DISDECO", the Delft Interactive Shell Design Code is described. The purpose of this project is to make the accumulated theoretical, numerical and practical knowledge of the last 20 years readily accessible to users interested in the analysis of buckling sensitive structures. With this open ended, hierarchical, interactive computer code the user can access from his work-station successive programs of increasing complexity.

Included are modules that contain Koiter's imperfection sensitivity theory extended to anisotropic shell structures under combined loading. The nonlinear Donnell-type shell equations in terms of the radial displacement W and the Airy stress function F are used. The spatial dependence is eliminated by Fourier decomposition. The resulting sets of algebraic equations form a standard matrix eigenvalue problem. Initial results indicate that an interactive use of these simple modules can greatly facilitate the search for an optimal lay up of composite shells under combined loading.

Introduction

In modern designs, which are often obtained by one of the structural optimization codes and which are made out of high strength materials (read advanced composites), it happens frequently that the stability behaviour dictates the choice of some of the critical dimensions of the structures.

This implies that one has to investigate the different loading cases quite accurately by carrying out extensive numerical calculations and/or experimental verifications.

Twenty five years ago it was so that numerical results were looked upon with a certain degree of distrust and they were only accepted if supported by some other facts. Now-a-days, as the older generation of engineers (the ones who have gotten their degrees before the advent of computers) is retiring and the younger ones with extensive training in the ever-so-popular finite element techniques take over, one begins to encounter in technical discussion a new mentality; the insight of how structures behave under loading of the older generation is being more and more replaced by the nearly religious faith of the younger ones in the predictions of their favorite computer codes.

Actually what one needs is not more of one and less of the other (and the reader is free to associate his preference with one or the other), but an optimal combination of both, namely insight into structural behaviour and familiarity with the appropriate numerical techniques. To provide the means for such an approach the development of DISDECO, the **Delft Interactive Shell Design Code** has been initiated. When finished this open ended, hierarchical, interactive computer code will provide for easy access to the theoretical knowledge, that has been accumulated by the many scientists who have been active in the field of shell stability, via the advanced interactive and computational facilities offered by the modern high-speed 32 bit personal workstations. Great care is being taken to present the results in a unified form so as to make it easy for the user to proceed step-by-step from the simpler approaches used by the early investigators to the more sophisticated analytical and numerical methods used presently.

Buckling and Imperfection Sensitivity of Anisotrpic Shells

The lowest hierarchical level consists of semi-analytical solutions for the buckling load of perfect anisotropic circular cylindircal shells under axial compressions, external lateral pressure, torsion and bending. Also included are modules that contain Koiter's imperfections sensitivity theory extended to anisotropic shell structures.

To illustrate this approach in the following the extension of the well known b-factor method [1,2] to anisotropic cylindrical shells loaded by axial compression, external pressure and torsion is presented. The Donnell type anisotropic shell equations [3] in terms of the radial displacement W and an Airy stress function F are used. The asymptotic perturbation expansions

$$W = W^{(0)} + \xi W^{(1)} + \xi^2 W^{(2)} + \dots \tag{1a}$$

$$F = F^{(0)} + \xi F^{(1)} + \xi^2 F^{(2)} + \dots \tag{1b}$$

$$\frac{\Lambda}{\Lambda_c} = 1 + a\xi + b\xi^2 + \dots \tag{1c}$$

generate 3 sets of equations governing the prebuckling, the buckling and the post-buckling states.

If the effect of boundary conditions and edge restraints is neglected then the following membrane solution satisfies the governing equations of the prebuckling state identically

$$W^{(0)} = h(W_v + W_p + W_t) \tag{2}$$

$$F^{(0)} = \frac{Eh^2}{cR} \left(-\frac{1}{2}\lambda y^2 - \frac{1}{2}\bar{p}x^2 - \bar{\tau}\, xy \right) \tag{3}$$

where

$$\lambda = \frac{\sigma}{\sigma_{c\ell}} \; ; \; \bar{p} = \frac{p}{\frac{h}{R}\sigma_{c\ell}} \; ; \; \bar{\tau} = \frac{\tau}{\sigma_{c\ell}} \; ; \; \sigma_{c\ell} = \frac{Eh}{cR} \; ; \; c = \sqrt{3(1-\nu^2)}$$

$$W_v = \frac{1}{c}\bar{A}^*_{12}\lambda \; ; \; W_p = \frac{1}{c}\bar{A}^*_{22}\bar{p} \; ; \; W_t = -\frac{1}{c}\bar{A}^*_{26}\bar{\tau}$$

The stiffness parameters $\bar{A}^*_{12}$, $\bar{A}^*_{22}$ and $\bar{A}^*_{26}$ are defined in Ref. [4].

In the linearized stability equations governing the buckling problem the applied loading consists of axial compression, internal or external pressure and clockwise or counter-clockwise torque. It is assumed to have a uniform spatial distribution and is divided into a fixed part and a variable part. The magnitude of the variable part is allowed to vary in proportion to a load parameter Λ. This leads to an eigenvalue problem for the critical load Λ_c. The user can select Λ_c to be the critical value of either the normalized axial load λ, or the normalized external pressure $\bar{p}$ or the normalized torque $\bar{\tau}$.

If one chooses $\Lambda_c = \lambda$ and assumes following Khot and Venkayya[5] that the buckling mode is represented by

$$W^{(1)} = h\xi \sin\frac{m\pi x}{L} \cos\frac{n}{R}(y - \tau_k x) \tag{4}$$

where m,n are integers and τ_k is Khot's parameter introduced to account for the torsion-bending coupling, then a straight forward Galerkin type solution of the linearized stability equations yields the following eigenvalue

$$\lambda_{mn\tau} = \frac{1}{2(\alpha_m^2 + \alpha_p^2)} \left(\bar{T}_1 + \bar{T}_2 + \frac{\bar{T}_3^2}{\bar{T}_2} + \frac{\bar{T}_4^2}{\bar{T}_6} \right) - \frac{2\bar{p}\beta_n^2}{(\alpha_m^2 + \alpha_p^2)} \tag{5}$$

This expression must be minimized with respect to m,n and τ_k. Notice that m and n are integers whereas τ_k is noninteger. The coefficiens $\bar{T}_1, \bar{T}_2$ etc and the parameters α_m^2, α_p^2 and β_n^2 are listed in Ref. [6].

Koiter [1] has postulated that the imperfection sensitivity of a structure is closely related to the initial post-buckling behaviour of the perfect structure represented by Eq. (1c). General expressions for the postbuckling coefficients a and b have been derived by Budiansky and Hutchinson [7]. If the shell-external loading combination possesses a unique asymmetric buckling mode associated with the lowest buckling load then it is easily verified that the first postbuckling coefficient a is identically equal to zero, and the initial postbuckling behviour of the shell is as indicated in Fig. 1.

If the initial imperfections are taken affine to the unique asymmetric buckling mode, then Koiter has shown that the buckling load of an imperfect shell λ_s, (defined as the maximum load the structure can support prior to buckling) is related to the imperfection amplitude $\bar{\xi}$ and the postbuckling coefficient b by

$$(1 - \rho_s)^{3/2} = \frac{3}{2} \sqrt{-3b} \, \rho_s \, |\bar{\xi}| \quad \text{for } b < 0 \tag{6}$$

where $\rho_s = \lambda_s/\lambda_c$ and λ_c is the normalized perfect shell buckling load. Notice that imperfection sensitive structures are characterized by negative values of b.

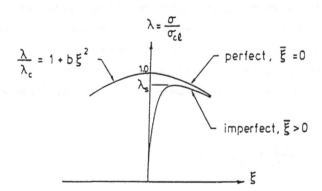

Fig. 1. Initial postbuckling behaviour of imperfection sensitive shells

In order to calculate the second postbuckling coefficient b it is necessary to solve for $W^{(2)}$ and $F^{(2)}$. The equations governing the 2nd-order state [4] admit separable solutions of the form

$$W^{(2)} = h \{ \sum_{i=1}^{\infty} A_i \sin \ell_i x + \frac{1}{2} \sum_{i=1}^{\infty} B_i \sin(\ell_{m_i} x - \ell_{2n} y)$$

$$+ \frac{1}{2} \sum_{i=1}^{\infty} C_i \sin(\ell_{p_i} x + \ell_{2n} y) \}$$

(7)

$$F^{(2)} = \frac{Eh^3}{2c} \{ \sum_{i=1}^{\infty} D_i \sin \ell_i x + \frac{1}{2} \sum_{i=1}^{\infty} E_i \sin(\ell_{m_i} - \ell_{2n} y)$$

$$+ \frac{1}{2} \sum_{i=1}^{\infty} F_i \sin(\ell_{p_i} + \ell_{2n} y) \}$$

(8)

where

$$\ell_i = \frac{i\pi}{L} \; ; \; \ell_{m_i} = \frac{i\pi}{L} + \frac{2n\tau_k}{R} \; ; \; \ell_{p_i} = \frac{i\pi}{L} - \frac{2n\tau_k}{R} \; ; \; \ell_{2n} = \frac{2n}{R}$$

The coefficients are readily determined by the Galerkin procedure. Finally the second postbuckling coefficient b is calculated by evaluating the integrals involved [7]. For orthotropic shells this operation leads to the following expression

$$b = -\frac{4c}{\pi} \frac{\beta_n^2}{\lambda_{mn}} \{ \sum_{j=1,3,...}^{\infty} (2\bar{F}B_j + E_j) \frac{1}{j}$$

(9)

$$- 2 \sum_{j=1,3,...}^{\infty} (2\bar{F}A_j + D_j) \frac{j}{j^2 - 4m^2} \}$$

The coefficients A_j, B_j, C_j, D_j and $\bar{F}$ are listed in Appendix C of Reference [8]. The series can be evaluated numerically to any degree of accuracy desired. This solution was first obtained by Hutchinson and Amazigo [2] using an asymmetric imperfection of the form

$$\bar{W} = h\bar{\xi} \sin\frac{m\pi x}{L} \cos\frac{ny}{R}$$

(10)

for stringer stiffened shells. Knowing b, one can use Eq. (6) to obtain the $\rho_s = \lambda_s/\lambda_c$ versus $\bar{\xi}$ curve of Fig. 2. Khot and Venkayya [5] presented a similar solution for anisotropic shells using an asymmetric imperfection of the form

$$\overline{W} = h \, \bar{\xi} \, \sin \frac{m\pi x}{L} \cos \frac{n}{R} (y - \tau_k x) \tag{11}$$

Notice that in this expression m and n are integers whereas Khot's parameter τ_k is noninteger.

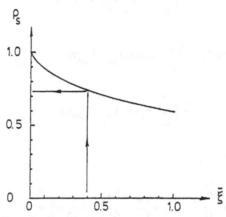

Fig. 2.   Imperfection sensitivity for a given asymmetric imperfection

Thus the computational modules of the lowest hierarchical level of "DISDECO" provide the user with a fast, interactive design tool to obtain both, initial estimates of the critical buckling loads of a large variety of possible shell configurations and an indication of expected degree of imperfection sensitivity of these critical buckling loads.

With the help of the graphic facilities available on modern desktop workstations the output can be displayed in the form of easily interpretable graphs. As can be seen from Fig. 3 the format of these plots often follow the layout of figures used in well known publications in the past by pioneers of the field of shell stability [2,3].

As "building blocks" of the hierarchical design and analysis system "DISDECO" the computational modules of the initial level make the first step towards acquiring of detailed understanding of the expected instability behaviour of different cylindrical shell configurations possible. This knowledge is a prerequisite for the development of discrete nonlinear computational models which can reliably predict the load carrying capability of the structure. It must be stressed that these initial predictions provide only a first indication of the expected nonlinear behaviour and all the results must be evaluated within the context of the fundamental assumptions involved in the theory used. Thus

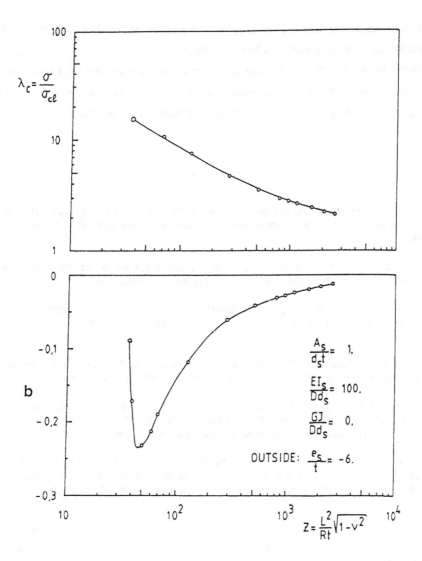

Fig. 3 Stringer stiffened shell under axial compression

the fact that membrane prebuckling was assumed and the effect of boundary conditions has been neglected may be misleading, especially for shorter shells.

To investigate the influence of these simplifying assumptions the user must switch to other more advanced computational modules available within "DISDECO", such as ANILISA [4] where a rigorous prebuckling analysis is used and the boundary conditions are enforced.

REFERENCES

1. Koiter, W.T., "On the Stability of Elastic Equilibrium", Ph.D. Thesis 1945, in Dutch, TH-Delft, The Netherlands, H.T. Paris, Amsterdam. (English translation issued as NASA TT F-10, pp. 833, 1967.)

2. Hutchinson, J.W. and Amazigo, J.C., "Imperfection-Sensitivity of Eccentrically Stiffened Cylindrical Shells", AIAA Journal, Vol. 5, No. 3, pp. 392-401, March 1967.

3. Tennyson, R.C. and Muggeridge, D.B., "Buckling of Laminated Anisotropic Imperfect Circular Cylinders under Axial Compression", Journal of Spacecraft, Vol. 10, no. 2, pp. 143-148, February 1973.

4. Arbocz, J. and Hol, J.M.A.M., "ANILISA - Computational Module for Koiter's Imperfection Sensitivity Theory", Int. J. Solids Structures, Vol. 26, No. 9/10, pp. 945-973, 1990.

5. Khot, N.S. and Venkayya, V.B., "Effect of Fiber Orientation on Initial Postbuckling Behavior and Imperfection Sensitivity of Composite Cylindrical Shells", Report AFFDL-TR-70-125, Air Force Flight Dynamics Laboratory, Wright-Patterson Air Force Base, Ohio, 1970.

6. Arbocz, J., "User's Guide for ANIPBIF - Program to calculate the buckling load of axially compressed anisottropic shells with a single axisymmetric imperfection", Memorandum M-624, Delft University of Technology, Faculty of Aerospace Engineering, The Netherlands, February 1990.

7. Budiansky, B. and Hutchinson, J.W., "Dynamic Buckling of Imperfection Sensitive Structures", Proceedings XI Intern. Congr. Appl. Mech. (H Görtler, Ed.), Springer Verlag, Berlin, pp. 636-651, 1964.

8. Arbocz, J., "The Effect of Initial Imperfections on Shell Stability", in: Thin Shell Structures, Theory, Experiment and Design (Eds. Y.C. Fung and E.E. Sechler) Prentice Hall, Englewood Cliffs, N.Y., pp. 205-245, 1974.

NONLINEAR DYNAMIC STABILITY OF ELASTO-PLASTIC SHELL STRUCTURES UNDER IMPULSIVE LOADS

W.B. Krätzig, Y. Basar, R. Quante
Institut für Statik und Dynamik
Ruhr-Universität Bochum, Germany

INTRODUCTION

The objective of this research report is the finite-element analysis of dynamic instability phenomena of arbitrary shell structures subjected to impulsive and monotonously increasing loads. A special attention is taken to the consideration of the elasto-plastic material behaviour which can affect the critical conditions, considerabely. The basis of the numerical investigations is the principle of virtual work for a finite rotation theory, more exact its transformation by a variational procedure into an incremental formulation. By means of a finite-element displacement discretization the incremental principle results into the tangential equation of motion to be used for the numerical tracing of arbitrary nonlinear motions caused, particularly, by the load classes mentioned above.

Numerical Analysis

The discretization of the incremental principle of virtual work by means of a finite element displacement model leads to the tangential equation of motion

$$\mathbf{M} \cdot \ddot{\mathbf{V}} + \mathbf{C} \cdot \dot{\mathbf{V}} + \mathbf{K}_T \cdot \dot{\mathbf{V}} = P - F_I \tag{1}$$

with   the mass matrix $\mathbf{M}$,
        the damping matrix $\mathbf{C}$,
        the tangential stiffness matrix $\mathbf{K}_T$,

the incremental nodal displacements,
velocities and accelerations $\tilde{v}$, $\dot{\tilde{v}}$, $\ddot{\tilde{v}}$,
the time dependent load vector P
and the internal nodal forces F_I .

Equation (1) represents the basis for the numerical tracing of arbitrary nonlinear dynamical processes. Herein K_T and F_I contain all geometrical and physical nonlinearities of the shell structure. The finite-element formulation is achieved by virtue of layered displacement models (Fig. 1) with which the change of material properties throughout the shell thickness can be considered at the element level numerically.

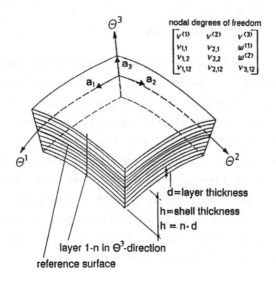

Figure 1. Layered Shell Element NACS 48

An elasto-plastic constitutive law based on a concept of internal variables allows to consider plastic flow in certain points of the structure. The yield condition

$$ F = (s_D^{ij} - \kappa^{ij}) (s_{Dij} - \kappa_{ij}) - k^2 \leq 0 \qquad (2) $$

with the deviatoric stresses s_D^{ij} ,
the kinematic hardening tensor κ^{ij} ,
the yield stress $k^2 = \sigma_F^2$

decides between elastic or elasto-plastic behaviour:

$\quad$ F < 0 - elastic

$\quad$ F = 0 - elasto-plastic.

The consistency condition

$$\frac{\partial F}{\partial s^{ij}} ds^{ij} + \frac{\partial F}{\partial \kappa^{ij}} d\kappa^{ij} = 0 \qquad (3)$$

has to be fulfilled for all stress increments. Additional to the yield condition a hardening law

$$d\kappa^{ij} = \eta \overset{o*ik}{a} \overset{o*jl}{a} d\gamma^P_{kl} , \qquad (4)$$

wherein η denotes a material function, $\overset{o*ik}{a}$ the metric tensor and $d\gamma^P_{kl}$ the plastic strain increment is used to describe the changement of the elastic domain. The kinematic hardening model leads to a translation of the yield surface in the stress state without an enlargement of the elastic region.

A decision about the stability of the structure can be made, in the case of impulsive loads, according a qualitative criterion proposed by Budiansky. Concluding, an example will be presented in the continuation of previous and well-established research results [1, 2, 3] in order to demonstrate the efficiency of the numerical models developed.

Dynamic Snap-through of a Cylindrical panel

The cylindrical panel [4] under uniform step loading is shown in Fig. 2. After its standard displacement discretization the nonlinear response, starting at time $t=t_o$, is computed by use of the finite-element method.

Therefore the nonlinear equation of motion is integrated step-wise. In every time-step iterative solution techniques are used in order to minimize the residual forces.

The time histories of the central deflection under uniform transverse step loading p^3 are presented in Fig. 3. Therein the last stable and the critical load intensity are considered for the purely elastic and the elasto-plastic material model. The dotted lines distinguish the elasto-plastic

Structure:

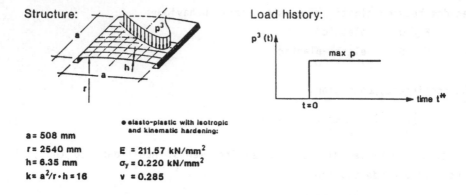

a= 508 mm
r= 2540 mm
h= 6.35 mm
k= a²/r·h = 16

● elasto-plastic with isotropic
and kinematic hardening:

$E = 211.57$ kN/mm²
$\sigma_y = 0.220$ kN/mm²
$\nu = 0.285$

Load history:

Figure 2. Structure - cylindrical panel

response from the elastic ones. Increasing to a critical load intensity
dynamic snap-through occurs for both constitutive laws. After the enforced
vibration has vanished the panel moves towards a static equilibrium state
in a snaped-through configuration. The quotient of the critical load
intensities of the elastic and the elasto-plastic model

$$p^3_{ep}/p^3_e = 0.221/0.275 = 0.804$$

demonstrates, that the plastic deformations of the structure reduce the
failure load to nearly 80%. A purely elastic computation would overestimate the load capacity, because this example represents a coupled elasto-
plastic stability problem.

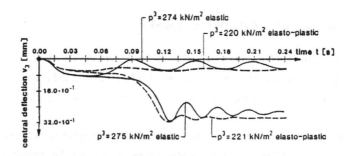

Figure 3. Time histories - central deflection v_3

Fig. 4 shows the difference of the vibration properties between the material models. The plastic dissipation capacity of the elasto-plastic constitutive law is responsible for the rapidly decreasing vibration of the
cylindrical panel.

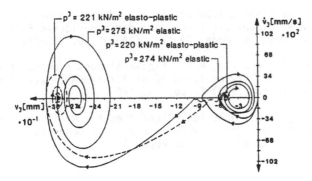

Figure 4. Phase diagramm

Conclusions

The contribution demonstrates, that the consideration of an elasto-plastic constitutive law influences the dynamic snap-through load essentially. Therefore a cylindrical panel as a part of a storage tank is investigated numerically. The numerical stability analysis requires the approximation of the material as well as the structural properties in an optimal manner. Satisfying these conditions the computer simulation supports the experimental investigations, which are necessary for the economical development of extremly optimizated constructions. The rapidly increased capacity of modern workstations and the wide range of software tools allow the numerical simulation of the bahaviour of complex structures. The engineer may not trust to the algorithms and their results without any criticism. His task will not be to believe in computer results but to review them.

Acknowledgement

This subject has been treated within a research project of the SFB 151: Financial support of the DFG-German Foundation is gratefully acknowledged.

REFERENCES

1. Basar, Y., Krätzig, W.B.: Introduction into Finite-Rotation Shell Theories and Their Operator Formulation. in: Krätzig, W.B.; Onate, E. (eds.): Computational Mechanics of Nonlinear Response of Shells. pp. 101-124, Springer-Verlag, Berlin, Heidelberg, New York 1990

2. Krätzig, W.B.: Eine einheitliche statische und dynamische Stabilitäts-
   theorie für Pfadverfolgungsalgorithmen in der numerischen Festkörperme-
   chanik. Z. Angew. Math. u. Mech. 69 (7), (1989), pp. 203-213

3. Basar, Y.; Eller, C.; Krätzig, W.B.: Finite element procedures for the
   nonlinear dynamic stability analysis of arbitrary shell structures.
   Computational Mechanics (1990) 6, pp. 157-166

4. Eller, C.; Krätzig, W.B.: Numerical Algorithms for Nonlinear Unstable
   Dynamic Shell Responses. Proc.: WCCMII, 2nd World Congress on Computational Mechanics, Stuttgart 1990 (to appear)

BUCKLING ANALYSIS OF SHELLS BY MEANS OF THE ANKA/PC CODE

Zenon WASZCZYSZYN, Maria RADWAŃSKA, Ewa PABISEK, Krzysztof RÓŻYCKI
Institute of Structural Mechanics and Computer Center,
Cracow University of Technology,
ul.Warszawska 24, 31-155 Kraków, POLAND

ABSTRACT

Basic equations of the displacement FEM are quoted. They are a base for the buckling analysis by means of the ANKA code implemented on PC computers. The following algorithms can be realised: i) Initial and linearized buckling, ii) Evaluation of nonlinearity of the fundamental equilibrium path, iii) Approximate analysis of nonlinear, prebuckling state, iv) Evaluation of the type of critical points, v) Incremental computation of nonlinear equilibrium paths. The isoparametric, degenerated, 8/9 node shell FE are used. The efficiency of the ANKA/PC code is shown on two numerical examples related to the buckling of cylindrical panels.

INTRODUCTION

Since 1979 the research on the nonlinear and stability FE analysis of structures has been developed at the Institute of Structural Mechanics and Computer Center of Cracow University of Technology. The computer code ANKA is a result of this research (cf.[1]). The code is oriented to research works and student training but can also be used by engineers for the analysis and design of structures.

The ANKA code is based on displacement version of FEM, assumptions of small strains and moderate large rotations as well as Total Lagrangian formulation. The static criterion of stability is used on assumptions of conservative and single parameter loads and linear elastic material. The code has been implemented on PC computers. Algorithms of different levels of complexity have been included into the code. This enables us to realise the strategy: 'from simple to refined models and methods'.

The ANKA code concers first of all the linear static analysis applying the following matrix equation:

$$K_0 \cdot [\bar{Q} \mid Q^*] = [\bar{P} \mid P^*] , \tag{1}$$

where: $\bar{P}$ - vector of fixed loads (e.g. dead load), P^* - reference vector of varying loads.

The next algorithms are related to the generalized eigenvalue problem:

$$[A + \lambda B]v = 0 , \tag{2}$$

where: λ - load parameter. After the matrices A, B are specified and related to the solution of Eq.(1) the following sequence of eigenvalues and eigenvectors can be computed

$$\{\lambda_i, v_i\} \quad \text{for} \quad i=1,\ldots,n . \tag{3}$$

They correspond to the buckling loads and buckling modes of perfect, initially deformed or laterally loaded structures depending on specification of the matrices A and B.

Information, obtained from the analysis of Eqs (1) and (2) enables us also to evaluate nonlinearities of the fundamental state of equilibrium and type of critical points. In case of linear elastic material the imperfection sensitivity of structure can be deduced on the base of the type of bifurcation point.

The next algorithm of ANKA is related to the incremental analysis of the following equations:

$$K_T \Delta Q - P^* \Delta\lambda = R ,$$
$$t^T \Delta Q + t_{N+1} \Delta\lambda = \Delta\tau , \tag{4}$$

where: K_T - tangential stiffness matrix, R - residual force vector, t - - control vector, τ - control parameter. In the constraint Eq.(4)$_2$ the components of vector $t \in \mathbb{R}^N$ and coefficient t_{N+1} make it possible to continue the step-by-step procedure under the control of load parameter $\tau \equiv \lambda$, selected displacement $\tau \equiv Q_j$ or arc-length parameter $\tau \equiv s$ - cf.[2].

The main goal of the analysis of Eqs (4) is the computation of the first critical point on nonlinear fundamental equilibrium paths. The ANKA code enables us to compute such paths also for structures with initial stresses induced by "load imperfections", with arbitrary geometrical imperfections or those associated with the buckling modes v_1. ANKA can be applied to compute the postcritical paths of equilibrium as well.

PROBLEMS CONSIDERED AND RELATED ALGORITHMS

Initial and linearized buckling

Depending on specification of the matrices A and B in Eq.(2) the following equations of the eigenvalue problem can be formulated:

$$\left[[K_0 + K_\sigma(\bar{s})] + \lambda [K_\sigma(s^*)] \right] v = 0 , \tag{5a}$$

$$\left[[K_0 + K_\sigma(\bar{s}) + K_{u1}(\bar{d})] + \lambda [K_\sigma(s^*) + K_{u1}(d^*)] \right] v = 0 , \tag{5b}$$

where: K_σ - initial stress matrix, K_{u1} - initial displacement matrix dependent linearly on displacement field , $\bar{s}, s^*$ - stress field vectors, $\bar{d}, d^*$ - displacement field vectors computed on the base of the solution of Eq.(1). In case of geometrical imperfections appropriate field of displacements is to be addded to $\bar{d}$.

Case a), corresponding to Eq.(5a), is called the *initial buckling* problem and case b) corresponds to the so called *linearized buckling* problem respectively.

The critical load parameter is usually referred to the lowest eigenvalue:

$$\lambda_{cr} = \min_i \lambda_i \qquad \text{for} \quad i = 1,\ldots,n \ll N \qquad (6)$$

where: N - number of DOF of the FE assemblage.

In case of shells of revolution a sequence of the eigenvalue problems is considered for subsequent, fixed numbers of the circumferential buckling waves j. In such case the critical load parameter equals:

$$\lambda_{cr} = \min_{i,j} \lambda_i^j \qquad \text{for} \quad j = 0,1,\ldots \qquad (6a)$$

Evaluation of nonlinearity of the fundamental equilibrium path

Nonlinearity of the fundamental equilibrium path can be associated with intrinsic features of the structure (e.g. deformation of a spherical cap under external pressure) or can be caused by imperfections of the perfect structure.

The simplest evaluation of deviation from the linear fundamental path is given by the following coefficient:

$$\beta = (\lambda_{cr}^a - \lambda_{cr}^b)/\lambda_{cr}^b , \qquad (7)$$

where: λ_{cr}^a, λ_{cr}^b - critical load parameters computed according to (6) from Eqs (5a) or (5b) respectively. In case of the linear prebuckling state (perfect structure) $\beta = 0$.

More precise evaluation is associated with the computation of the parameter $\overset{..}{\zeta} \in [0,1]$ - cf.[3]:

$$\overset{..}{\zeta} = a^T a/W \qquad (8)$$

where: $a = X^T P^T$, $X = [v_1,\ldots,v_n]$, $X^T K_0 X = I$, $W = Q^{*T} P^*$. Components of vector a are coefficients of the linear combination of vectors v_i, i.e. $\overset{..}{Q} = Xa$. The value of parameter $\overset{..}{\zeta} = \overset{..}{Q}^T K_0 \overset{..}{Q}/W$ is the work of the 'parallel' loads $\overset{..}{P} = K_0 \overset{..}{Q}$ related to the work W of all loads $P^* = \overset{..}{P} + \overset{\perp}{P}$. In case of the linear prebuckling state $a = 0$, $\overset{..}{\zeta} = 0$ and for increasing nonlinearity $\overset{..}{\zeta} \to 1$.

Approximate nonlinear, prebuckling equilibrium state.

On the base of the initial modal superposition method (cf.[4]) the following formula can be derived:

$$Q(\lambda) = \sum_{i=1}^{n} \frac{\lambda_i \lambda}{\lambda_i - \lambda} v_i v_i^T P^* \qquad \text{for} \quad \lambda < \lambda_i . \qquad (9)$$

Evaluation of the type of critical points

Critical state of equilibrium is associated with the zero eigenvalue $\varkappa_1 = 0$ related to the standard eigenproblem $(K_T - \varkappa_1 I)v_1 = 0$. Taking into account the equation of adjacent equilibrium $K_T \Delta Q = P^* \Delta \lambda$ the following criterion can be formulated in order to evaluate the type of critical points:

$$v_1^{T} P^* \begin{cases} = 0 & \text{- bifurcation point B,} \\ \neq 0 & \text{- limit point L .} \end{cases} \tag{10}$$

In case of point B two sets of linear equations (4) can be solved substituting the following tangent stiffness matrices

$$K_T^{\pm} = K_o + K_\sigma(\lambda_{cr}) + K_u(\lambda_{cr} Q^* \pm \alpha \hat{v}) , \tag{11}$$

where: K_u - full initial displacement matrix, $\alpha < 1$, $\hat{v} = v_1 / |v_1|$. The computed values $\Delta\lambda^+$, $\Delta\lambda^-$ determine the following types of point B:

a) asymmetric point B: $\Delta\lambda^+ \cdot \Delta\lambda^- < 0$,
b) symmetric stable point B: $\Delta\lambda^+ \approx \Delta\lambda^- > 0$,   (12)
c) symmetric unstable point B: $\Delta\lambda^+ \approx \Delta\lambda^- < 0$.

Nonlinear equilibrium paths

The set of incremental equations (4) can be rewritten in the following form, suitable for the Newton-Raphson method:

$$^mK^i \, \Delta\Delta^mQ^i - P^* \Delta\Delta^m\lambda^i = {}^mR^i ,$$
$$^mt_i^{i^{T}} \Delta\Delta^mQ^i + {}^mt_{N+1}^{i} \, \Delta\Delta^m\lambda^i = \alpha^i \Delta^m\tau . \tag{13}$$

The subincrement $\Delta\Delta^mQ^j$, $\Delta\Delta^m\lambda^i$ are cumulated to compute the current increments for step m :

$$\Delta^mQ^i = \sum_{j=1}^{i} \Delta\Delta^mQ^j , \quad \Delta^m\lambda = \sum_{j=1}^{i} \Delta\Delta^m\lambda^j , \tag{14}$$

and the predictor-corrector parameter equals $\alpha^1 = 1$, $\alpha^i = 0$ for $i > 1$. The iteration process is stopped after the following convergence criteria are fulfilled - cf.[5]:

$$\left| (dc - dp)/d1 \right| < \varepsilon_Q , \quad \left| Rc - R1 \right| < \varepsilon_R , \tag{15}$$

where: $dc = \|\Delta^mQ^i\|$, $dp = \|\Delta^mQ^{i-1}\|$, $d1 = \|\Delta^mQ^1\|$, $Rc = \|^mR^i\|$, $R1 = \|^mR^1\|$.
The modified Newton-Raphson method can also be used in ANKA. In such an option the stiffness matrix K_T is updated only at the beginning of iteration process, i.e. $^mK^i = K_T(^{m-1}Q + \Delta^mQ^1)$.

At each step m of the incremental procedure the stability of equilibrium state as well as type of critical points can be evaluated without the eigenanalysis. This can easily be achieved by checking the signs of the stability determinant $\det{}^m K$ and the determinant of the main matrix of Eqs (13) - $\det{}^m \tilde{K}$. Instead of (10) the following criterion of critical state can be used - cf.[1]:

$$\det K_T = 0 \quad \text{and} \quad \det \tilde{K} \begin{cases} = 0 & \text{- bifurcation points B ,} \\ \neq 0 & \text{- limit point L .} \end{cases} \qquad (16)$$

THE ANKA/PC CODE AND USED FINITE ELEMENTS

The ANKA code has been written in FORTRAN and C and implemented on PC computers under MS-DOS. The code is composed of 3 moduli: i) input data processing, ii) computations, iii) graphical postprocessing. The moduli contact through disk files.

The program can analyse a set of about 4000 linear equations. Symmetric part of the global matrix is stored according to the sky-line concept and devided into blocks. Two matrix blocks needed to perform the LDL^T decomposition and the back subtitution are processed in the main memory at the same time.

The generalized eigenvalue problem is solved by means of subspace iterations with a possibity of the matrix spectrum shifting. The code can compute up to 10 first eigenvalues and eigenvectors.

All information about finite elements is stored in "packs", i.e. appropriate disk records which are decoded during the computation of the FE matrices and assembling of the global matrices. Such an approach is of value especially at the incremental analysis of nonlinear problems.

Finite elements of different levels of complexity have been incorporated into the ANKA library of FE - cf.[1]. The cone FE, called SRSK1 (Shell of Revolution, Straight meridian, Kirchhoff hypothesis) is the simplest one. For every number j of the circumferential buckling waves the number of DOF of a node equals $N_k = 4$.

Much more complicated are isoparametric, degenerated shell FE, called in ANKA as SQDR1 and SQDR2 (Shell Quadrilateral, Double curved, Reissner hypothesis). The element SQDR1 is the 8-node Seredipy type FE of number DOF equals $N_e = 40$. The element SQDR2 is 9-node Heterosis type FE of $N_e = 42$. Reduced integration is used in SQDR1 and selective integration in SQDR2, as it was discussed in [1,6].

NUMERICAL EXAMPLES

Only two numerical examples, which show possibilities of the ANKA/PC code, are discussed below. Other examples from [1] as well as new ones will be presented at the conference.

A shallow, axially compressed cylindrical panel

The panel has all edges supported in tangentially movable way (Fig.1a). A plate, which corresponds to the panel projection on the (x,y) plane has been additionally solved. Besides the edge compression p_y the imperfections are introduced as a concentrated, lateral force $\ni_P = P_z$ or an initial sinusoidal deflection with the amplitude $\epsilon_e = w_s$.

16 SQDR1 finite elements have been used. For perfect structures the

following critical buckling loads have been computed: i) for plate - p_{cr}^a = = p_{cr}^b = 1136.4 kN/m, ii) for panel - p_{cr}^a = 1481.2 kN/m, p_{cr}^b = 1481.6 kN/m. The critical load p_{cr} =1512 kN/m for the perfect panel was computed in [7].The symmetric stable bifurcation point for the plate and asymmetric point for the panel have been numerically confirmed using the criterium (12) - cf. Fig. 1b.

For both load imperfection $\ni_p$ as well as for the initial deflection $\ni_w$

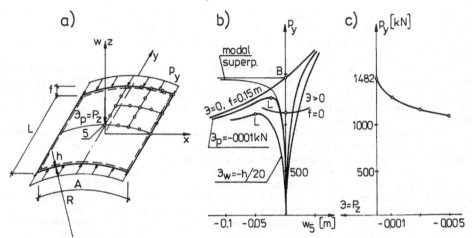

$L = A = 10.0m$, $h = 0.1m$, $f = 0.15m$, $R = 83.3m$, $E = 3.4 \cdot 10^7 kN/m^2$, $\nu = 0.2$

Fig. 1. Axially compressed shallow cylindrical panel

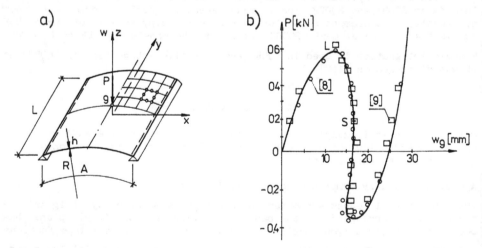

$L = A = 504mm$, $R = 2540mm$, $h = 6.35mm$, $E = 3.105 kN/mm^2$, $\nu = 0.2$

Fig. 2. Cylindrical panel under a concentrated load

the limit points has been computed for $\ni < 0$. The sensitivity on the panel to imperfections is shown in Fig.1c.

The approximate curves computed from formula (9) are also shown in Fig.1b.

Shallow cylindrical panel under lateral load

In order to check the accuracy and efficiency of ANKA the cylindrical panel under a concentrated load has been computed (Fig.2a). A quarter of the panel has been analysed using the arc-type control parameter.

In Fig.2b the computed equilibrium path as well as results from [8,9] are shown. Besides the limit point L also the snap-back point S is visible on the plane (P, w_g).

FINAL REMARKS

The ANKA code is under continuous development. A version of the code has been used to the analysis of reinforced concrete plates and shells [10].

Another version, so called small ANKA, is also being developed. This version is suitable for checking new finite elements and teaching students on the structural stability.

The version of the code under UNIX is to be developed from the viewpoint of the more general CAD system for designing shell structures.

REFERENCES

1. Waszczyszyn, Z., Cichoń, Cz., Radwańska, M., The Finite Element Method for Stability of Structures (in Polish), Arkady, Warsaw, 1990.

2. Waszczyszyn, Z., Numerical problems of nonlinear stability analysis of elastic structures, Comp. & Stru., 1983, 17, 13-24.

3. Wieczorek, M., Numerical analysis of structures sensitive to buckling (in Polish), Dodatek do Biuletynu WAT, 1986, 11 (441).

4. Nagy, D.A., Modal representation of geometrically nonlinear behavior by the finite element methods, Comp. & Stru., 1979, 10, 683-8.

5. Owen, D.R.J., Hinton, E., Finite Element in Plasticity, Pineridge Press, Swansea, 1980.

6. Hinton, E., Owen, D.R.J., Finite Element Software for Plates and Shells, Pineridge Press, Swansea, 1984.

7. Basar, V., Harte, R., Kratzig, W.B., General theory, computational concepts and applications of thin shell structures. In Lecture Notes to the CISM Course, Udine, September 1985.

8. Crisfield, A.M., Solution procedures for nonlinear structural problems. In Recent Advances in Nonlinear Computational Mechanics, Eds.E.Hinton, D.R.J.Owen, C.Taylor, Pineridge Press, Swansea, 1982, pp.1-29.

9. Meek, J.L., Tan, H.S., Instability analysis of thin plates and arbitrary shells using a faceted shell element with Loof nodes, Comp. Meth. Appl. Mech. Eng., 1986, 57, 143-70.

10. Pamin, J., Radwańska, M., Waszczyszyn, Z., A simplified FE analysis of reinforced concrete plates and shells. In Computer Aided Analysis and Design of Concrete Structures, Eds. N.Bicanic, H.Mang, Pineridge Press, Swansea, 1989, Vol.1, pp.329-40.

BUCKLING OF THIN CYLINDRICAL SHELLS SUBMITTED TO EXTERNAL PRESSURE

by M. DJERROUD, I. CHAHROUR and J.M. REYNOUARD
Concretes and Structures Laboratory INSA LYON (France)

ABSTRACT

The problem under discussion is the efficiency of different discretization procedures used for the stability and the geometrically non linear analysis of shell structures. A curved element based on Marguerre's shallow shell theory and a flat element based on plate theory are compared to isoparametric Assumed Natural coordinate Strain (ANS) elements. The numerical results of the different analytical approaches are compared to experimental data on buckling and post-buckling of thin cylindrical shells submitted to external pressure. It appears that the convergence is very sensitive to the choice of discretization procedure and that this choice is case dependent.

1 . INTRODUCTION

Thin and slender structures are being increasingly used for various reasons. However, along with this tendancy towards the use of optimum structural forms it could be expected that buckling will become a significant design constraint.

To predict the buckling loads of shell type structures, it is necessary to use good mechanical and geometrical models. More than anything else the numerical results of the finite element method tend to stress the importance of the choice of discretization procedure, particulary when it comes to buckling and non-linear analysis.

Nevertheless it is questionable whether the assumption of a linear prebuckling path which restricts investigation of instability up to the points of bifurcation on the linearized load-displacement path is justified.

2 . LINEAR STABILITY CRITERION

The stability of a structure is investigated using the method of adjacent equilibrium which is also called Euler's method. Only static loads are considered. In the Total Lagrangian Description (T.L.D.), the criterion is :

$$\delta u \ K_t \ u = 0 \ ;$$

where K_t is the tangent stiffness matrix ;

u is the incremental displacement field. (1)

When the structure is elastic and the loading is conservative, the Euler's criterion reduces to the Trefftz's energy criterion

$$\delta (\delta^2 \ \Pi) = 0 \qquad (2)$$

Where Π is the total potential energy.

The stability is called linear when the problem is linearized at the initial unloaded reference configuration. The tangent stiffness matrix is decomposed into four parts, i.e.

$$K_t = Ko + Ku + Kuu + K\sigma \qquad (3)$$

In this sum,

Ko is the infinitesimal matrix
Ku + Kuu forms the initial displacement matrix, linear and quadratic
contributions respectively
$K\sigma$ is the initial stress matrix

If λ_c is the critical load level leading to instability and if the initial displacements can be considered as very small, i.e. the Ku matrix can be neglected, the stability criterion reduces to the classical form

$$det \ (ko + \lambda_c \ K\sigma) = 0 \qquad (4)$$

3 . HISTORICAL DEVELOPMENT OF SHELL ELEMENTS

Since the begining of the finite element method, a continuing effort has been made to develop shell elements that satisfy a number of requirements hard to combine:

- the element should show good convergence ;
- the element should be well behaved, i.e. reliable results should be
  obtained in all situations ;
- the element should be easy to use and to implement.

Although not strictly necessary, it is preferable to have an element based on a displacement type formulation for ease of implementation.

In the history of the development of shell elements, the three following approaches are distinguished :

1) Flat elements

Flat elements are often used to represent curved surfaces. The reason for this is that it simplifies the formulation and also that it eliminates problems with strain energy under rigid body displacement. As long as the constitutive equations for the shell wall do not introduce membrane-bending coupling, it is possible to superimpose independent bending and membrane elements. The development of an effective but simple bending element, valid for thick to thin plates, is essential for the safty of a flat element. The main difficulties to overcome are related to the rank of the element stiffness matrices and to the absence of shear locking. In stability analysis of shell structures, it is necessary to describe initial geometry and therefore a high number of flat elements must be used.

2) Degenerated three-dimensional elements :

This approach was originated by Ahmad and Zienkiewicz [1]. By degeneration of three dimensional elements, a so-called thick shell element is obtained. In this element type, separate interpolation functions are used for displacement of the mid-surface and rotation of the normal to the surface. The mismatch between the two leads to the transverse shear strains. As the element gets thinner, these transverse shear strains should get smaller, thus approaching the thin shell formulation.

However it has been shown that in many cases the element become too stiff with decreasing thickness. This so-called "locking" phenomenon can be avoided by selection of an appropriate reduced or selective integration scheme.

3) Elements based on shell theory :

This type of element if of sufficiently high order, usually shows good performance. However, due to the neccessity of C1 compatibility between elements, it leads to rather unwieldy degrees of freedom [2].

4 . THE APPROACHES USED

Because of the complexity of the formulation of elements based on shell theory, two approaches have been choosen which can be considered as improvements or extensions of the approaches by degenerated elements and flat elements :

- The first approach derives from the degenerated solid concept introduced by Ahmad [1]. It uses the Continuum-Based (CB) approach. Standard shell hypotheses are embedded directly into the linearize CB equations. Following the Assumed Natural coordinate Strain (ANS) elements introduced by Park [3], two new triangular elements (ANST 3 and ANST 6) are developed

- The second approach is based on the Marguerre shallow shell theory which incorporates the effects of the curvature of the shell but avoids locking and spurious kinematic modes. The Constant Strain Triangle (CST) element is superimposed on the Discrete Shear Triangle (DST) element. The resulting element is called Discrete Shear Triangle in Marguerre theory (DSTM).

5 . THE ANS ELEMENTS

These are based on the approach proposed by Park [3] and the kinematical Hencky-Mindlin's hypothesis. Generally, two problems are encountred during the modelisation of the C° curved structures. The first one is the geometrical representation of the structure and is solved by making a simple geometrical approximation (isoparametric transformation). The second one is the membrane and transverse shear locking phenomenon. The generally used solutions for this problem consists of treating it with reduced integration or with the mode decomposition technique. However, the problems associated with these solutions can be cured but lead to many other problems such ase restriction of application and/ or of a numerical nature. These difficulties have led to the choosing of a method based on Park's approach which constitutes a preventive solution. Indeed, it avoids the problems before they appear.

Two new shell elements based on this approach are formulated : The first one is a three node linear triangular element (ANST 3) and the second one is a six node quadratic triangular element (Fig. 1)

1) Geometry and displacement field :

The coordinates and the displacements in the global system are interpolated by the two-dimensional shape functions as follow

$$(x, \, y, \, z) = \sum_{i=1}^{N} N_i \, (\xi, \, \eta) \, (x_i, \, y_i, \, z_i) \quad \text{and} \quad (u) = \sum_{i=1}^{N} N_i \, U_i \quad (5)$$

At each point of the element, we define a covariant and a cartesian basis. The unit vectors of the local cartesian coordinates system at each point of the elements are adopted as :

$$e_x = a_\xi \quad e_z = a_\zeta \quad \text{and} \quad e_y = e_z \wedge e_x \quad (6)$$

The transformation matrix related the covariant basis (a) to the cartesian one (e) is given by :

$$a = T^{-1} . e$$
$$\text{where } a = \langle a_\xi \ a_\eta, \ a_\zeta \rangle$$

and

$$T = \frac{1}{|a_\xi \wedge a_\eta|} \begin{bmatrix} e_y a_\eta & -e_y a_\xi & 0 \\ -e_x a_\xi & e_x a_\xi & 0 \\ 0 & 0 & |a_\xi \wedge a_\eta| \end{bmatrix}$$

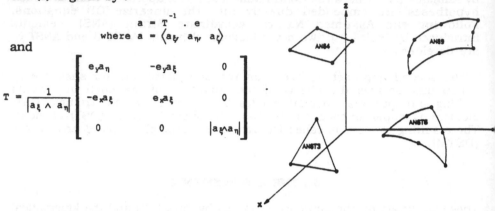

Fig. 1 : The ANS elements

2) Strain field

This approach is characterised by :

1. The way in which we obtain the product form of the strain-displacement relations. Hence, in the construction, first we interpolate the displacements that are fixed in the inertial coordinates system. The unit normals, a_ζ , vary along the ξ and η lines. We then obtain their derivatives along the natural-coordinates lines. Third, we project the interpolated quantities and the derivatives into the appropriate natural-coordinates lines to obtain the necessary covariant derivatives. By combining them, we obtain the desired covariant natural-coordinates strains.

2. At each integration point the necessary interpolations are performed along the two natural-coordinates lines which pass throught the integration point.

3. The third point is the way the natural coordinates in-plane strain and twist, $k_{\xi\eta}$ and $\varepsilon_{\xi\eta}$, are interpolated. We sample these strains at the Barlow points, which are then tensorially transformed and interpolated at each integration point.

4. The inertial-coordinates strains at any points in the element are finally obtained by tensorial transformation of the natural coordinate strains.

5. The membrane, the bending and transverse shear stiffness matrix are not coupled. The strain field is given by :

$$\varepsilon_{\xi\xi} = a_\xi^T \frac{\partial U}{\partial s_\xi} + \zeta \, a_\xi^T \frac{\partial \widehat{U}}{\partial s_\xi} + \frac{1}{2}\left[\left(a_\xi^T \frac{\partial U}{\partial s_\xi}\right)^2 + \left(a_\eta^T \frac{\partial U}{\partial s_\xi}\right)^2 + \left(a_\zeta^T \frac{\partial U}{\partial s_\xi}\right)^2\right]$$

$$\varepsilon_{\eta\eta} = a_\eta^T \frac{\partial U}{\partial s_\eta} + \zeta a_\eta^T \frac{\partial \widehat{U}}{\partial s_\eta} + \frac{1}{2}\left[\left(a_\xi^T \frac{\partial U}{\partial s_\eta}\right)^2 + \left(a_\eta^T \frac{\partial U}{\partial s_\eta}\right)^2 + \left(a_\zeta^T \frac{\partial U}{\partial s_\eta}\right)^2\right]$$

195

$$\varepsilon_{\xi\eta} = a_\eta^T \frac{\partial U}{\partial s_\xi} + a_\xi^T \frac{\partial U}{\partial s_\eta} + \zeta \left(a_\eta^T \frac{\partial \widehat{U}}{\partial s_\xi} + a_\xi^T \frac{\partial \widehat{U}}{\partial s_\eta} \right) + \left[\left(a_\xi^T \frac{\partial U}{\partial s_\xi} a_\xi^T \frac{\partial U}{\partial s_\eta} \right) + \left(a_\eta^T \frac{\partial U}{\partial s_\xi} a_\eta^T \frac{\partial U}{\partial s_\eta} \right)^2 + \left(a_\zeta^T \frac{\partial U}{\partial s_\xi} a_\zeta^T \frac{\partial U}{\partial s_\eta} \right) \right]$$

$$\varepsilon_{\xi\zeta} = a_\zeta^T \frac{\partial U}{\partial s_\xi} + a_\xi^T \widehat{U} + \left[\left(a_\xi^T \frac{\partial U}{\partial s_\xi} \right) \left(a_\xi^T \widehat{U} \right) + \left(a_\eta^T \frac{\partial U}{\partial s_\xi} \right) \left(a_\eta^T \widehat{U} \right) \right]$$

$$\varepsilon_{\eta\zeta} = a_\zeta^T \frac{\partial U}{\partial s_\eta} + a_\eta^T \widehat{U} + \left[\left(a_\xi^T \frac{\partial U}{\partial s_\eta} \right) \left(a_\xi^T \widehat{U} \right) + \left(a_\eta^T \frac{\partial U}{\partial s_\eta} \right) \left(a_\eta^T \widehat{U} \right) \right]$$

with (8)

$$a_\xi^T = \frac{1}{A_\xi} \langle X_{,\xi} \quad Y_{,\xi} \quad Z_{,\xi} \rangle$$

$$a_\xi = \frac{a_\xi \wedge a_\eta}{|a_\xi \wedge a_\eta|}$$

$$a_\eta^T = \frac{1}{A_\eta} \langle X_{,\eta} \quad Y_{,\eta} \quad Z_{,\eta} \rangle$$

$\langle U \rangle = \langle u \quad v \quad w \rangle$ (Displacement field)
$\langle \widehat{U} \rangle = \langle \widehat{u} \quad \widehat{v} \quad \widehat{w} \rangle$ (Rotation field)

$$\frac{\partial U}{\partial s_\xi} = \frac{1}{A_\xi} \frac{\partial U}{\partial \xi}$$

These strains are then implemented taking into account the deformation state. Thus, the strains along $\xi=cte$ or $\eta=cte$ line must be constant for the ANST3 case and linear for ANST6 case.

Following these steps the interpolation of the rotations used in the development of transverse shear strain for a η constant line are obtained as follows :

* For the ANST3 element   * For the ANST6 element

$$\overline{N}_1 = 1/2 - \eta$$
$$\overline{N}_2 = 1/2$$
$$\overline{N}_3 = \eta$$

$$\overline{N}_1 = 2\eta^2 + 4\xi\eta - \xi - 3\eta + 2/3$$
$$\overline{N}_2 = -4\xi\eta + 2/3$$
$$\overline{N}_3 = \xi - 1/3$$
$$\overline{N}_4 = 4\xi\eta$$
$$\overline{N}_5 = 2\eta^2 - \eta$$
$$\overline{N}_6 = -4\eta^2 - 4\xi\eta + 4\eta$$

(9)

3) Tangent and initial-stress stiffness matrix

The description of the notion and of the equilibrium state are made in Total Lagrangian Description (T.L.D.). The tangent stiffness matrix is given by :

(10)

$$\partial q . K_T . q = \int_{^\circ V} [C_{ijkl} . \dot{\varepsilon}_{ij} . \partial \dot{\varepsilon}_{ij} + \sigma_{ij} . \partial \dot{\varepsilon}_{ij}^*] . d^\circ V$$

Where $\overset{\bullet}{\varepsilon}_{ij}$ and $\overset{\bullet\ast}{\varepsilon}_{ij}$ are the linear and the non-linear parts of the incremental Green Strains, σ_{ij} is the Kirchhoff stress tensor, ^{o}V the initial volume.

The initial stress stiffness matrix is given by :

$$K_{\sigma} = \int_{^{o}V} G_{il} \cdot N_{lk} \cdot G_{kj} \cdot d^{o}V \qquad (11)$$

where G_{ij} is the vector of quadratic displacement contribution and N_{ij} is the membrane stresses in the element.

6 . The DSTM ELEMENT

The (DST+CST) element is a flat element resulting from the superposition of the Constant Strain Triangle and the Discrete Shear Triangle (DST) proposed by Batoz [4]. The introduction of the Marguerre shallow shell theory to incorporate the effects of the initial curvature gives the DSTM element. The effect of coupling membrane and bending creates the membrane locking phenomenon which is treated by the mode decomposition technique [5].

The formulation of the DST element is based on a generalization of the discrete Kirchhoff technique to include the transverse shear effects. DST has a proper rank and is free to shear locking. It conicides withe the DKT (Discrete Kirchhoff Triangle) element if the transverse shear effects are not significant.

The DSTM element now possesses five degrees of freedom at each node. The sixth degree of freedom associated with the shell normal rotation (not usually required by the theory) is the drilling rotation proposed by Bergan and used in Marguerre theory by Frey. Fig. 2 shows the stages of the construction of the DSTM element

CST+Curvature   DST   Drilling rotation   DSTM

Fig. 2 : Construction of the DSTM element

For the construction of the initial stress matrix, the element uses a cubic Hermite interpolation of the slope of the transverse displacement defined only at the element's boundary.

The description of motion and of the equilibrium state are made in A.U.L.D. (Approximate Updated Lagrangien Description). Its assocation with Marguerre's theory referenced in corotational axes leads to the corototional Lagrangian Description in Marguerre's theory (C.L.D.M.).

For solving the nonlinear process, the Newton-Raphson method is associated with the load, the displacement or the arc-lenght control.

7 . NUMERICAL RESULTS

1- Bi-hinged arch cylindrical shell

Figure 3 describes a bi-hinged shell used to test the ability of the shell elements (ANS, DST + CST, DSTM) to model the buckling of deep and shallow shells subjected to external pressure. Finite element results for the Euler's critical pressure are shown in Fig.4.

It appears that the incorporation of the initial curvature by the Marguerre's theory improves the convergence of the flat element. The linear triangular element ANST 3 gives the same results than the linear quadrilateral element ANS4 developed by PARK.

The quadratic triangular element ANST9 converges more rapidly than the ANS9 element and the two linear ANS elements.

2- Buckling of cylindrical shell due to external pressure

An end clamped thin cylindrical shell subjected to a uniform external pressure is analyzed to compare linear buckling to geometrically non lineare response.

The geometry and the material properties are shown in Fig. 5.
Results of linear buckling are shown in Tab 1 and the load-displacement response is shown in Fig 6.

It can be seen that good agreement exists between numerical and experimental results in linear stability. The load-displacement response and the buckling load obtained with the ANS and the DSTM elements correlate well with the experimental results.

8. CONCLUSIONS AND RECOMMENDATIONS

For buckling analysis of thin shells, the ANST6 and the DSTM elements are efficient and reliable. The first one converge rapidly and it is more performant than ANS elements using a linear interpolation for the in-plane dixplacements. The DSTM element, by incorporating the initial curvature, improves the convergence of the flat element (DST+CST).

It must be stressed that the investigation reported here is far from

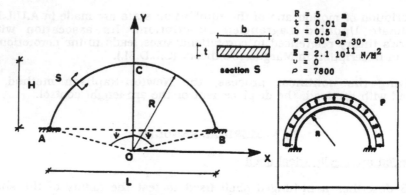

Fig. 3 : Bi-hinged arch

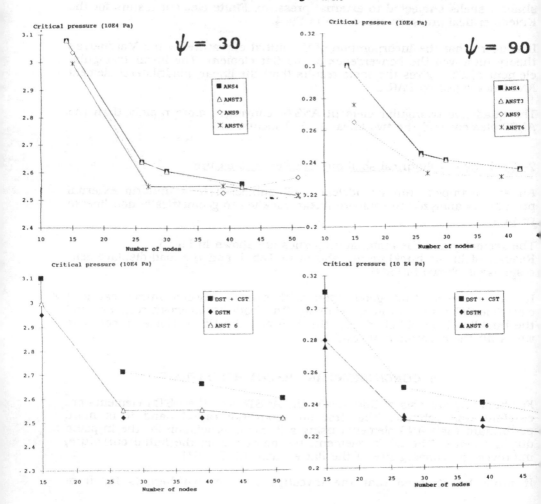

Fig . 4 : Critical pressure of the bi-hinged arch

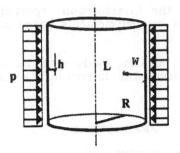

L=150 mm
R=75 mm
h=0.16 mm
E=166473 N/mm²
ν=0.34

Fig. 5 :   Cylindrical shell

	Critical pressure 10E2 Pa
Experience	224
ANS 4	244
ANS 9	240
ANST 3	245
ANST 6	239
DST+CST	241
DSTM	238

Tab. 1: Critical, pressure of the cylindrical shell

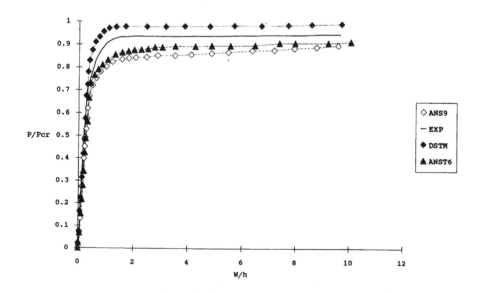

Fig. 6 : Load-displacement respose of the cylindrical shell

It must be stressed that the investigation reported here is far from exhaustive. For example, no curved elements based on shell theory were included in the study.

The difference between the buckling loads obtained with linear and non linear analysis tend to stress the importance of the concept of the incremental stability.

REFERENCES

[1] : AHMAD S, IRONS B.M, ZIENKIEWICZ
"Curved thick shell and membrane elements with particular reference to axisymmetric problems" Proc. conf. Matrix Meth. Struct. Wright-Patterson. Air Force Base, Ohio, AFFDL- TR - 68 - 15O, 1968, PP 539-572.

[2] : DHATT G-S.
"Instability of thin shells by the finite element method", Proc. IASS symp., Vienna 1970, Vol 1, pp 1-36.

[3] : PARK K.C.
"Improved strain interpolation for curved C° elements" Int. J. Num. Meth. Eng., 1986, Vol 22, pp 281-288.

[4] : BATOZ J.L. and LARDEUR P.
"A discrete shear triangular nine d.o.f. element for the analysis of thick to very thin plates"- Int. J. Num. Meth. Eng., 1989, Vol 28, pp 533-560.

[5] : STOLARSKI H., BELYTSHKO T., CARPENTIER N.
"Bending and shear mode decomposition in C° structural elements".
J. Struct. Mech., 1983, Vol 11, n°2, pp 153-176.

BUCKLING OF LAMINATED PLATES AND CYLINDRICAL SHELLS SUBJECTED TO COMBINED THERMAL AND MECHANICAL LOADS

H. KOSSIRA and M. HAUPT
Institut für Flugzeugbau und Leichtbau,
Technische Universität Braunschweig, Fed. Rep. Germany

ABSTRACT

The topic of this paper is the behaviour of laminated shells which are subjected to temperature fields and mechanical loading. The numerical investigations refer to the nonlinear deformations and buckling of shells with different lamina. Special types of nonuniform temperature distributions are considered. The numerical results show that the influence of the ply angle, radius of curvature and imperfections on the nonlinear behaviour are significant for laminated shells.

INTRODUCTION

Future developments in aircraft- and spacetechnologies need a combined thermal and structural analysis of structures. Designing hypersonic aircrafts or spacecrafts, especially reentry vehicles, the thermal influence on structures cannot be ignored. In this paper the behaviour of shells is investigated, which are used as covering skins of aircrafts. Aerodynamic heating of the surface will lead to a generally nonuniform temperature distribution within the shell. For example high temperature gradients will appear in the presence of aerodynamic shocks. Heatsinks, e.g. stringers and ribs, also induce temperature gradients. Therefore the investigated shells are loaded with uniform and nonuniform temperature distributions partly combined with mechanical loads.

GOVERNING EQUATIONS

Kinematic Description and Strain-Displacement Equations

The description of the deformation state of the shell is based on Kirchhoff-Love-hypothesis. The normal vector of the shell midsurface remains straight and perpendicular during the deformation. The deformation state $\mathbf{u}^*$ of any point within the shell can be described as :

$$(1) \qquad \mathbf{u}^* = \mathbf{u} + \vartheta^3\, \mathbf{w} \; .$$

$\mathbf{u}$ is the vector of the midplane displacement and $\mathbf{w}$ the rotation vector of the normal, ϑ^3 indicates the coordinate in thickness direction.

From the definition of the symmetric Green-Lagrange strain tensor γ_{ij} its tangential components can be written in the form :

$$(2) \qquad \gamma_{\alpha\beta} = \alpha_{\alpha\beta} + \beta_{\alpha\beta}\, \vartheta^3 \; .$$

$\alpha_{\alpha\beta}$ is the first strain tensor of the midplane, $\beta_{\alpha\beta}$ is the second symmetric strain tensor, which is linearised under the assumption of moderate rotations of the shell [1] :

$$
\begin{aligned}
\alpha_{\alpha\beta} &= \tfrac{1}{2}\left(\varphi_{\alpha\beta} + \varphi_{\beta\alpha} + \varphi_{\lambda\alpha}\,\varphi_{\rho\beta}\,a^{\lambda\rho} + \varphi_{3\alpha}\,\varphi_{3\beta}\right) \ , \\
(3) \qquad \beta_{\alpha\beta} &= -\tfrac{1}{2}\left(\varphi_{3\alpha|\beta} + \varphi_{3\beta|\alpha} + \varphi_{\lambda\alpha}\,b^{\lambda}_{\beta} + \varphi_{\lambda\beta}\,b^{\lambda}_{\alpha}\right) \ ,
\end{aligned}
$$

with the deformation gradients $\varphi^{\beta}_{.\alpha} = (u^{\beta}|_{\alpha} - u^{3}\,b^{\beta}_{\alpha})$, $\varphi^{3}_{.\alpha} = (u^{3}_{,\alpha} + u^{\beta}\,b_{\alpha\beta})$

u^{i} are the components of the midplane displacement vector $\mathbf{u}$, b^{α}_{β} is the tensor of curvature and $|_{\alpha}$ the covariant derivative. The strain components perpendicular to the midsurface $\gamma_{3\alpha}$ and γ_{33} vanish under the assumptions of the Kirchhoff-Love-theory.

Stress Resultants
It is usual to define the membrane forces $n^{\alpha\beta}$ and the bending moments $m^{\alpha\beta}$ by integration of the second Piola-Kirchhoff stress tensor $s^{\alpha\beta}$ weightened with the coordinate ϑ^{3} :

$$
(4) \qquad n^{\alpha\beta} = \int_{h} s^{\alpha\beta}\,d\vartheta^{3} \ , \qquad m^{\alpha\beta} = \int_{h} s^{\alpha\beta}\,\vartheta^{3}\,d\vartheta^{3} \ .
$$

Constitutive Equations
The constitutive equations couple these forces with the two strain tensors – assuming linear elastic material, including the thermal effects – in the following form [2] :

$$
\begin{aligned}
(5) \qquad n^{\alpha\beta} &= C_{D}^{\alpha\beta\lambda\mu}\,\alpha_{\lambda\mu} + C_{K}^{\lambda\mu\alpha\beta}\,\beta_{\lambda\mu} - \overset{T}{n}{}^{\alpha\beta} \ , \\
m^{\alpha\beta} &= C_{K}^{\alpha\beta\lambda\mu}\,\alpha_{\lambda\mu} + C_{B}^{\alpha\beta\lambda\mu}\,\beta_{\lambda\mu} - \overset{T}{m}{}^{\alpha\beta} \ .
\end{aligned}
$$

The tensors of elasticity result from the integration of the lokal tensor of elasticity $E^{\alpha\beta\lambda\mu}$ corresponding to the plane stress conditions :

$$
(6) \qquad \left[\,C_{D}^{\alpha\beta\lambda\mu} \,|\, C_{K}^{\alpha\beta\lambda\mu} \,|\, C_{B}^{\alpha\beta\lambda\mu}\,\right]^{T} = \int_{h} E^{\alpha\beta\lambda\mu}\,\left[\,1\,|\,\vartheta^{3}\,|\,(\vartheta^{3})^{2}\,\right]^{T}\,d\vartheta^{3}
$$

where the indices D, K, B indicate the tensors of extensional, coupling and bending stiffness. The relations for the thermal induced forces are derived from a similar integration, where additionally the tensor of thermal expansion $\overset{T}{\gamma}_{\lambda\mu}$ and the temperature difference Θ with respect to a stress-free reference temperature appears :

$$
(7) \qquad \left[\,\overset{T}{n}{}^{\alpha\beta} \,|\, \overset{T}{m}{}^{\alpha\beta}\,\right]^{T} = \int_{h} E^{\alpha\beta\lambda\mu}\,\overset{T}{\gamma}_{\lambda\mu}\,\Theta\,\left[\,1\,|\,\vartheta^{3}\,\right]^{T}\,d\vartheta^{3} \ .
$$

In this paper it is assumed, that the temperature varies linearly across the thickness of the shell

$$
(8) \qquad \Theta(\vartheta^{3}) = \Theta_{0} + \Theta_{1}\,\vartheta^{3} \ .
$$

This is valid for the most part of applications. For temperature independent material properties the thermal induced strain tensors become linear functions of the temperature coefficients Θ_{0} and Θ_{1}

$$
\begin{aligned}
(9) \qquad \overset{T}{\alpha}_{\alpha\beta} &= -\left(Q^{D}_{\alpha\beta\lambda\mu}\,\overset{T}{n}{}^{\lambda\mu}_{\Theta_{0}} + Q^{K}_{\lambda\mu\alpha\beta}\,\overset{T}{m}{}^{\lambda\mu}_{\Theta_{0}}\right)\Theta_{0} - \left(Q^{D}_{\alpha\beta\lambda\mu}\,\overset{T}{n}{}^{\lambda\mu}_{\Theta_{1}} + Q^{K}_{\lambda\mu\alpha\beta}\,\overset{T}{m}{}^{\lambda\mu}_{\Theta_{1}}\right)\Theta_{1} \ , \\
\overset{T}{\beta}_{\alpha\beta} &= -\left(Q^{K}_{\alpha\beta\lambda\mu}\,\overset{T}{n}{}^{\lambda\mu}_{\Theta_{0}} + Q^{B}_{\alpha\beta\lambda\mu}\,\overset{T}{m}{}^{\lambda\mu}_{\Theta_{0}}\right)\Theta_{0} - \left(Q^{K}_{\alpha\beta\lambda\mu}\,\overset{T}{n}{}^{\lambda\mu}_{\Theta_{1}} + Q^{B}_{\alpha\beta\lambda\mu}\,\overset{T}{m}{}^{\lambda\mu}_{\Theta_{1}}\right)\Theta_{1} \ .
\end{aligned}
$$

The expressions $\overset{T}{h}{}^{\alpha\beta}_{,\Theta_0}, \overset{T}{h}{}^{\alpha\beta}_{,\Theta_1} = \overset{T}{m}{}^{\alpha\beta}_{,\Theta_0}$ and $\overset{T}{m}{}^{\alpha\beta}_{,\Theta_1}$ are results from :

$$(10) \qquad \left[\; \overset{T}{h}{}^{\alpha\beta}_{,\Theta_0} \mid \overset{T}{h}{}^{\alpha\beta}_{,\Theta_1} = \overset{T}{m}{}^{\alpha\beta}_{,\Theta_0} \mid \overset{T}{m}{}^{\alpha\beta}_{,\Theta_1} \;\right]^T = \int_h E^{\alpha\beta\lambda\mu}\,\overset{T}{\gamma}_{\lambda\mu}\,\left[\; 1 \mid \vartheta^3 \mid (\vartheta^3)^2 \;\right]^T \, d\vartheta^3 \;,$$

$Q^i_{\alpha\beta\lambda\mu}$ are the inverse stiffness tensors.

Mixed Work Principle

A mixed work principle is used to express the boundary value problem in the weak formulation. The variables of state are combined in the vectors :

$$(11) \qquad \begin{aligned} \boldsymbol{\epsilon}^T &= [\; \alpha_{\alpha\beta}\,,\, \beta_{\alpha\beta} \;], & \overset{T}{\tilde{\boldsymbol{\epsilon}}}{}^T &= [\; \tilde{\alpha}_{\alpha\beta}\,,\, \tilde{\beta}_{\alpha\beta} \;], & \mathbf{u}^T &= [\; u_1\,,\, u_2\,,\, u_3 \;], \\ \boldsymbol{\sigma}^T &= [\; n^{\alpha\beta}\,,\, m^{\alpha\beta} \;], & \overset{T}{\boldsymbol{\sigma}}{}^T &= [\; \overset{T}{h}{}^{\alpha\beta}\,,\, \overset{T}{m}{}^{\alpha\beta} \;], & \bar{\mathbf{p}}^T &= [\; \bar{p}^1\,,\, \bar{p}^2\,,\, \bar{p}^3 \;], \end{aligned}$$

with $\bar{p}_i$ being the vector components of the mechanical load. Thus the field equations including the nonlinear operator matrices $\mathbf{D}_k, \mathbf{D}_e$ can be expressed as follows :

$$(12) \qquad \begin{aligned} \text{Equilibrum :} & \quad \mathbf{D}_e\,\boldsymbol{\sigma} + \bar{\mathbf{p}} = 0 \;, \\ \text{Kinematic :} & \quad \boldsymbol{\epsilon} = \mathbf{D}_k\,\mathbf{u} \;, \\ \text{Constitutive Relations :} & \quad \boldsymbol{\sigma} = \mathbf{C}\,\boldsymbol{\epsilon} - \overset{T}{\boldsymbol{\sigma}} \;, \\ & \quad \boldsymbol{\epsilon} = \mathbf{Q}\,\boldsymbol{\sigma} + \overset{T}{\boldsymbol{\epsilon}} \;. \end{aligned}$$

The boundary conditions on C consist of prescribed forces $\bar{\boldsymbol{\Sigma}}$ on C_Σ and prescribed displacements $\bar{\mathbf{U}}$ on C_u. The applied mixed variational principle which has forces, moments and displacements as independent unknown variables can be written in the following formulation :

$$(13) \qquad \begin{aligned} I^* = &\int_F \left(\tfrac{1}{2}\boldsymbol{\sigma}^T\,\mathbf{Q}\,\boldsymbol{\sigma} + \boldsymbol{\sigma}^T\,\overset{T}{\boldsymbol{\epsilon}}\right) dF - \int_F \boldsymbol{\sigma}^T\,\mathbf{D}_k\,\mathbf{u}\,dF \\ & + \oint_{C_\Sigma} \bar{\boldsymbol{\Sigma}}^T\,\mathbf{u}\,ds + \oint_{C_u} \boldsymbol{\Sigma}^T\,(\mathbf{u} - \bar{\mathbf{u}})\,ds + \int_F \bar{\mathbf{p}}^T\,\mathbf{u}\,dF = \text{stat} \;. \end{aligned}$$

FINITE ELEMENT DISCRETISATION

In practice a closed treatment of the work principle is not possible. Therefore the finite element method is applied here, to yield an algebraic system of equations. The contribution of a finite element to the work principle is obtained by integrating over the element area. Inside the element the degrees of freedom – assembled in the state vector $\mathbf{z}$ – by means of shapefunctions – arranged in the matrix $\boldsymbol{\Phi}_z$ – and the nodal degrees of freedom $\hat{\mathbf{z}}$ are approximated :

$$(14) \qquad \mathbf{z} = \boldsymbol{\Phi}_z\,\hat{\mathbf{z}} \quad \text{with} \quad \mathbf{z}^T = [\; \mathbf{u}^T \mid \mathbf{n}^T \mid \mathbf{m}^T \;] \;.$$

This approximation permits the integral evaluation in an analytical or numerical way. The resulting element matrices are :

$$(15) \qquad \mathbf{K}^e_{L,N} = \int_{F_e} \boldsymbol{\Phi}_z^T\,\mathbf{A}_{L,N}\boldsymbol{\Phi}_z\,dF \;.$$

The matrices $\mathbf{K}_L$ und $\mathbf{K}_N$ describe the linear elastic and the geometric nonlinear behaviour of the finite shell element. The matrices $\mathbf{A}_L$ and $\mathbf{A}_N$ consist of the linear and nonlinear operators of the work principle. For more details see [1]. The thermal load vector is given by the second term of the first integral of the mixed work principle. Using an interpolation of temperature coefficients Θ_0, Θ_1 corresponding to (14) the temperature induced thermal strains $\overset{\approx}{\alpha}{}^T$ and $\overset{\approx}{\beta}{}^T$ can also be approximated and assuming temperature independent material properties the therefrom resulting load vector is :

(16) $\qquad \overset{\star}{b}_T = \int_{F_e} \Phi_z^T\, \Phi_\Theta\, dF\, \overset{\star}{p}_T \quad \text{with} \quad {}^T\overset{\star}{p}{}^T = [\,0\mid 0\mid 0\mid \overset{\approx}{\alpha}{}^T \mid \overset{\approx}{\beta}{}^T\,]\,.$

${}^T\overset{\star}{p}{}^T$ is the vector of thermal nodal strains. Bilinear shape functions have been chosen for the analysis.

COMPOSITES

The basic assumption for the constitutive relations of laminates is the classical lamina theory [1,3]. The orthotropic properties of a single unidirectional layer are usually given with respect to the principle material directions. With layer data (E_1, E_2 elastic modulus parallel and perpendicular to the fibre direction, ν_{12}, ν_{21} orthotropic Possion's ratios, G_{12} in-plane shearmodulus, α_1, α_2 coefficients of thermal expansion parallel and perpendicular to the fibre orientation) the stress-strain-relation relative to the layer oriented axes may be written as :

(17)
$$\sigma^{11} = \frac{E_1}{1-\nu_{12}\nu_{21}}\left((\gamma_{11}-\alpha_1\Theta)+\nu_{12}(\gamma_{22}-\alpha_2\Theta)\right),$$
$$\sigma^{22} = \frac{E_2}{1-\nu_{12}\nu_{21}}\left((\gamma_{22}-\alpha_2\Theta)+\nu_{21}(\gamma_{11}-\alpha_1\Theta)\right),\qquad \sigma^{12} = 2G\,\gamma_{12}\,.$$

The stiffness coefficients $E^{\alpha\beta\lambda\nu}$ and the tensor of thermal expansion $\bar{\gamma}_{\alpha\beta}$ are results of appropriate tensor transformation relations [1,3]. Assuming constant stiffness $E_k^{\alpha\beta\lambda\nu}$ in the single layer with the thickness t_k the integrals (6) can be rewritten as :

(18) $\qquad \left[\,C_D^{\alpha\beta\lambda\mu}\mid C_K^{\alpha\beta\lambda\mu}\mid C_B^{\alpha\beta\lambda\mu}\,\right]^T = \sum_k E_k^{\alpha\beta\lambda\mu}\,t_k\,\left[\,1\mid\vartheta_k^{3m}\mid(\vartheta_k^{3m})^2+\tfrac{1}{12}t_k^2\,\right]^T.$

ϑ_k^{3m} is the ϑ^3 coordinate of the midplane of layer k. The integrals (10) are transformed analogously [4]:

(19) $\qquad \left[\,{}^T\!n_{,\Theta_0}^{\alpha\beta}\mid{}^T\!n_{,\Theta_1}^{\alpha\beta}={}^T\!m_{,\Theta_0}^{\alpha\beta}\mid{}^T\!m_{,\Theta_1}^{\alpha\beta}\,\right]^T = \sum_k \left(E_k^{\alpha\beta\lambda\mu}\,{}^T\!\gamma_{\lambda\mu\,k}\,t_k\,\left[\,1\mid\vartheta_k^{3m}\mid(\vartheta_k^{3m})^2+\tfrac{1}{12}t_k^2\,\right]^T\right).$

ALGORITHMS FOR NONLINEAR ANALYSIS

In structural analysis the characterisation of the load carrying behaviour is the essential viewpoint. The analysis of thinwalled structures under thermal loads shows that geometric nonlinearities occur as a result of the temperature induced strains, i.e. besides the tension problem, bifurcation- and snap-through-problems exist. The nonlinearities influence the load carrying behaviour remarkably. The finite element approach leads to a system of nonlinear equations, which express the equilibrium :

(20) $\qquad\qquad\qquad [K_L + K_N(z)]\,z = \overset{\star}{b}\,.$

The system is handled with an incremental procedure. The resulting nonlinear relations are solved with a modified Newton-Raphson iteration with arc-length constraints [5,6,7]. To detect points of neutral equilibrum – bifurcation and limit points – during the computation of nonlinear solution paths the following equation can be used :

(21) $\qquad\qquad\qquad [K_L + K_N(z)]\,\Delta z = 0\,.$

This equation indicates the existence of multiple states at the same load level and can be formulated as a linear eigenvalue problem :

$$(22) \qquad [\mathbf{K}_L + \lambda \, \mathbf{K}_N(\mathbf{z})] \, \mathbf{z}_E = \mathbf{0} \ .$$

For $\lambda = 1$ a point of neutral stability is reached. In the case of a bifurcation point the eigenvector $\mathbf{z}_E$ gives the eigenmode of the secondary solution path of the intersection.

Using the results of a linear analysis to compute the nonlinear matrix $\mathbf{K}_N$ leads to an eigenvalue problem similar to that of the classical linear buckling analysis. The smallest eigenvalue gives an approximation for the buckling load.

NUMERICAL EXAMPLES

Linear Analysis of Plates

To test the accuracy of the computer code FiPPS a linear bending problem was treated which had been studied before by WU und TAUCHERT using a Fourier-approximation [8]. The rectangular composite plate with the dimensions $a \times b$ consists of a glas-epoxy laminate with the material properties :

$$\begin{array}{lll} E_1 = 53,8 \, GPa & E_2 = 17,9 \, GPa & G_{12} = 8,62 \, GPa \\ \nu_{12} = 0,25 & \alpha_1 = 6,3 \cdot 10^{-6} \, K^{-1} & \alpha_2 = 20,5 \cdot 10^{-6} \, K^{-1} \end{array}$$

First $[+\alpha/-\alpha]_m$ antisymmetric angle-ply laminates are analysed. The plate is simply supported; the choice of boundary conditions is expressed as :

$$\vartheta_1 = 0, a : u_1 = u_3 = 0, \ n^{12} = m^{11} = 0; \quad \vartheta_2 = 0, b : u_2 = u_3 = 0, \ n^{12} = m^{22} = 0 \ .$$

The temperature distribution $\Theta = \Theta_1 \, \vartheta^3$ induces the out-of-plane deformations as shown in Fig. 1 for a two-ply laminate with a ply angle of $\alpha = 15°$. The out-of-plane deflection at the plate centre u_3^M is plotted versus the ply angle for different numbers of layers in Fig. 2. The deflection u_3^M is scaled by the bending stiffness $C_B^{1111^*}$ and the thermal moment $\widetilde{m}_{,\Theta_1}^{11^*}$ of a $[0/90]$ laminate

$$(23) \qquad u_3^* = u_3^M \, \frac{C_B^{1111^*} \, \pi^4}{16 \, \widetilde{m}_{,\Theta_1}^{11^*} \, \Theta_1 \, b^2} \ .$$

The whole plate has been modelled by 32×32 finite elements. Already a 16×16 mesh shows an error under 1%, i.e. good agreement with the results of reference [8] is indicated. As illustrated in Fig. 2 the influence of the antisymmetry and the therewith connected bending-membran-couplings of a two-ply laminate is significant in comparison with other laminates. The local minimum of u_3^* at $\alpha = 45°$ of all ply-numbers is caused by the opposite effects of the high thermal expansions α_2 and the high elastic modulus E_1 of the $90°$ turned neighbour layers.

Second a $[0/90]_m$ cross-ply laminate is considered. The temperature load is defined by a uniform temperature rise $\Theta = \Theta_0$ and the boundary conditions (simply-supported) are :

$$\vartheta_1 = 0, a : u_2 = u_3 = 0, \ n^{11} = m^{11} = 0; \quad \vartheta_2 = 0, b : u_1 = u_3 = 0, \ n^{22} = m^{22} = 0 \ .$$

Fig. 3 shows the deflection of a square plate, calculated with a 16×16 mesh. The out-of-plane deflection disappears along the two diagonals. The diagonals on which u_3 is zero are antisymmetry axes of the deformation pattern. The ply-number $2n$ and the varied plate aspect ratio a/b influence the centre deflection as shown in Fig. 4. The magnitude of the dimensionless displacement $^*u_3^M$ decreases as the number of layers $2n$ increases. For $n = \infty$ the coupling stiffness and the thermal moments $^T m^{\alpha\beta}$ vanish and no transversal deflection occurs. It can be concluded that the membrane-bending coupling has a significant effect upon the thermal deformation of the two-ply laminate, while on the other hand the coupling is relatively unimportant for plates having a large number of plies.

Linear Buckling Analysis of Plates

This example serves to compare the own linear buckling anlyses with the results from CHEN und CHEN [9]. The in-plane dimensions of the plates are $a \times b$, the material properties are given by :

$$\frac{E_1}{E_2} = 40 , \quad \frac{G_{12}}{E_2} = 0,5 , \quad \nu_{12} = 0,25 , \quad \alpha_1 = \alpha_2 = 1,0 \cdot 10^{-6} .$$

The laminate is antisymmetric $[+45/-45]_3$ with a relative thickness $a/t = 100$. For a uniform temperature rise, $\Theta = \Theta_0$, bifurcation occurs at the critical buckling temperature Θ_{crit}. The results for Θ_{crit} are obtained by a linear buckling analysis. The plotted buckling temperature Θ_{crit} versus the plate aspect ratio a/b is shown in Fig. 5 for two boundary conditions, simply supported

$$\vartheta_1 = 0, a : u_2 = u_3 = 0, n^{11} = m^{11} = 0; \quad \vartheta_2 = 0, b : u_1 = u_3 = 0, n^{22} = m^{22} = 0$$

and clamped

$$\vartheta_1 = 0, a : u_1 = u_2 = u_3 = 0, n^{12} = m^{12} = 0; \quad \vartheta_2 = 0, b : u_1 = u_3 = u_3 = 0, n^{12} = m^{12} = 0$$

The results corresponds well with [9]. As expected, the clamped plates have higher critical temperature values than the simply-supported ones and the critical temperature increases with the aspect ratio a/b.

Nonlinear Analysis of CFRP-Shells

Curved shells tend to nonlinear behaviour already at low load levels and the nonlinear deformation influences the buckling behaviour. For a better understanding of these interactions the full nonlinear pre- and postbuckling behaviour of flat and cylindrical CFRP shells is investigated. The dimensions of the studied shells are shown in Fig. 6. The ply data ($t_i = 0.275mm$) of the fibre-matrix system T300/914C are :

$$E_1 = 139400 \; N/mm^2 \quad E_2 = 8990 \; N/mm^2 \quad G_{12} = 4640 \; N/mm^2$$
$$\nu_{12} = 0,35 \quad \alpha_1 = 0,23 \cdot 10^{-6} K^{-1} \quad \alpha_2 = 29,0 \cdot 10^{-6} K^{-1} .$$

In connection with the two different boundary conditions simply supported :

$$\vartheta_1 = 0, a : u_1 = u_2 = u_3 = 0, m^{11} = 0; \quad \vartheta_2 = 0, b : u_1 = u_2 = u_3 = 0, m^{22} = 0$$

and clamped

$$\vartheta_1 = 0, a : u_1 = u_2 = u_3 = 0, m^{12} = 0; \quad \vartheta_2 = 0, b : u_1 = u_3 = u_3 = 0, m^{12} = 0$$

the plate was subjected to three different temperature distributions as illustrated in Fig. 7. In general the laminates were symmetric $[+\alpha/-\alpha]_s$ or antisymmetric $[+\alpha/-\alpha]_2$. For plates Fig. 8 shows the relationship between the critical maximum temperature Θ_{crit}^{max}, at which bifurcation occurs, and the ply angle α. As expected the clamped plates have larger critical values than the simply supported ones. The influence of antisymmetric laminates is low. Such laminates have slightly higher buckling values than the symmetric ones. For antisymmetric stacking the bending stiffness C^{1111} is equal to C^{2222} and the thermal moment $\bar{m}^{11}$ is equal to $\bar{m}^{22}$, whereas these are different for symmetric laminates.

In the case of constant temperature fields $\Theta_{01} = 0$ the curve of buckling loads is symmetric to the $\alpha = 45°$ axis; the reason is trivial. The gradient of the buckling temperature with increasing ply angle is caused by the decrease of the maximum thermal forces. Parallel to this tendency a change in the buckling mode is observed (Fig. 8). The buckling pattern is effected by the orientation of the stiffness and the thermal stresses. If the angle α is small respectively great the buckling patter consists of two halfwaves in the direction of minimum stiffness and maximum thermal force. In the moderate angle area the loads and the stiffness along the principle directions are nearly the same. For this reason a halfwave occurs in both direction.

Adequate buckling modes are also observed in the cases of nonuniform temperature fields (Fig. 9). The reason for the small differences in buckling temperature for all temperature fields at small ply angles are the nearly identical critical pressure loads along the $\vartheta_1 = a/2$ axis. For great angles up to $\alpha = 90°$ the critical forces are along the $\vartheta_2 = const$ axis and for $\Theta_{01} = \Theta_{00}$ respectively $\Theta_{00} = 0$ the critical temperature is obout 30% respectively 100% higher than that for $\Theta_{01} = 0$.

Fig. 10 shows the effect of the radius of curvature on thermal buckling. The existence of the curvature stabilizes the shellstructure and the buckling temperature increases if the radius of curvature decreases. The difference between the two different boundary conditions becomes relatively considered smaller, when the radius decreases. The curvature is of high importance in the presence of thermal loads, because it leads to transverse deflections in the prebuckling state influencing the buckling behaviour. In Fig. 11 for a radius of $R = 5000mm$ is plotted the buckling temperature versus the ply angle α. It shows the results of the linear buckling analysis and the nonlinear calculation.

There is no great difference in buckling temperatures between the two boundary conditions. The results of the linear buckling analysis are in good agreement with the results of the nonlinear calculation for angles up to $\alpha = 40°$, because the thermal induced deflections in the prebuckling state are not significant. In the range $\alpha = 45° - 60°$ there is no buckling in the considered temperature range. An interesting nonlinear behaviour can be noticed at angles $\alpha = 60° - 90°$. Here the results from the linear buckling analysis differ significantly from those of the nonlinear calculation. These differences can be attributed to the considerable prebuckling transverse deflections. The prebuckling behaviour is nonlinear and the linear eigenvalue analysis is at its limits.

Fig. 12 - 14 illustrate the nonlinear behaviour of shells, which are clamped and have a symmetric laminate. The maximum temperature is plotted versus the centre deflection u_3^M for different temperature fields. The extreme points of the curves indicate a bifurcation points. Further more there are some deformation patterns shown.

Referring to Fig. 14, in the case of great ply angles a snap through behaviour can be observed similar to an imperfect bifurcation. At higher temperatur levels a bifurcation point exits. The interpretation of the snap through effect as an imperfect bifurcation is supported by Fig. 15. This Figure shows the bifurcation behaviour of the corresponding simply supported shell ($\alpha = 60°$). The calculations started with antisymmetric imperfections of different magnitude, passed the bifurcation points and show on the second solution path a snap through behaviour. The second solution path stabilizes and reaches a deformation pattern with five halfwaves. This pattern is similar to the second eigenvector of the linear buckling analysis, while the deformation behind the first snap through corresponds with the first eigenvector. In the nonlinear calculation these two buckling modes can be reconstructed. The thermal load produces with its expansion in every direction predeformations which have the same effect as imperfections.

The temperature induced deformation behaviour does not change significantly with increasing shell thickness. Fig. 16 shows the temperature-deflection curves of symmetric laminates with $n = 4, 8, 12$.

Nonlinear Behaviour of CFRC-Shells
Carbonfibre reinforced carbon (CFRC) with its special properties is applicable at high temperatures. The previously mentioned clamped shell configuration is now considered with following material coefficients :

$$E_1 = 165000 \ N/mm^2 \quad E_2 = 6000 \ N/mm^2 \quad \nu_{12} = 0,23$$
$$G_{12} = 5670 \ N/mm^2 \quad \alpha_1 = -0,30 \cdot 10^{-6} K^{-1} \quad \alpha_2 = 4,30 \cdot 10^{-6} K^{-1}$$

Fig. 17 illustrates in a great extent a linear shell behaviour for small ply angles α. Caused by the negative thermal expansion α_1 the deflection u_3^M has a negative amplitude. The critical temperatures are relatively high compared with the results of CFRP shells, as the thermal expansion coefficients are an order of magnitude smaller. This results in a buckling pattern with more halfwaves. For great ply angles the primary bifurcation disappears and the nonlinear behaviour can be compared with this of CFRP. The differences in the deformation pattern are caused by the slightly different stiffnesses. The reasonable higher temperature level results from the smaller extension coefficients.

Thermal and Mechanical Loading of CFRP-Shells
For the investigation of the structural behaviour of CFRP-shells with respect to hot loadbearing structures the interest is focused on the superposition of thermal and mechanical loads. As an example a CRFP-shell with a ply angle of $\alpha = 45°$, exposed to temperature fields of $\Theta_{01} = 0, 10, 20, 30, 40, 50$ and a shear deformation γ, have been investigated. Fig. 18 shows the centre deflection u_3^M as a function of the shear angle γ for the different temperature loads. The non heated shell reacts with typical bifurcation behaviour and a major buckling deformation with a negative amplitude along the diagonal of tension. Additional thermal loading produces a prebuckling deformation with positive amplitude, which is nearly constant in the prebuckling area. An increase in the shear deformation angle leads to the buckling form of the non heated shell but with a positive amplitude snapping through into negative u_3^M at a certain level of shear load. It is obvious that the investigated low-level temperature loads have no greater influence on the main bifurcation.

CONCLUSION

Plates and cylindrical shells made of multi-layer composites exposed to thermal and mechanical loads have been investigated. This has been done with a computer code which based on the nonlinear theory of shells considering thermal induced strains. The used finite element formulation based on a mixed work principle.

The bifurcation behaviour of CFRP-plates with constant temperature across the thickness is similar to that of plates under mechanical loads. The critical load level depends on the boundary condition. A variation of the ply angle or of the temperature distribution changes the critical buckling load by a maximum factor of 2.

The nonlinear behaviour of shells is significantly dependent on the fibre orientation. The critical temperature increases evidently in the scope of the ply angle $\alpha = 45° - 60°$. The increase depends very much on the temperature distribution. For great angles of α the primary bifurcation behaviour changes to an imperfect one. The second path of solution can be reached after a bifurcation at elevated temperatures by snapping through. For such configurations the linear stability analysis can not be used.

Shells fabricated from CFRP and CFRC have a similar behaviour. The critical temperatures of comparible CFRC-shells are considerably higher, due to the smaller expansion coefficients.

ACKNOWLEDGEMENT

The authors gratefully acknowledge the financial support of their investigations by the Deutsche Forschungsgemeinschaft (Sonderforschnungsbereich 257).

REFERENCES

[1] WOLF, K. : Untersuchungen zum Beul- und Nachbeulverhalten schubbeanspruchter Teil-schalen aus kohlefaserverstärktem Kunststoff; Ph.D.Thesis; TU Braunschweig, 1988

[2] HETNARSKI, R.B. : Thermal Stresses I; North-Holland; Amsterdam, New York, Oxford, Tokyo; 1986

[3] TUNKER, H. : Über das Tragverhalten schubbeanspruchter quadratischer Platten aus Faserverbundwerkstoffen bei großen Deformationen; Ph.D.Thesis; TU Braunschweig, 1982

[4] HAUPT, M., KOSSIRA, H. : Thermisches Beulen von Platten und Zylinderschalen aus faserverstärkten Werkstoffen; in: Proc of the Jahrestagung der Deutschen Gesellschaft für Luft- und Raumfahrt 1990; Friedrichshafen; Jahrbuch 1990 I, 1990, 1327–1336

[5] WOLF, K.; KOSSIRA, H. : Untersuchungen zum Nachbeulverhalten schubbeanspruchter CFK-Teilschalen; in: Proc of the Jahrestagung der Deutschen Gesellschaft für Luft- und Raumfahrt 1987; Berlin; Jahrbuch 1987 I, 1987, 365–373

[6] WOLF, K.; KOSSIRA, H. : Buckling and Postbuckling Behaviour of Curved CFRP Lami-nated Shear Panels; in: Proc. of the 16th ICAS Congress, Jerusalem, 1988, 920–930

[7] WOLF, P.; KOSSIRA, H. : Zur Berechnung des nichtlinearen Tragverhaltens von Struktu-ren aus Faserverbundwerkstoffen; in: Entwurf und Anwendungen von Faserverbund-Struk-turen; DGLR-Report 87-02, 1987, Bonn

[8] WU, C.H.; TAUCHERT, T.R. : Thermoelastic Analysis of Laminated Plates 2 : Antisym-metric Cross-Ply and Angle-Ply Laminates; J. Therm.Stresses, 1980, **3**, 365–378

[9] CHEN, L.-W.; CHEN, L.-Y. : Thermal Buckling Analysis of laminated cylindrical Plates by the Finite Element Method; Computers&Structures, 1990, **34**, 71–78

FIGURES

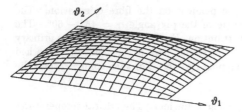

Fig.1: Deformation state of a plate with $\alpha = 15°$; $\Theta = \Theta_1 \vartheta^3$

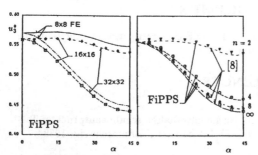

Fig.2: Comparison of the results of [8] and the finite element analysis (FiPPS)

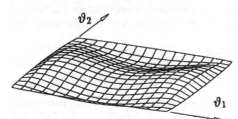

Fig.3: Deformation state of a square plate; [0/90] laminate; $\Theta = \Theta_0$

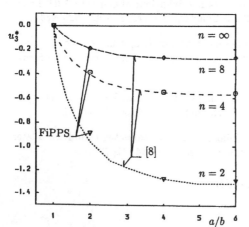

Fig.4: Comparison of the results of [8] and the finite element analysis (FiPPS)

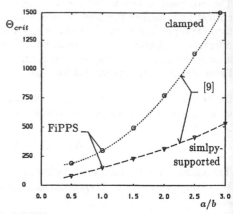

Fig.5: Comparison of the results of [9] and the finite element analysis (FiPPS)

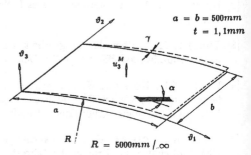

Fig.6: Geometry of the cylindrical shell

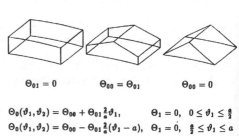

Fig.7: Temperature distributions (see (8))

$$\Theta_0(\vartheta_1, \vartheta_2) = \Theta_{00} + \Theta_{01}\frac{2}{a}\vartheta_1, \qquad \Theta_1 = 0, \quad 0 \le \vartheta_1 \le \frac{a}{2}$$

$$\Theta_0(\vartheta_1, \vartheta_2) = \Theta_{00} - \Theta_{01}\frac{2}{a}(\vartheta_1 - a), \qquad \Theta_1 = 0, \quad \frac{a}{2} \le \vartheta_1 \le a$$

211

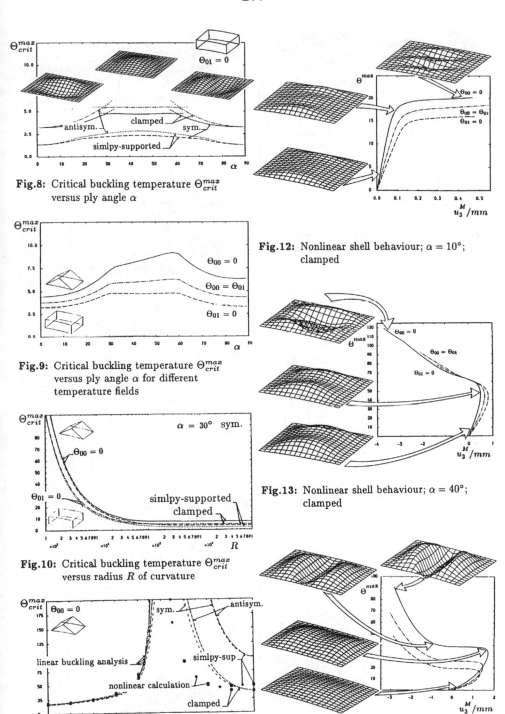

Fig.8: Critical buckling temperature Θ_{crit}^{max} versus ply angle α

Fig.9: Critical buckling temperature Θ_{crit}^{max} versus ply angle α for different temperature fields

Fig.10: Critical buckling temperature Θ_{crit}^{max} versus radius R of curvature

Fig.11: Critical buckling temperature Θ_{crit}^{max} versus ply angle α ($R = 5000mm$)

Fig.12: Nonlinear shell behaviour; $\alpha = 10°$; clamped

Fig.13: Nonlinear shell behaviour; $\alpha = 40°$; clamped

Fig.14: Nonlinear shell behaviour; $\alpha = 90°$; clamped

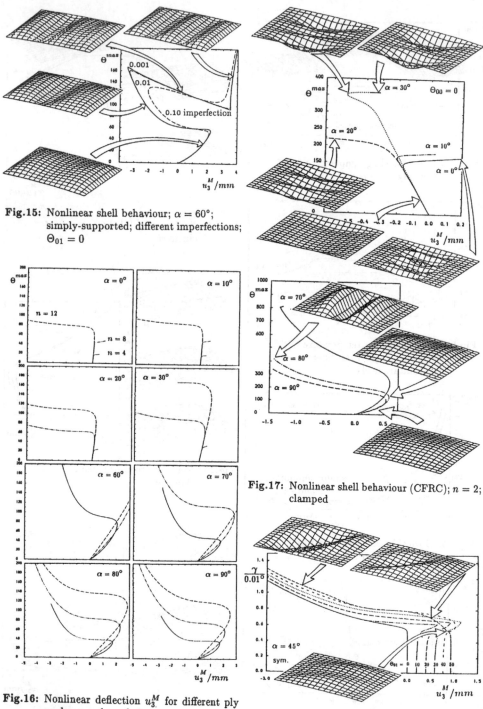

Fig.15: Nonlinear shell behaviour; $\alpha = 60°$; simply-supported; different imperfections; $\Theta_{01} = 0$

Fig.16: Nonlinear deflection u_3^M for different ply angles; number of ply as parameter; $\Theta_{00} = 0$

Fig.17: Nonlinear shell behaviour (CFRC); $n = 2$; clamped

Fig.18: Deformation of a CFRP shell mechanically and temperature loaded

INSTABILITY OF SHELLS OF REVOLUTION USING ALREF:
STUDIES FOR WIND LOADED SHELLS

FERNANDO G. FLORES and LUIS A. GODOY
Departamento de Estructuras - Universidad Nacional de Córdoba
Casilla de Correo 916, 5000 Córdoba, ARGENTINA

ABSTRACT

The object of this paper is twofold: first, to report on the development of
a computer code ALREF for stress and stability analysis of shells of
revolution, in which the general theory of elastic stability is employed
and a semianalytical finite element formulation used in the discretization.
Second, to explain the techniques employed in ALREF for bifurcation under
non-axisymmetric loads. This is illustrated by means of examples of
cylinders and cooling towers under wind load.

1. INTRODUCTION

The subject of stress and stability analyisis of shells of revolution has
attracted the interest of a number of researchers during the past 20 years
and as a consequence of that, there are several computer codes that can
handle such geometry. A review of existing codes may be found, for
instance, in Ref.[1]. Most codes are limited to the evaluation of
bifurcation loads from a linear fundamental path, and any non linearity is
considered via continuation methods.

The general theory of elastic stability [2,3], on the other hand, has
provided a strong framework to consider stability problems, in which the
singularity occurring at a critical state is carefully considered. It thus
allows to classify critical states; to approximate all postcritical paths
emerging from a critical point, and performing sensitivity analysis. In the
computer code ALREF, such facilities have been imcorporated in conjunction
with the finite element method; and details are given in this paper on the
application to the stability analysis under non axisymmetric pressures.

2. THE COMPUTER CODE ALREF

2.1. Generalities and fields of application

ALREF is a computer code oriented to the stress and stability analysis of shells of revolution. There are many codes already available which deal with this problem, but a particular feature of ALREF is that it is fully based on the theory of stability of discrete systems. In stability theory, special attention is given to critical points, which are studied in detail in order to obtain a classification based on the nature of the critical point itself; to investigate the stability of all paths emerging from the critical point; to approximate post critical paths; and to obtain sensitivity with respect to imperfections. An account of the theory may be found in the book by Thompson and Hunt [3], and in Ref.[4].

The basic assumptions on which ALREF has been developed are:

1. The constitutive material is linear elastic
2. The shell has axisymmetric geometry
3. Kirchhoff -Love assumptions of thin shell theory are adopted
4. For performing non linear analysis, moderate rotations are considered.
5. Mode interaction in stability analysis is not considered.

The tasks that ALREF can perform may be listed as follows

1. Linear and non linear stress analysis for axisymmetric loads (continuation methods)
2. Linear stress analysis for non axisymmetric loads
3. Eigenvalue analysis for linear fundamental paths under non axisymmetric loads (subspace iteration).
4. Eigenvalue analysis for non linear fundamental paths under axisymmetric loads.
5. Non linear, non axisymmetric post buckling path (perturbation analysis).
6. Imperfection sensitivity due to load or geometric imperfections (perturbation analysis).

2.2. Finite element discretization

The discretization of the shell is achieved by means of a finite element formulation. Fig.1 shows the element chosen, with the geometry and nodal degrees of freedom assumed. Special atention was given to the need to model complex, branched shells of revolution, and the element developed is fully capable of modelling intersections of several shells without any difficulties. The geometry is defined by a cubic interpolation based on coordenates and tangents of the nodes; while the displacement field is approximated by cubic functions for in-plane displacements and quintic for out-of-plane displacements. The element does not have blocking problems and has good convergence rate [5]

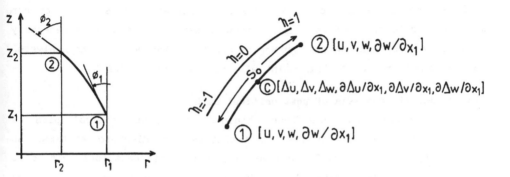

Figure 1. Finite element geometry and nodal displacements

2.3. The fundamental path

The most general techniques, known as continuation methods, and based on the Newton-Raphson method has been implemented to follow a non linear fundamental path. The algorithms of advance along a path considered have been those of prescribed load; prescribed displacement; spherical arc length; normal tangent plane; and prescribed external work [5]. The first algorithm is not adequate to follow a post buckling path; the last one presents problems in zones with significant loss of stiffness. The other three are available in the present version of ALREF, and have shown to have good convergence conditions and lead to reliable results.

2.4. Distinct Critical Points

As stated in the hipothesis, Section 2.1, only distinc critical points are considered in the present version of ALREF. For the evaluation of critical points, and eigenvalue is stated in the form

$$(K_T + \Delta\Lambda \ K_T') \ x = 0 \tag{1}$$

where K_T is the tangent stiffness matrix at the equilibrium state (Q^E, Λ^E); K_T' is the derivative of the tangent matrix with respect to the load; $\Delta\Lambda = \Lambda^C - \Lambda^E$; and x is the mode of bifurcation normalised so that $x_1 = 1$. Details on how eq.(1) is obtained may be found in [6].

For the evaluation of critical states on a linear fundamental path, eq.(1) is written at $(Q^E, \Lambda^E) = (0,0)$. If the fundamental path is not linear, the bifurcation state is seeked together with the advance along the path, until convergence is reached.

The eigenvalue problem for axisymmetric loads is not coupled in the different harmonics considered as possible bifurcation modes. But for non axisymmetric load, there is coupling between harmonics; and it is necessary to specify which and how many harmonics are to be included in the analysis. Special consideration to this point will be given in section 3.

2.5. Asymptotic analysis of post critical path

Once a critical point has been obtained, the post critical path is approximated using a perturbation technique for discrete systems, as employed by Thompson [3]. To achieve this, some basic equations related to the total potential energy V of the system should be recalled. Since $V = V(Q_i, \Lambda)$ where Q_i are the degrees of freedom of the discretised structure, and Λ the load parameter, the equilibrium conditions stem from stationary of V with respect to Q_i in the form

$$V_i = \frac{\partial V}{\partial Q_i} = 0 \tag{2}$$

The condition for the existence of a critical point is given by

$$V_{ij} \, x_j = 0 \tag{3}$$

To follow the post critical path, the variables Q_i and Λ are expanded in Taylor series of the perturbation parameter s. The parameter s is chosen as some of the Q_i (in particular $s \equiv Q_1$). Substitution of the Taylor expansion in (2), and by succesive derivation with respect to s one obtains a set of linear perturbation equations that are solved in a sequential order. In all systems, the matrix associated to the unknowns is allways the same: matrix V_{ij}. This is a regular degenerated problem [9] since V_{ij} is singular; and a contraction mechanism [3] is employed in the solution.

A limit point occurs when the loads f_i have component on the kernel of V_{ij}, (x), and the first derivative of the load is zero. If the loads do not have component on the critical mode we are in the presence of a bifurcation, with two solutions, each one associated to the two paths that intersect at the bifurcation state. These two solutions correspond to the contracted form of the second perturbation equation, which is quadratic in $\Lambda^{(1)} = d\Lambda/ds = d\Lambda/dQ_1$

$$A \; \Lambda^{(1)^2} + 2 \; B \; \Lambda^{(1)} + C = 0 \tag{4}$$

with

$$
\begin{aligned}
A &= V_{ijk} \; x_i \; y_j \; y_k \\
B &= V_{ijk} \; x_i \; x_j \; y_k \\
C &= V_{ijk} \; x_i \; x_j \; x_k
\end{aligned}
\tag{5}
$$

and

$$V_{ij} \; y_j = -f_i \qquad i,j=2,n \qquad y_1=0 \tag{6}$$

the derivatives of the degrees of freedom Q_i result in

$$Q_i^{(1)} = x_i + y_i \; \Lambda^{(1)} \qquad i=1,n \tag{7}$$

The higher order derivatives are obtained from the following linear perturbation equations and their contracted forms

2.3. Imperfection sensitivity

To investigate the influence of imperfections (either load, geometric or material imperfections) on the maximum loads that the system may attain. A variable ϵ represents the amplitude of the imperfection and is included in the formulation through V. In this case, it is convenient to apply perturbation on two sets of equations: equilibrium equations and those of critical state. The solution may be obtained using a component Q_1 as perturbation parameter s, and emploing the contraction mechanism. Notice that ϵ itself cannot be used as perturbation parameter since it would lead to a singular perturbation problem.

2.7. Final remarks

For shells of revolution under axisymmetric loads the analysis may be carried out on a personal computer due to the small number of d.o.f. and to the banded form of the system of equations. Most of the computer time is devoted to following the non linear path and the evaluation of the critical state.

Bifurcation occurs for a specific harmonic n in the semi analytical finite element formulation; while the derivatives of the secondary path only include harmonics which are multiples of n.

Numerical results obtained with ALREF have been published, for instance, in [5,7,8]

3. BIFURCATION LOADS OF WIND LOADED SHELLS

Special atention is given in this Section to the application of ALREF to the analysis of wind loaded shells. The main problem here is that the critical mode has components in all harmonics, with the result that both the non linear fundamental path and the bifurcation analysis are coupled in the harmonics.

3.1. Harmonics to be coupled

Simplified analysis for linear fundamental path is usually carried out using one of the following criteria:

(i) The worst loaded meridian, in which an axisymmetric pressure is considered, with value equal to the maximum pressure due to wind load.

(ii) The worst stressed meridian, in which the meridian with largest stressed is identified, and bifurcation is computed this state as axisymmetric fundamental state. Notice that this criteria requieres a double computation.

In the solution of a problem coupled in harmonics, it is necesary to stablish how many and which harmonics to be included in the eigenvalue analysis. This problem has been addressed before by Wang and Billington [10] and by Kundurpi et al [11] for cantilevered cylindrical shells. Consecutive harmonics are considered in both cases. Wang and Billington found that the first two harmonics j=0 and j=1 have a small influence in the bifurcation mode, and suggested the use of only 5 consecutive harmonics from a trial and error process, until a minimun in bifurcation load was reached.

In the present criteria, incorporated in ALREF, the harmonics included in the coupled analysis are those which individually produce a lower critical load under axisymmetric conditions. The number of harmonics to be coupled in the analysis depends on the actual pressure distribution in circunferencial direction, but it seems that 8 harmonics lead to sufficiently accurate results.

3.2. Numerical results

3.2.1 <u>Cantilever</u> <u>cylinder</u> <u>under</u> <u>wind</u> <u>load</u>: The geometry, material and pressure for this problem is shown in Fig.2. For this cylindrical shell with length/radius = 2 and radius/thickness = 100 with no internal suction, the simplified criterion (i) yields the lowest critical load 2.88 for harmonic 5; whereas for criterion (ii) the lowest critical load is 3.47 again for harmonic 5. With 10 coupled harmonics, the present criterion leads to 6.08 and the completer set of results is included in Table 1.; thus both simplified criteria produces extremely conservative loads (48% and 58% of the non axisymmetric load). The results of Wang and Billington for the present case lead to 6.6208; this difference of 10% with our results may be due to the incorrect selection of harmonics considered by them.

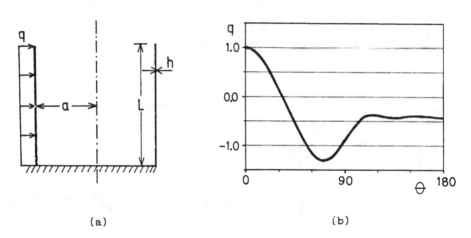

(a)   (b)

Figure 2. Cantilever Cylinder under wind load. (a) geometry $E=10^6$ $\nu=0.3$. (b) wind pressure variation along the parallel

Further results are compared with experiments by Kundurpi et al on clamped-free cylinders in a wind tunnel. For geometries with L/a = 2 and a/h = 376, Fig.3 shows the geometry, material and pressure distribution, which includes internal suction. The experimental value obtained is $\Lambda_c=0.98$, while the same authors calculated a numerical result for 5 harmonics of $\Lambda=1.15$. With 6 elements and 10 harmonics (4 to 13) in ALREF, a value of $\Lambda_c=1.00$ is obtained, which shows only 2% difference with experiments. For 6 harmonics (5 to 10), the critical load $\Lambda_c=1.088$ is only 10% higher than the experimental load. The simplified criteria (i) leads to

Λ_c=0.665; while (ii) to Λ_c=0.657; both are approximately 2/3 of the experimental load.

TABLE 1

Critical loads of a cantilever cylinder under wind loads for diferent selection of harmonics

Number of har	left shifted har.	Λ	Proposed criter. har.	Λ	rigth shifted har.	Λ
4	3- 6	14,56	4- 7	11,99	5- 8	14,18
5	3- 7	8,98	4- 8	8,55	5- 9	10,27
6	3- 8	7,25	4- 9	7,37	5-10	8,81
8	2- 9	6,48	3-10	6,26	4-11	6,67
10	2-11	6,09	3-12	6,08	4-13	6,52
12	2-13	6,01	3-14	6,05	4-15	6,47

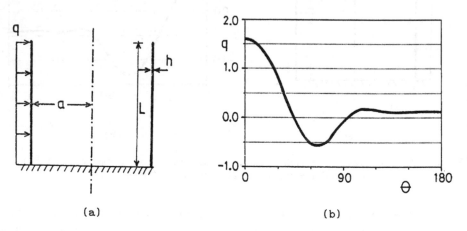

(a)   (b)

Figure 3. Cantilever Cylinder under wind load. (a) geometry $E=10^6$ $\nu=0.3$. (b) wind pressure variation along the parallel

3.2.2 Cooling tower under wind load. Fig.4 illustrates the geometry and loads; for this case using ALREF, the self weigth is considered as a fixed load and the wind pressure is incresed until buckling occurs.

The simplified criteria (i) leads to Λ_c=5.28 and n_{cr}=8; while (ii) to Λ_c=12.47, n_{cr}=7 and θ=65°. The present criteria incorporated in ALREF leads to $\Lambda_c=8.40$. Notice that criteria (ii) is not conservative in this case, while (i) yields Λ_c about 60% of the coupled analysis.

Abel et al [12] employed a 2-dimensional, doubly curved, element with non linear facilities, and found that for various cooling tower geometries the fundamental path was escentially linear before bifurcation. However, the significant differences detected between linearised bifurcation and experiments would indicate a strong imperfection sensitivity. Thus, the simplified criteria (i) is often used as an attempt to produce lower bounds in buckling of cooling towers.

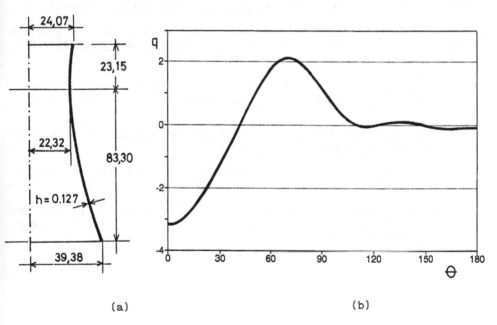

(a)   (b)

Figure 4. Cooling tower under self weigth and wind load. (a) geometry.
(b) wind pressure variation along the parallel

4. REFERENCES

1. Bushnell, D., Stress, stability and vibration of complex shells of revoution. Comp. & Struct., 1974, 4, 399-435.

2. Koiter, W.T., Over the stabiliteit van het elastich evenwicht, Delft, H.J.Paris, Amsterdam, 1945; (English translation) NASA Rep. TTF-10, 1967.

3. Thompson, J.M.T. and Hunt, G.W., A General Theory of Elastic Stability, J.Wiley & Sons, London, 1973.

4   Flores, F.G., _Elastic Stability and Geometric Non-linear Analysis of Shells of Revolution by the Finite Element Method with Applications to Pressure Vessels._ Ph.D. Thesis, National University of Cordoba, (in Spanish), 1990.

5. Flores, F.G. and Godoy, L.A., Finite element applications to the internal pressure loadings on spherical and other shells of revolution. In _Finite Element Applications to Thin Walled Structures,_ Ed. John Bull, Elsevier Science Publishers, Barking, 1990, 259-96.

6. Flores, F.G. and Godoy, L.A., Elastic post buckling analysis via finite element and perturbation techniques: Part I, formulation, _Int. J. Num. Meth. Engng.,_ (in press).

7. Flores, F.G. and Godoy, L.A., Elastic post buckling analysis via finite element and perturbation techniques: Part II, Application to shells of revolution. _Int. J. Num. Meth. Engng.,_ (submitted for publication).

8. Flores, F.G. and Godoy, L.A., Post-buckling of elastic cone-cylinder and sphere-cylinder complex shells. _Int. J. Pres. Ves. & Piping,_ **45**, 1991, 237-58.

9. Godoy, L., Flores, F., Raichman, S. and Mirasso, A., _Perturbation Techniques in Non-linear Analysis via Finite Elements,_ AMCA, Cordoba, (in Spanish), 1990.

10. Wang, Y. and Billington, D.P., Buckling of cylindrical shells by wind pressure. _ASCE J. Engng. Mech. Div.,_ 1974, 1005-23.

11. Kundurpi, P.S., Gopalacharyulu, S. and Johns, D., Stability of cantilevered shells under wind load. _ASCE J. Engng. Mech. Div.,_ 1975, 517-30.

12. Abel, J.F., Chang, S.C. and Hanna, S.L., Comparison of complete and simplified elastic buckling analysis for cooling towers shells. _Engng. Struct.,_ 1986, **8**, 25-28.

STABILITY OF WIND-LOADED SHELLS OF CYLINDRICAL TANKS WITH UNRESTRAINED UPPER EDGE

HELMUT SAAL
WENDELIN SCHRÜFER
Universität Stuttgart, Institut für Stahlbau und Holzbau
Pfaffenwaldring 7, D-7000 Stuttgart 80

ABSTRACT

The unrestrained upper edge occurs during the erection stage and for the catch basin perhaps even as a permanent situation. The investigation shows how in this case the buckling loads depend upon the boundary conditions at the lower edge. As long as the axial tensile stresses due to the wind load do not exceed the compressive stresses due to dead weight, the radial displacements at the upper edge are rather small, and the buckling load may be obtained from an analysis with axially fixed lower edge of the tank shell. Beyond this load level the lower edge uplifts, the radial displacements increase rapidly, and buckling occurs at a load which is very close to that at which the maximum tensile stresses from wind load cancel the compressive stresses due to dead weight loading.

INTRODUCTION

Because of their large r/t-ratios tank shells are prone to buckle under the action of compressive stresses. The dominant contribution is from circumferential compression due to wind loading. In order to simplify the stability analysis for this nonuniform external pressure situation (see figure 1) DIN18800, part 4 [1] according to [2] introduces a constant equivalent pressure. The buckling loads for this axisymmetric loading may be obtained from an analysis or taken from [1] for various boundary conditions.

During erection the upper edge of the tank shell is free while at the lower edge the in plane displacements are fixed by the bottom plate. In the vertical direction downward displacements are prevented by the foundation at the lower edge. The upward displacement of the lower edge is unrestrained in the standard case with no anchorage. With the boundary conditions depending on the direction of the

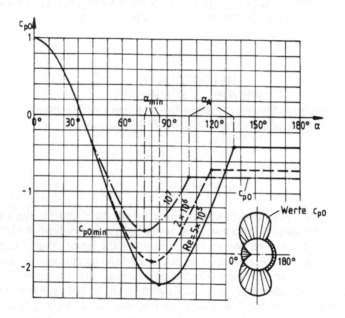

Figure 1: Drag coefficients for wind loaded cylinders according to DIN 1055

displacements the usual eigen-value-analysis is no longer applicable for calculating a buckling load. The buckling loads for the two limiting cases "axial displacements at the lower edge fully restrained" and "axial displacements free" differ for a tank with the dimensions $r/t = 1300$ and $l/r = 1$ by a factor of about 100.

The boundary conditions described above for the tank shell during erection apply to the catch basin as a permanent situation.

DESIGN OF STIFFENERS FOR THE CATCH BASIN OF A CYLINDRICAL TANK

Figure 2 shows a tank and a catch basin which had to be added to the tank for the purpose of water protection. With such a catch basin the location and cross section of the ring stiffeners is determined by the stability considerations. The stages of erection are shown in figure 3. The design of the ring stiffeners (position and cross section) was according to DASt-Richtlinie 013 [3] which apart from safety factors and reduction factors is equivalent to [1] : The equivalent pressure was in no case smaller than 97% of the stagnation pressure. A coefficient 0.6 was applied to the stagnation pressure at the upper edge of the catch basin to get the uniform internal suction due to wind loading.

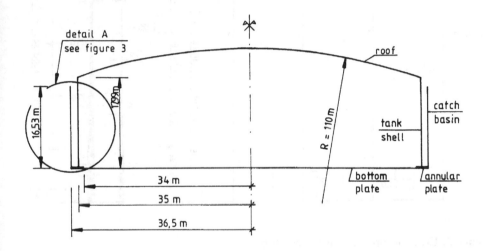

Figure 2: Geometry of tank and catch basin

The boundary conditions which according to [1] may be freely chosen were assumed to be an unrestrained upper edge, and the three displacement components u, v, w fully restrained at the lower edge. For this configuration the buckling pressures were obtained from linear theory. Subsequent to their multiplication by a reduction factor 0.7 [3] they were required to give a safety factor 1.1 [4] in comparison with the actual load. For the 3 situations with stiffeners shown in figure 3, this calculation gave safety factors between 1.11 and 1.15.

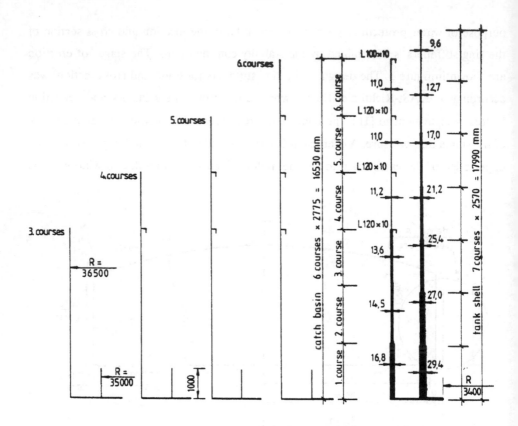

Figure 3: Erection progress of the catch basin

FINITE-ELEMENT-MODEL FOR THE NONLINEAR ANALYSIS

Because of the assumptions of invariable boundary conditions at the lower edge and of axisymmetric loading a finite-element-analysis was provided to check the results of the above analysis. Figure 4 shows the finite-element-model with the catch basin, the annular plate and part of the lowest course of the tank shell. With the finite-element-program ANSYS shell elements of type STIF 63 werde used for modelling these parts. Gap elements which only transfer compressive forces were used to represent the variable boundary conditions. The ring stiffeners were represented by beam elements type STIF 4. Both of these element types are also contained in figure 4 which is a plot for the final erection stage. Symmetry displacement constraints were introduced at the

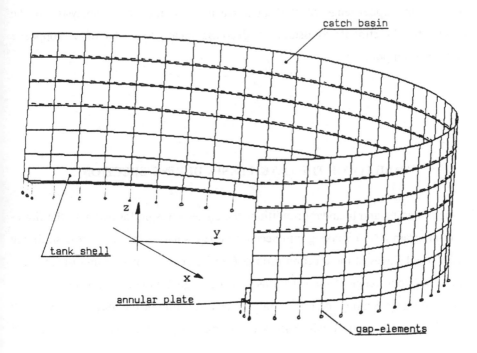

- - - - beam elements representing ring stiffeners

Figure 4: Finite-element-model of the 6 course catch basin

nodes on the x-z-plane. The in plane displacements of the nodes at the inner edge of the annular plate were assumed to be fully restrained by the bottom plate. According to [5] a material-safety-factor of 1.1 was applied to Youngs-modulus.

The dead load of the roof and the parts of the tank shell not contained in the model were taken into account by increasing the specific weight of the lowest course of the tank shell. For the catch basin including the ring stiffeners and for the annular plate the specific weight was 78.5kN/m. The wind suction acting on the roof was assumed to conform with [4] and represented by vertical forces along the upper edge of the lowest course of the tank shell. The circumferential distribution of the wind load acting on the catch basin (see figure 1) was taken from [6] for a Reynolds number 10^7 which is the closest to the actual value. The vertical distribution of this wind loading was given by a stagnation pressure of 0.5kN/m below and 0.8kN/m above the height of 8.265m above ground. The suction between catch basin and tank shell which is due to the wind was assumed constant over the height with a coefficient 0.6 applied to the stagnation pressure at the upper edge of the catch basin as in the above analysis. In the circumferential direction it was assumed to decrease linearly to a value of zero opposite to the stagnation point.

STABILITY OF THE CATCH BASIN DURING ERECTION

With these loading and boundary conditions the analysis was performed with the finite-element-program ANSYS taking into account large displacements and stress stiffening as well. For all stages of the erection process (see figure 3) the maximum displacement occurs at the upper edge of the catch basin at the stagnation point. This displacement is plotted in figure 5 along the horizontal axis as a function of the load factor γ which is plotted along the vertical axis. Both axes are on a logarithmic scale. With the load

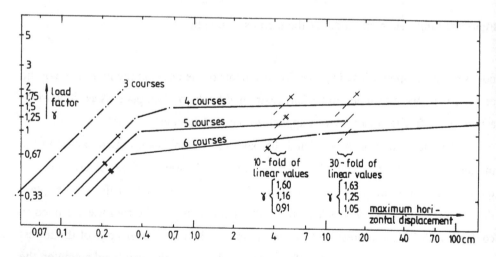

Figure 5: Load-displacement curves for 4 erection stages

deflection curves in figure 5 the relevant number of courses according to figure 3 is given. Obviously up to a load factor of 2 there is no risk of instability for the 3 course erection stage. However, with four and more courses there is a distinct knee in the load-deflection curve. The up to then very small displacements increase progressively. The load-deflection curves terminate at the last load step where convergence was achieved. Because of the discrete load steps of 0.25 between $\gamma = 1.0$ and $\gamma = 2.0$ the buckling load may be somewhat higher than indicated by the curves. A displacement criterion would seem reasonable if only because of the small distance between catch basin and tank shell. The load factors where the maximum displacement attains the 10-fold and 30-fold respectively of the value according to linear theory are marked in figure 5. Independent of the curves they are associated with maximum displacements of about 4cm and 14cm respectively. Taking the latter one as relevant criterion one arrives at a load safety factor for the 6 course erection stage which corresponds to that for the above mentioned buckling analysis if one there also takes into account the material safety factor. However, the safety factors for the 4 and 5 course erection stages are considerably larger. The progressive increase of the maximum deflections in figure 5 is associated with an uplift at the lower edge of the catch basin.

STABILITY OF THE TANK SHELL DURING ERECTION

The situation for the tank shell during erection differs form that for the catch basin: there is no beneficial action comparable to that of the tank shell for the catch basin, and the wall thicknesses of the tank shell are larger than those for the catch basin on the other hand. Therefore the erection stages of the tank shell in figure 3 were also investigated. The finite-element-model corresponds to that in figure 4. However, it only consists of the annular plate extending from the shell to the inside of the tank, the tank shell with each course represented by one ring of elements and of the gap elements. The outer part of the annular plate and the catch basin are missing compared to figure 4. The wind loading corresponds to that for the catch basin apart from the change of the stagnation pressure from 0.5kN/m to 0.8kN/m being at the height of 7.710m above ground and the absence of wind suction acting on the roof of the tank. The dead weight of tank shell and annular plate were covered by their specific weight.

Figure 6a shows the results for the finite-element-model where an uplift is allowed by using gap elements whereas the results in figure 6b are for a finite-element-model with vertical rigid support which allows no uplift. In both cases the figures

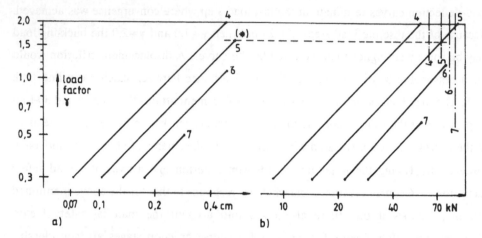

Figure 6: a) maximum inward radial displacement at the stagnation point for the model witch gap elements
b) maximum tensile force at the stagnation point for the model with rigid support

indicate the number of courses and the lines terminate where buckling occurs at the next load step (see below). The asterik in figure 6a indicates that at this load step an uplift occured at the lower edge of the tank shell with 5 courses. A comparison with figure 6b where the vertical tensile force due to wind load is used as abscissa and the compression forces due to dead weight are indicated by the dash-dotted vertical lines, shows that buckling can be contributed to the tensile forces exceeding the compression due to dead weight.

In order to get an idea how the size of the tank shell influences the buckling situation, a half-size and a double-size (although somewhat unrealistic) tank were investigated. Together with the dimensions of the tank shown in figure 3 their dimensions are given in table 1. The results of this investigation are shown in figure 7. The termination of the load-deflection curves indicates that buckling occured at the next load step which corresponds to a load factor which is by 0.1 (0.2 at $\gamma = 0.3$) larger. With the double-size tank shell buckling occurs without any uplift. However, with the half-size tank shell uplift occurs with the 3 and 4 course erection stages and is followed by buckling of the shell at an adjacent load level. The 5, 6 and 7 course erection stages buckle at even lower load levels without preceding uplift.

TABLE 1

Dimensions of the tank shell

	system 2	system h	system d
tank shell radius = outside radius of annular plate	35 000	17 500	70 000
inside radius of annular plate	34 000	16 500	69 000
wall thickness of annular plate	17.0	8.5	34.0
wall thickness of tank shell course 1 2 3 4 5 6 7	9.6 12.7 17.0 21.2 25.4 27.0 29.4	4.8 6.4 8.5 10.6 12.7 13.5 14.7	19.2 25.4 34.0 42.4 50.8 54.0 58.8
	all dimensions in mm		

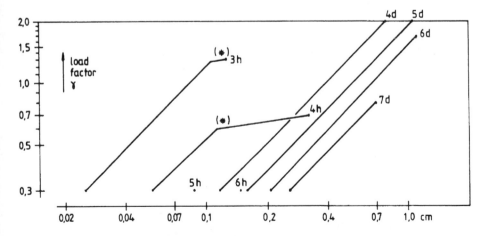

Figure 7: maximum inward radial displacement at the stagnation point for the half size (h) and the double-size (d) tank shell

REFERENCES

1. DIN18800, Teil 4, (11.90): Stahlbauten; Stabilitätsfälle, Schalenbeulen.

2. Resinger, F. and Greiner, R., Buckling of Wind-Loaded Cylindrical Shells - Application to Unstiffened and Ring-Stiffened Tanks. In: Buckling of Shells. Ed.: E. Ramm. Springer-Verlag. 1982.

3. DASt-Richtlinie 013: Beulsicherheitsnachweise für Schalen. Deutscher Ausschuß für Stahlbau. Juli 1980.

4. DIN4119, Teil 2, (02.80): Oberirdische zylindrische Flachboden-Tankbauwerke aus metallischen Werkstoffen; Berechnung.

5. DIN18800, Teil 1, (11.90): Stahlbauten; Bemessung und Konstruktion.

6. DIN1055, Teil 4, (08.86): Windlasten bei nicht schwingungsanfälligen Bauwerken.

ELASTIC AND INELASTIC BUCKLING OF RING–STIFFENED CIRCULAR CYLINDRICAL SHELLS SUBJECTED TO EXTERNAL PRESSURE

W. WUNDERLICH, Z. LU
Lehrstuhl für Statik, Technische Universität München

H. OBRECHT
Lehrstuhl für Baumechanik–Statik, Universität Dortmund

ABSTRACT

The paper deals with the results of systematic numerical studies on the nonlinear load–carrying and buckling behavior of ring–stiffened circular cylindrical shells under external pressure. In particular it focuses attention on the conditions under which stiffeners with rectangular cross–sections lead to either global or local (interstiffener) buckling and on the influence of elastic–plastic material behavior. The numerical results are presented in the form of nondimensional diagrams in which the buckling loads of stiffened shells are given directly as functions of a global stiffening parameter containing all relevant information on both the shell's geometry and the type and arrangement of the stiffeners. They indicate the buckling mode as well as the amount by which the load–carrying capacity of a stiffened shell is larger than that of the corresponding unstiffened one. This makes them particularly useful for practical design purposes.

INTRODUCTION

An accurate assessment of the safety and economy of a thin–walled shell structure requires a sufficiently detailed knowledge of its nonlinear load–carrying and buckling behavior. This is particularly true of ring–stiffened shells, because the stiffeners lead to strong stiffness variations along the axis and hence to a highly non–homogeneous stress and displacement state. In addition, the deformation and buckling behavior of ring–stiffened shells is governed by widely different characteristic dimensions: the size of the stiffener cross–section, the length of the unstiffened shell between the stiffeners, and the overall dimensions of the shell. Thus, depending on the geometrical and material parameters and on the type of loading, the buckling behavior may be dominated either by overall buckling or by interstiffener shell buckling. Furthermore, in addition to buckling in these individual modes one may also have circumstances in which they interact nonlinearly. This may lead to an increase in imperfection–sensitivity and thus to a further reduction in the shell's load–carrying capacity.

To gain more detailed information on the influence of ring–stiffeners on the load–carrying behavior of shells of revolution, systematic numerical studies on the elastic and elastic–plastic buckling behavior of ring–stiffened circular cylinders under external pressure have been carried out. The numerical results thus obtained are essentially free from simplifying assumptions and therefore may – within the limitations inherent in the discrete analysis – be considered to be accurate. In addition, they constitute a comprehensive and theoretically sound basis for the derivation of practical design guidelines. The analyses performed so far show that the numerical results can be presented in nondimensional diagrams which directly relate the buckling loads of stiffened shells to a global stiffening parameter containing all relevant information on both the shell's geometry and arrangement of the stiffeners. Without the need to evaluate additional quantities, they indicate the type of buckling as well as the amount by which the load–carrying capacity of a stiffened shell is larger than that of the corresponding unstiffened one. This makes them particularly useful for practical design purposes.

NUMERICAL SOLUTION PROCEDURE

As described in more detail in [1–4], the numerical results presented here are based on a semi–analytic treatment of the field equations governing the nonlinear static behavior of shells of revolution. It differs from standard two–dimensional finite element approaches in that Fourier series are used to represent the circumferential variation of the geometrical parameters, external loads, boundary conditions, initial geometric imperfections and state variables, while ring elements – one for each Fourier harmonic – are used to perform a discretization with respect to the meridional coordinate. The associated stiffness matrices are obtained via an asymptotically exact numerical integration of the governing differential equations over a finite interval of the meridional coordinate and the resulting transfer matrices. To be able to deal with the various Fourier harmonics separately, all terms resulting from geometrically and/or physically nonlinear influences are treated as pseudo–loads. This results in nonlinear implicit systems of algebraic equations which are then solved via a Newton–type procedure combined with a modified conjugate gradient technique.

To reduce the numerical effort and to avoid certain shortcomings inherent in other computational models, the ring–stiffeners have been treated as integral curved beam elements. They are considered to be attached to the outer and/or inner shell surface, while the coefficients of their respective stiffness matrices are referred to the shell's middle surface. Thus, stiffener excentricities are automatically taken into account, and both the shell and the stiffeners have the same degrees–of–freedom referred to the same nodal circles. As a result, the incorporation of the latter into a discrete shell model is a matter of routine. More important, however, is the fact that the small bandwidth of the global stiffness matrix is preserved, whereas when stiffeners are modelled as an assembly of shell ring elements it may increase considerably.

ELASTIC BUCKLING OF UNSTIFFENED CYLINDRICAL SHELLS

In order to obtain a comprehensive and compatible set of reference values, preliminary calculations have been carried out to determine the elastic bifurcation pressures of unstiffened, simply supported circular cylinders for various parameters.

The numerical results are summarized in Fig.1. It gives the elastic bifurcation pressure $p_{cr,0}^{el}$, normalized by Young's modulus and the radius–to thickness ratio, as a function of l/r and r/t. For a given value of the length–to–radius ratio the differences between the buckling pressures of thick and thin shells are considerably more pronounced when l/r is small. Thus, bifurcation pressures obtained for long and thin shells should not be extrapolated directly to the case of short and thick ones, and vice versa.

ELASTIC BUCKLING OF RING–STIFFENED CYLINDRICAL SHELLS

The results presented in this and the subsequent chapter apply to simply supported circular cylinders with ring–stiffeners having a rectangular cross–section. For consistency, the l/r– and r/t–values were taken to be the same as for the unstiffened shells, and seven equidistant stiffener arrangements as well as a number of different rectangular cross–sections were considered. As could be expected, the results showed that in the elastic range different stiffener shapes will lead to essentially the same buckling pressure as long as their respective in–plane bending moments of inertia are identical. Hence, both in the elastic and elastic–plastic analyses rectangular cross–sections with a constant height–to–width ratio of 10:1 were considered, and the only parameter that was changed systematically was the in–plane bending moment of inertia I_r.

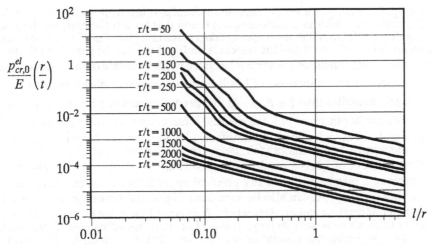

Figure 1. Elastic bifurcation pressures of unstiffened circular cylindrical shells

The numerical results given in Fig. 2 show the dependence of the buckling pressure p_{cr}^{el}, normalized by the bifurcation pressure $p_{cr,0}^{el}$ of the respective unstiffened shell

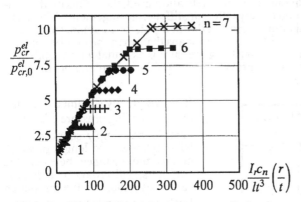

Figure 2. Elastic bifurcation pressures of circular cylindrical shells with symmetric rectangular ring-stiffeners ($l/r = 3$, $r/t = 200$)

(Fig. 1), on the number of symmetrically arranged internal and external ring stiffeners and their total bending stiffness. The quantity on the abscissa is a global stiffening parameter which contains all relevant information on both the shell's geometry and the degree of stiffening. Crudely speaking, it represents a weighted ratio relating the combined in-plane bending stiffness of all stiffeners to a measure of the shell's total bending stiffness. Apart from I_r and the shell parameters r, l, and t, it also contains the factor

$$c_n = \sum_{i=1}^{n} \sin \frac{i\pi}{(n + 1)} , \qquad (1)$$

where n denotes the number of stiffeners. The product $I_r c_n$ characterizes the combined effective bending stiffness of all stiffeners. Note that when n varies from 1 to 6, c_n takes on the following values

n	1	2	3	4	5	6
c_n	1	1.732	2.414	3.078	3.732	4.381

Thus, except for n = 1, c_n is always smaller than n. This indicates that an increase in the number of stiffeners does not lead to an equivalent increase in the total effective bending stiffness $I_r c_n$. This, of course, also affects the shell's load-carrying behavior.

Considering, for example, a simply supported circular cylindrical shell with $l/r = 3$, $r/t = 200$ and four equidistant stiffeners, it is found that for a sufficiently small value of I_r – e.g. $I_r = 75 \, lt^3 \, (t/r)/3.078$ – buckling takes place in a global displacement mode which involves both the shell and the stiffeners. In Fig. 2 this can be seen from the fact that the associated bifurcation pressure p_{cr}^{el} is given by the ordinate of the convex curve on the left. Moreover, an increase of I_r will lead to a corresponding increase of p_{cr}^{el} as long as p_{cr}^{el} is smaller than $5.8 \, p_{cr,0}$. At that point the buckling behavior changes, because for a further increase of I_r the smallest bifurcation pressure p_{cr}^{el} is associated not with a global but with a local (interstiffener) buckling mode. Moreover, as indicated by the almost horizontal branch, the corresponding bifurcation pressures are largely independent of I_r. As a result, the largest value of p_{cr}^{el} that is attainable with a given value of n may be regarded as a practical upper limit of the respective shell's load–carrying capacity. As can also be seen from Fig. 2, the maximum value of p_{cr}^{el} strongly depends on the number of stiffeners. A single stiffener, for example, will lead to an upper limit which is twice as large as that of the corresponding unstiffened shell, whereas for n = 7 one only obtains an increase by a factor of about 10. This means that the addition of stiffeners will always lead to a rise in the absolute load–carrying capacity, but each time n is increased, the relative increase of the upper limit p_{cr}^{el} becomes smaller. In other words, with each added stiffener the relative efficiency of the stiffeners diminishes.

It is also interesting to note that due to the normalization used, the points on the convex curve in Fig.2 may correspond to very different stiffener arrangements. Those immediately above the horizontal branch associated with n = 2, for example, correspond to shells with either 3, 4, 5, 6 or 7 stiffeners. This means that when the number of stiffeners exceeds a certain minimum, the buckling pressure is essentially independent of how many more stiffeners are added, provided the combined in–plane bending stiffness $I_r \, c_n$ has a certain fixed value.

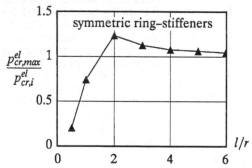

Figure 3. Maximum (interstiffener) buckling pressures of ring–stiffened circular cylindrical shells (n = 7, r/t = 150)

It is frequently assumed that the elastic buckling pressure $p_{cr,i}^{el}$ of an unstiffened, simply supported circular cylinder, the length of which equals the distance between the stiffeners, will always be a conservative estimate of the maximum buckling pressure $p_{cr,max}^{el}$ of the respective stiffened shell. The curve in Fig. 3, which gives the dependence of the ratio $p_{cr,max}^{el} / p_{cr,i}^{el}$ on l/r, indicates that for sufficiently long shells ($l/r > 1.5$) this assumption is justified, whereas for relatively short ones it may lead to a drastic overestimation of the load–carrying capacity. In particular, for $l/r = 0.5$ one finds that $p_{cr,max}^{el}$ is less than 25 percent of $p_{cr,i}^{el}$. This again confirms the earlier observation that the load–carrying behavior of relatively short cylinders differs considerably from that of long ones, and that the straightforward extrapolation of results from one range of parameters to the other should be avoided.

When the horizontal branches in Fig. 2 are omitted and the convex curves corresponding to an l/r–ratio of 3 and r/t–ratios of 50, 100, 150, 200, 250, 500, 100, 1500, 2000 and 2500 are plotted together, one obtains the left diagram in the center of Fig. 4.

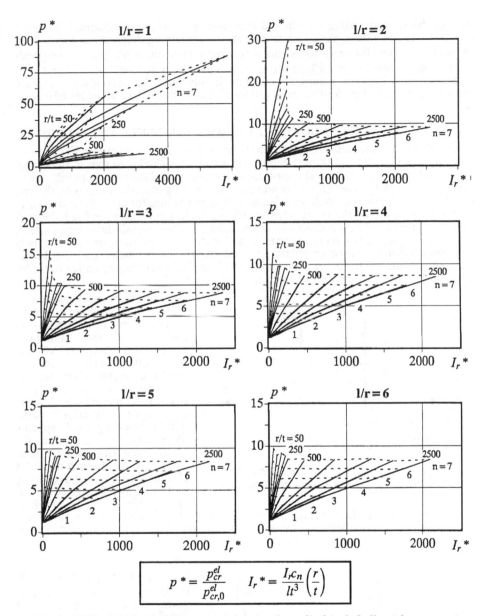

$$p^* = \frac{p^{el}_{cr}}{p^{el}_{cr,0}} \qquad I_r^* = \frac{I_r c_n}{lt^3}\left(\frac{r}{t}\right)$$

Figure 4. Elastic bifurcation pressures for circular cylindrical shells with symmetric rectangular ring–stiffeners (r/t=50,100,...,2500)

There, dashed lines have been used to denote the points at which the horizontal branches corresponding to n = 1, 2, 3, 4, 5, 6 and 7 intersect the respective convex curves (see Fig. 2). As can be seen, for r/t–values larger than about 250, the dashed lines are essentially horizontal for all values of n, whereas for r/t–values less than about 150 they rise sharply. Thus, except for relatively thick shells, the maximum increase in buckling pressure that can be achieved by a given number of stiffeners is approximately equal for all values of r/t. In the case of l/r = 3 and n = 7, for example, the highest buckling pressure of a stiffened shell will be approximately 9 times larger than that of the corresponding unstiffened one. Similar results are obtained for other length–to–radius ratios, as can be seen from the remaining diagrams in Fig.4. They show that – except for very short shells – the maximum value of $p_{cr}^{el}/p_{cr,0}$ corresponding to a given number of stiffeners does not depend very strongly on l/r. For n = 7 and sufficiently large r/t–values, for example, it only varies between about 8 and 10, even though the radius–to–thickness ratio varies over more than an order of magnitude.

ELASTIC–PLASTIC BUCKLING OF RING–STIFFENED CYLINDERS

When elastic–plastic deformations occur during loading, the buckling behavior of ring-stiffened circular cylindrical shells is in some ways similar to that in the elastic range, but it also shows several important differences. The latter are primarily due to the strong coupling between the continuously decreasing material moduli on the one hand and the geometric nonlinearities as well as the nonuniformity of the stress and displacement state on the other. This usually results in a considerable, load–dependent, stress redistribution and an associated change in the overall deflection pattern. In particular, depending on the geometric and material parameters, one may not only have bifurcation into a nonaxisymmetric mode, but also axisymmetric collapse. Moreover, the nonuniformity in the axial variation of the field quantities as well as the influence of the boundary regions is much more pronounced.

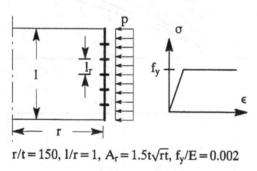

$r/t = 150,\ l/r = 1,\ A_r = 1.5t\sqrt{rt},\ f_y/E = 0.002$

Figure 5a. Ring–stiffened cylindrical shell under external pressure – Geometry and material data.

An important consequence of the strongly nonlinear interaction of all of these factors – in particular of the continuous nonuniform decrease of the shell's overall stiffness and the resulting considerable increase in the magnitude of the respective displacements – is that the definition of a practically useful 'plastic buckling load' appears to be less obvious than in the elastic case.This can be seen from Figs. 5a–d which show the elastic–plastic load–carrying and buckling behavior of a moderately thin and relatively short shell with five equidistant rectangular ring stiffeners. For simplicity the material was assumed to be elastic–ideally plastic and the yield stress f_y was taken to be 0.2 percent of Young's modulus E (see Fig. 5a).

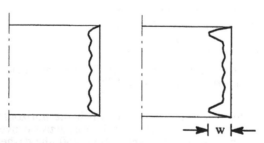

Figure 5b. Elastic pre-buckling deflections

Figure 5c. Elastic–plastic prebuckling deflections

The corresponding elastic bifurcation pressure p_{cr}^{el} may be obtained from the upper left diagram in Fig. 4, and the axial distribution of the prebuckling deflections are shown in Fig. 5b. As can be seen, the latter are quite regular with the stiffeners exhibiting more–or–less the same uniform contraction. In the elastic–plastic range, on the other hand, the deflection pattern is considerably less uniform (see Fig. 5c).

In particular, the displacement amplitudes are much larger near the supports and, moreover, the stiffeners also experience a certain amount of twisting. The relationship between the external pressure p – normalized by the yield stress f_y and the radius–to–thickness ratio r/t – and the maximum lateral deflection w has been plotted in Fig. 5d. It shows that as soon as p is large enough to result in significant plastic deformation, the slope of the curve in Fig. 5d – which may be regarded as a measure of the shell's local stiffness – also decreases rapidly. Thus, over a comparatively short range of p and/or w the load–deflection curve approaches an almost horizontal asymptote. This indicates that when p is sufficiently large the initially high (elastic) stiffness has essentially disappeared and the shell therefore, cannot carry a higher load in a quasistatic manner. Consequently, the respective value of p represents an upper limit of the shell's load–carrying capacity. Note that the associated deflection pattern is axisymmetric, so that when p were to actually approach this limiting value, axisymmetric collapse would occur following a rapid growth of the inward deflections near the boundaries.

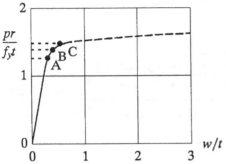

Figure 5d. Load–displacement relationship

To circumvent the effort involved in an accurate evaluation of the maximum value of p, a number of heuristic 'plastic limit pressures' have been defined. Their main purpose is to yield reasonable lower bound estimates of the collapse pressure without – supposedly – involving any complex calculations. One such limit is the pressure p_y, which is considered to be reached when in the most highly stressed region the plastic zone has spread from the outer surfaces to the middle surface. In the load–deflection curve in Fig. 5d the respective point is denoted by A. An alternative limiting pressure has been proposed by ECCS. It is considered to be reached when at the midpoint between two stiffeners the circumferential membrane stress of the shell first exceeds the material's yield stress f_y. In Fig. 5d this particular point has been denoted by B. In the present case B lies above A and thus is closer to the horizontal part of the load–deflection curve. On the other hand, point A is more conservative.

Unfortunately, neither the pressures corresponding to points A, B and the horizontal asymptote, represent the actual limit of the shell's load–carrying capacity. It is given by the pressure p_{cr}^{pl} at which bifurcation into a nonaxisymmetric mode take place. E.g. Fig. 5d the respective point on the load–deflection curve is denoted by C. The axial distribution of the corresponding buckling mode shows, that in the elastic–plastic range buckling occurs in a global mode which involves both the shell and the stiffeners, whereas in the corresponding elastic case the buckling displacements are essentially confined to the shell sections between the stiffeners.

As can be seen from Fig. 5d, the actual upper limit of the shell's load–carrying capacity (Point C) is larger than the approximate limit pressures corresponding to points A and B. Thus, the latter may be regarded as conservative. This, however, is not always the case. Rather, for certain values of the geometric and material parameters, point C may also lie considerably below points A and B. This often happens when the elastic and elastic–plastic bifurcation pressures p_{cr}^{el} and p_{cr}^{pl} are reasonably close. Thus, the intuitively attractive idea of deriving approximate plastic limit pressures from the stress levels at particular points of the shell's middle surface may actually lead to an overestimation of the buckling pressure precisely in those cases in which it is expected to give a sufficiently realistic representation of the negative effects of plastic deformations on the shell's load–carrying behavior.

In Figs. 6a and b numerical results for p_{cr}^{pl}, normalized by the respective elastic value p_{cr}^{el}, are plotted as functions of l/r and an appropriate global stiffening parameter

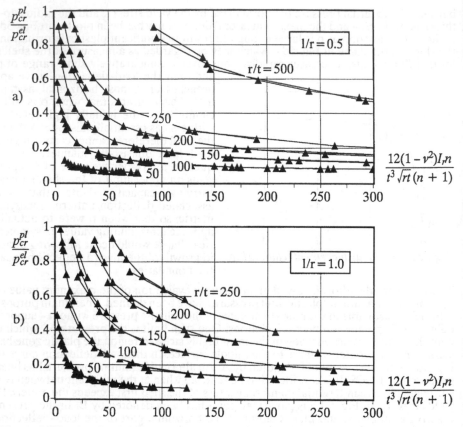

Figure 6. Elastic–plastic bifurcation pressures of ring–stiffened circular cylindrical shells

which again accounts for the bending moment of inertia I_r and the number of stiffeners. They show that for sufficiently thin shells with large r/t–values no plastic deformations occur at all, so that the respective values of $p^{pl}_{cr} / p^{el}_{cr}$ one equal to one. For increasingly thicker shells, on the other hand, p^{pl}_{cr} becomes progressively smaller than p^{el}_{cr}, and for $r/t = 50$, for example, p^{pl}_{cr} may even be as small as 5 percent of p^{el}_{cr}. This general tendency holds for both l/r–values, but it is more pronounced for the shorter shells having an l/r–ratio of 0.5. This again demonstrates the familiar result that thicker and shorter – and thus geometrically stiffer – shells are more affected by plastic deformations and the associated decrease in material stiffness than thinner and longer ones. Moreover, Figs. 6a and b also indicate that the rather complex nonlinear interaction between the governing parameters cannot be described in a sufficiently simple manner by a single 'plastic reduction factor'.

In Fig. 7 similar numerical results are plotted in an alternative form which more closely resembles that in Fig. 2. There, the plastic buckling pressures, normalized by the yield stress f_y and the r/t–ratio are given as a functions of a different global stiffening parameter which only involves the cross-sectional area A_r of the stiffener as well as the shell's radius r, and its thickness t. Crudely speaking, the parameter on the abscissa may be regarded as a global stretching stiffness involving the stiffener and part of the shell to which it is attached. Interestingly, for a given value of r/t (and l_r/r) the numerical results obtained for both three and seven stiffeners lie on almost the same curves, which,

241

moreover, – as in the case of elastic shells – approach horizontal asymptotes when the stiffening parameter becomes sufficiently large. Again, the buckling modes associated with the maximum values of p_{cr}^{pl} are essentially confined to the shell sections between the stiffeners. Nevertheless, the respective stiffener deformations are somewhat larger than those in the corresponding elastic case. In addition, Fig. 7 also shows that interstiffener buckling – and hence a pronounced upper limit of the load–carrying capacity – is more likely when r/t is large, whereas for reasonably thick shells global buckling is the rule. In such cases, an increase in the cross–sectional area A_r will lead to a proportionate almost linear increase in the shell's load–carrying capacity.

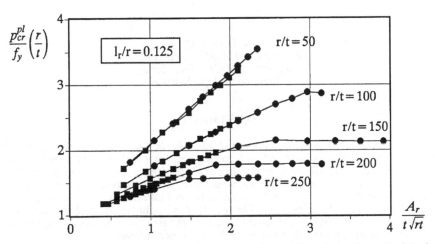

Figure 7. Elastic–plastic bifurcation pressures of ring–stiffened circular cylindrical shells (■ 3 stiffeners, ● 7 stiffeners)

REFERENCES

1. Wunderlich, W., Rensch, H. J., Obrecht, H.: Analysis of elastic–plastic buckling and imperfection–sensitivity of shells of revolution. In: Buckling of Shells, E. Ramm (ed.), Springer–Verlag, Berlin (1982), 137–174.
2. Wunderlich, W., Cramer, H., Obrecht, H.: Application of ring elements in the nonlinear analysis of shells of revolution under nonaxisymmetric loading. Comp. Meth. in Appl. Mech. Eng. (1985), 51, 259–275.
3. Wunderlich, W., Obrecht, H., Schnabel, F.: Nonlinear behavior of externally pressurized toriconical shells – analysis and design criteria. In: Stability of Plate and Shell Structures. P. Dubas, D. Vandepitte (eds.), ECCS Brussels (1987), 373–386.
4. Wunderlich, W., Obrecht, H., Springer, H., Lu, Z.: A Semi–analytical approach to the nonlinear analysis of shells of revolution. In: Analytical and Computational Models for Shells. A.K. Noor, T. Belytschko, J.C. Simo (eds.), ASME, New York (1989), 509–536.

ACKNOWLEDGEMENT

The authors gratefully acknowledge financial support of the "Arbeitsgemeinschaft Industrieller Forschungsvereinigungen e.V. (AIF)" and the "Deutsche Gesellschaft für Chemisches Apparatewesen, Chemische Technik und Biotechnologie e.V. (DECHEMA)" under grant 7707.

STRENGTH AND RELIABILITY OF STRINGER-STIFFENED CYLINDERS IN OFFSHORE STRUCTURES

P J Dowling
Professor and Head of Department
Department of Civil Engineering
Imperial College, London

ABSTRACT

This paper summarizes research conducted over the last ten years on stringer-stiffened cylinders at the Department of Civil Engineering, Imperial College. The research includes experimental studies, development of non-linear finite element formulations and their use in experimental correlations and parametric studies, simplified analytical methods for strength prediction, probabilistic analyses for reliability evaluation and assessment of buckling recommendations in codes of practice. The results of these studies, as well as similar work on ring-stiffened cylinders, have been used to improve the structural design of cylindrical shell components in semi-submersible and tension leg offshore platforms.

INTRODUCTION

Research into thin-walled plated structures at Imperial College was initiated by work on box beams and ship hulls well over twenty years ago. A major programme on box-girder bridge structures was undertaken in the 1970's following the collapse of a number of these structures during erection. The culmination of this effort was the publication of the UK Bridge Code (BS5400) in 1982. Much of the background work required for the drafting of this code can be found in experimental, analytical and design orientated studies carried out at the Department of Civil Engineering. Since the late 1970's, with research objectives switching from bridge to offshore structures in the light of North Sea developments, a similar programme has been directed at shell buckling behaviour in offshore rigs. A summary of part of this research focusing on stringer-stiffened cylinders is given in this paper.

From a practical point of view, the scope of work can be sub-divided into three main areas. First, buckling strength prediction at the design stage has to be addressed. This can be aided by experimental and analytical/numerical results that determine the influence of various parameters on strength and likely failure mode under various types of loading. Comparisons with available design guidance is used to quantify the area of applicability and the degree of conservatism of different code formulations, and to identify areas in need of further research.

Secondly, the in-service performance of structures has to be examined. In the context of stiffened cylinders in offshore applications, a topic of major importance is the assessment of residual buckling strength following accidental loading due to collision of supply boats and other vessels with the main platform. Work in this area comprised of experimental and analytical studies that led to the development of a simple approach to evaluate buckling strength reduction due to collision damage.

Finally, the extensive experimental and analytical data produced in the course of these studies was used to introduce probabilistic concepts in the design of stiffened cylinders. These methods can improve structural performance in future designs and have also been applied in assessing structural reliability of typical stringer-stiffened cylinder geometries.

BUCKLING STRENGTH

The first experimental study was undertaken on a large-scale stringer-stiffened cylindrical model under axial compression (Figure 1). This model was extensively instrumented during the test, in addition to full imperfection scanning and residual strain measurements during various stages of fabrication. The results gave valuable information for the geometry under consideration [1] but it was clear from the outset that the extensive experimental corroboration needed in shell buckling problems could not be achieved with large scale models due to the costs involved.

Fig. 1: Large-scale stringer stiffened cylinder under axial compression

Thus, effort was devoted to the development of a manufacturing method for small-scale models that would, however, represent to some degree manufacturing practice for offshore components, at least in adopting welding for the assembly of the stiffened cylinder. Reference [2] describes the technique used for the fabrication of all small-scale welded steel models. Table 1 presents typical geometries for the various types of models manufactured, whilst Figure 2 shows the test rig used in small-scale testing under axial compression. Specially ordered steel sheets were used with a material curve that resembles that of normal steel with a yield stress of about 350 N/mm^2. In all models, geometric imperfections were carefully controlled by extensive jigging and recorded in detail after completion of fabrication. The final imperfections were well within acceptable tolerances stipulated by design codes. However, in order to eliminate the effect of residual stresses, which could not be properly scaled, the models were heat treated before testing. This factor alone suggests caution in relying on the results of small-scale modelling in the development of experimental databases for comparisons with design formulations. Nevertheless, bearing in mind that the primary reason for small-scale testing is the reduction in cost, the models provided a useful benchmark on which numerical and analytical tools were validated before being used in the generation of design data through wider parametric studies.

Insofar as stringer-stiffened cylinders are concerned, twenty-four tests were carried out under the following types of loading:
- (i)   axial compression
- (ii)   axial compression and bending
- (iii)   external pressure
- (iv)   axial compression and external pressure

Fig. 2: Set-up for small-scale axial compression test

Table 1. Typical model dimensions

Stringer stiffened

Radius	Thickness	Length	No. of stringers	Panel width	Stringer dimensions
160	0·84	64	40	25.1	6.72 × 0.84
160	0·84	128	40	25.1	6.72 × 0·84
160	0.84	192	40	25.1	6.72 × 0.84
160	0.84	64	20	50.2	6.72 × 0.84

Ring stiffened

Radius	Thickness	Stiffener spacing	No. of bays	Ring dimensions
160	0.60	24	5	12.8 × 1.6
160	0.60	24	5	9.6 × 0.6
160	0.60	24	5	6 × 0.6(f); 5 × 0.6(w); T-bar
160	0.60	24	5	5 × 0.6

Orthogonally stiffened

Radius	Thickness	Bay length	No. of stringers	No. of rings	Dimensions
160	0.60	64	20	2	4.8 × 0.6
160	0.60	107	40	2	4.8 × 0.6

All values in mm

The first study consisted of six stringer-stiffened models divided in two sets [1, 3]. Cylinders in the first set were loaded under axial compression, whilst those in the second were loaded by a combination of axial compression and bending (by introducing a load eccentricity of 0.25R-0.40R). Uniaxially loaded models failed at a high proportion of squash load in a controlled manner with local panel collapse mechanisms forming between stiffeners in all cases. Eccentrically loaded models showed first failure at the same local stresses as uniaxially loaded models and broadly similar failure modes. There was, however, some redistribution of load at collapse of the first panels of the closely stiffened models ($Z_s = (1-v^2)s^2/Rt = 4.3$) which reached a stress approximately 10% higher than the axially loaded equivalents.

The second series comprised twelve small-scale models [4, 5] with identical cross section properties ($Z_s = 4.3$), the variables being the length of the cylinder (short, medium and long with $Z = (1-v^2)L^2/Rt = 27.7$, 110.9 and 249.6 respectively) and the pressure/axial loading ratio (axial compression, external pressure and two combined loading cases). The shortest models provided an upper-bound for the cross-section under consideration, with elastic buckling effects being negligible. Longer models, however, exhibited substantial reductions especially in combined and pressure loading cases. It was apparent that a linear interaction diagram, joining the two limiting capacities under axial compression and pure radial pressure would be very conservative in all but the longest set of cylinders.

Finally, the last experimental study concentrated on six cylinders of more slender cross-section ($Z = 231.1$ and $Z_s = 5.7$ or 22.8) stiffened by weak flat-bar stringers (depth to thickness ratio of six) in order to investigate overall buckling modes [6]. Three proportions of interactive loading were used and the results are summarized in Table 2. It is interesting to note that although the strength under axial compression is substantially higher for the narrow panelled cylinder (E vs. A), the strengths under combined and pressure loading are similar. It appears that although the small stringers used are able to increase the axial buckling capacity, even a moderate amount of destabilizing pressure is sufficient to reduce the strength to approximately the same level under combined loading.

Table 2: Experimental results for slender models

Cylinder	A	B	C	E	F	G
Z	231.1	231.1	231.1	231.1	231.1	231.1
Z_s	22.8	22.8	22.8	5.7	5.7	5.7
w_{omax} (out)	0.44	0.54	0.49	0.47	0.40	0.66
w_{omax} (in)	0.54	0.50	0.49	0.45	0.36	0.44
σ_x (N/mm^2)	284	26	246	400	26	240
σ_θ (N/mm^2)	0	56	37	0	60	39
Failure mode	Local	Local	Local	Overall	Overall	Overall

The influence of both material and geometric non-linearities on the buckling behaviour of all the models was evident in the tests. This led to the development of non-linear finite element analysis tools suitable for stiffened shell structures [7, 8]. The resulting FE package was used extensively in reproducing numerically test behaviour and, following this validation exercise, in wider parametric studies. In particular, a comprehensive parametric study was carried out on a range of geometries and loadings which produced local buckling failure [6]. The finite element model is shown in Fig. 3(a). The range of applicability was identified by finite element runs using large models that could cater for overall buckling (Fig. 3(b)). Fully non-linear incremental collapse analyses were undertaken on stiffened imperfect curved panels with the aim of obtaining interaction diagrams (Fig. 4) for assessing design guidance.

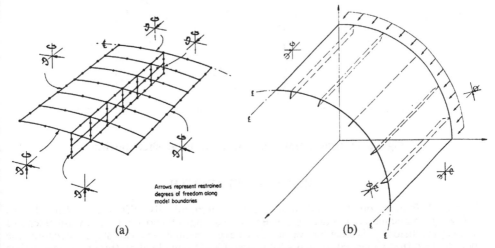

Arrows represent restrained degrees of freedom along model boundaries

(a) (b)

Figure 3: Finite element models for stringer-stiffened cylinders

Comparisons with available design guidance was also undertaken within this work [6]. Axially loaded cylinders are, in general, treated satisfactorily, although the distinction between local and overall modes is not always predicted accurately. In this respect, the ECCS Recommendations [9] offered the best comparisons by distinguishing correctly between modes. Furthermore, the ECCS proposal of modifying the knockdown factor for 'poorly' fabricated models was also found to give reasonable results.

In contrast, predictions for pressure-loaded cylinders [10, 11] exhibit a high degree of conservatism. The discrepancy arises mainly for short heavily-stiffened cylinders that develop considerable strength due to overall bending of the shell/stiffener unit. This is an area where further research is needed in order to improve available design guidance. Finally, of the various simple interaction equations, the one used in the DnV Notes [11] appears to give reasonable results although not always conservative.

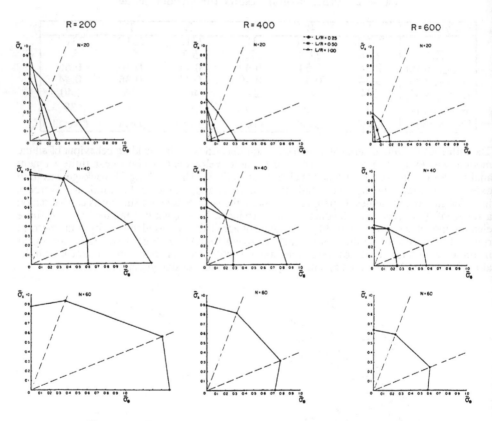

Figure 4: Interaction diagrams for local shell buckling

RESIDUAL STRENGTH

This work focused on the residual strength of longitudinally stiffened cylindrical bays following a particular type of damage, when the deck edge of a boat impacts a vertical leg in a sideways collision. This situation was idealized experimentally in a two-stage procedure. The models were first subjected to quasi-static lateral knife-edge loading (Figure 5) to simulate denting due to accidental collision. The model ends were stiffened by heavy end rings to give flexurally fixed conditions. In addition, in some models the ends were further restrained in the longitudinal direction by massive blocks to simulate the continuity effect of adjacent bays. Both single- and three-bay small-scale models were manufactured using the techniques developed in the previous experimental programme. However, orthogonally stiffened cylinders required further development of the small-scale fabrication capability and necessitated the manufacturing of complex jigs made of numerous pieces [2].

To model analytically the behaviour of the experimental models and to be able to predict the response of other geometries, a plastic mechanism approach was developed [12] to follow the denting process, by treating the stringer-stiffened shell as a series of T-section strips, comprising of the stringer and associated effective shell plating, spanning across ring stiffeners. Springs attached at the ends of the beams provide partial restraint to axial pull-in. The analysis was initially developed for single bay dents but has been extended to treat multi-bay dents where the rings also deform. Comparisons with experimental results have verified the assumptions used in the mechanism analysis and have demonstrated that the model is capable of predicting the response for the entire the load history [13, 14].

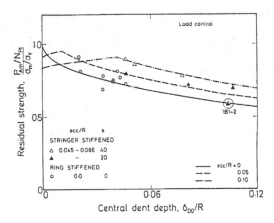

Fig. 5: Set-up for denting tests   **Fig. 6: Residual strength of damaged shells**

Residual compressive strength was first determined experimentally by testing the damaged models under axial compression. It was then verified analytically using a combination of elastic and plastic techniques on an effective cylinder cross-section. It may be intuitively supposed that very little compressive load will be carried in the dent area, and this was confirmed experimentally using strain gauge readings. Thus, a simplified elastic analysis of the response was developed [15] in which the axial stiffness of the dented material was neglected. Figure 6 presents typical results for residual strength obtained by this method and their correlation with test results on both stringer- and ring-stiffened cylinders [16]. The agreement is very good, although the analysis is slightly conservative for the stockier models because they are sufficiently ductile to allow circumferential redistribution of stresses. The principal parameter affecting residual strength is the maximum depth of the dent imperfection but the importance of smaller imperfections in the area adjacent to the dent provided they are sympathetic to the critical buckling mode of the shell has also been identified. The manner in which the axial compression has been applied is important and two methods, load control and displacement control have been compared [16]. Actual boundary conditions in any particular structure would determine which method is best suited for analysis.

In general, the strength losses due to damage were less severe than might have been expected and the proposed methods can be used to take into account this type of accidental loading during initial design and may also help determine whether an existing damaged unit should be repaired and/or remain in operation.

PROBABILISTIC ANALYSIS

Following the collection of test results and the generation of numerical/design data, application of probabilistic concepts was introduced in the research [17]. The need for probabilistic analysis arises from the underlying uncertainties that exist in geometric and material parameters, as well as in loading and structural modelling aspects. The results can be used in initial design to ensure that the structure meets a target reliability level and is not dominated by uncertainties associated with a single parameter (e.g. material variability) and in assessing the influence of tolerance specifications in offshore codes. It is important to note that the use of these methods is not a substitute for deterministic methods of structural analysis, such as those described above. However, as a result of their application some modifications to existing deterministic methods will be introduced, since, by comparison, probabilistic methods utilize available information in a more rational and systematic manner.

The first study was concerned with the development of a structural reliability approach, in which strength modelling was directly based on finite element analysis. The results of the parametric study on stringer-stiffened cylinders buckling in a local mode [6] were incorporated in a multi-stage reliability approach [18] that catered explicitly for the variability in material properties and the amplitude of initial geometric imperfections in a critical mode. Figure 7 shows typical results of the sensitivity factors obtained for a range of stringer-stiffened cylinder geometries under axial compression assuming a constant notional reliability level [19]. As can be seen, the effect of adding stringers to the design is to move from a region where the reliability is influenced by uncertainty in imperfection amplitude to a region where material variability becomes dominant. A balance in the sensitivity factors is obtained for designs with about 36-40 stringers.

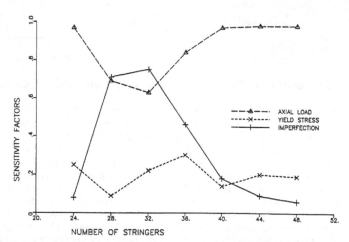

Fig.7: Reliability-based design of stringer-stiffened cylinders

In buckling problems, an important source of uncertainty that has large influence on both strength and reliability is the randomness of imperfection amplitude at any given point on the cylinder surface as well as its spatial distribution. The preceding experimental studies [3, 5, 6] had produced a large data bank of imperfection measurements on cylinders manufactured in a similar manner. Following data reduction using Fourier analysis, these measurements were re-analysed in order to study their statistical properties [20]. Figure 8 presents typical mean value results on imperfection amplitudes from a database containing Fourier coefficients of twenty four stringer-stiffened cylinders. The statistical analysis revealed that due to the common manufacturing process several trends exist in the imperfection patterns. Thus, guidelines were developed in selecting and combining imperfection modes. Furthermore, simulation methods were employed to determine the probabilistic properties of characteristic imperfection models required for buckling strength analysis. These models were used to study the imperfection sensitivity of a particular geometry and to demonstrate the potential of imperfection surveys in full-scale structures [21].

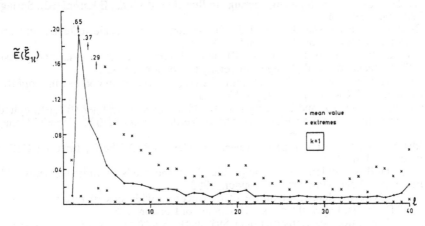

Fig. 8: Mean imperfection modal amplitudes

The probabilistic modelling of variables influencing buckling behaviour can be significantly improved by data collection and analysis. The need for initial imperfection measurements on full-scale structures is one obvious example and measuring methods should be sufficiently developed to allow this without substantially hindering the fabrication and assembly process. In addition, the spatial variation of material properties can be of significance in elasto-plastic buckling response due to the possible combination of 'soft' zones with high imperfection levels. The definition of compressive yield stress and the measurement of residual stresses should also be addressed in order to improve modelling uncertainty parameters in future comparisons of analytical and design orientated formulations with experimental results.

CONCLUDING REMARKS

This summary paper aims to demonstrate the broad range of techniques and methods used to address design aspects of stringer-stiffened cylinders used in offshore construction. Emphasis has been placed on results that can be used in the drafting of shell buckling recommendations. Appropriate references provide a means for the reader to follow up specific topics, since space only allows a brief introduction to be made of the various topics investigated. At present, significant effort is directed at the drafting of Eurocode 3 and shell buckling guidance should be developed in accordance with the general format and spirit of existing parts. The current ECCS Recommendations provide a useful starting point from which further advances can be achieved and it is hoped that past and future research will be appropriately used to corroborate the design formulations put forward in various topics.

ACKNOWLEDGEMENTS

The author would like to thank the research workers and laboratory technicians involved in the various programmes. The axial compression tests were carried out at Imperial College, while the interactive and pressure tests were conducted at the University of Surrey. The work was sponsored by the UK Science and Engineering Research Council, the UK Department of Energy and Det norske Veritas.

REFERENCES

1. Dowling P J, Harding J E, Agelidis N and Fahy W, 'Buckling of orthogonally stiffened cylindrical shells used in offshore engineering', in *Buckling of Shells*, E Ramm (ed), Springer, Berlin, 1982, pp. 239-273.
2. Scott N D, Harding J E and Dowling P J, 'Fabrication of small-scale stiffened cylindrical shells', *J. Strain Analysis*, Vol. 22, 2, 1987, pp.97-106.
3. Fahy W, 'Collapse of longitudinally stiffened cylinders subject to axial and pressure loading', PhD Thesis, Imperial College, University of London, 1984.
4. Agelidis N and Dowling P J, 'The strength of stringer stiffened cylinders', in *Stability of Metal Structures*, CTICM, Paris, 1984, pp. 169-178.
5. Agelidis N, Harding J E and Dowling P J, 'Buckling tests on stringer-stiffened cylinder models subjected to load combination', Det norske Veritas Report 82-098, Imperial College, 1982.
6. Agelidis N, 'Collapse of stringer-stiffened cylinders', PhD Thesis, Imperial College, University of London, 1984.
7. Trueb U, 'Stability problems of elastic plastic plates and shells by finite elements', PhD Thesis, Imperial College, University of London, 1983.
8. Bates D N, 'The mechanics of thin-walled structures with special reference to finite rotations', PhD Thesis, Imperial College, University of London, 1987.
9. European Convention for Constructional Steelwork, *Buckling of steel shells: European recommendations*, 1988.
10. American Society of Mechanical Engineers, *Boiler and pressure vessel code*, New York, 1980.
11. Det norske Veritas, *Buckling strength analysis of mobile offshore units*, Classification note 30.1, 1987.
12. Ronalds B F, 'Mechanics of dented orthogonally stiffened cylinders', PhD Thesis, Imperial College, University of London, 1985.
13. Ronalds B F and Dowling P J, 'A denting mechanism for orthogonally stiffened cylinders', *Int. J. Mech. Sci.*, Vol. 29, 10/11, 1987, pp. 743-759.
14. Ronalds B F and Dowling P J, 'Stiffening of steel cylindrical shells against accidental lateral impact', *Proc. Instn. Civ. Engrs.*, Pt. 2, 83, Dec. 1987, pp. 799-814.
15. Ronalds B F and Dowling P J, 'Residual compressive strength of damaged orthogonally stiffened cylinders', in *Stability of Plate and Shell Structures*, P Dubas and D Vandepitte (eds), Proc. of International Colloquium, Ghent, 1987, pp. 503-512.
16. Ronalds B F and Dowling P J, 'Compressive strength of stiffened cylindrical shells with large imperfections', in *Buckling of Structures*, I Elishakoff et al (eds), Elsevier 1988, pp. 313-334.
17. Chryssanthopoulos M K, 'Probabilistic buckling analysis of stringer-stiffened cylinders', PhD Thesis, Imperial College, University of London, 1988.
18. Chryssanthopoulos M K, Baker M J and Dowling P J 'A reliability approach to the local buckling of stringer-stiffened cylinders', in *Proc. of 5th Offshore Mechanics and Arctic Engineering Symposium*, Tokyo, 1986.
19. Chryssanthopoulos, M K, Baker, M J and Dowling P J 'Reliability-based design of stringer-stiffened cylinders under axial compression', in *Proc. of 6th Offshore Mechanics and Arctic Engineering Symposium*, Houston, 1987.
20. Chryssanthopoulos M K, Baker M J and Dowling P J 'Statistical analysis of imperfections in stiffened cylinders', *J. of Struct. Engg.*, ASCE, 117 (7), 1991, pp. 1979-1997.
21. Chryssanthopoulos M K, Baker M J and Dowling P J 'Imperfection modelling for buckling analysis of stiffened cylinders', *J. of Struct. Engg.*, ASCE, 117(7), 1991, pp. 1998-2017.

THE ULTIMATE STRENGTH OF CYLINDRICAL SHELL STRUCTURES UNDER EARTHQUAKES

HIROSHI AKIYAMA
Department of Architecture
Faculty of Engineering, University of Tokyo,
7-3-1, Hongo, Bunkyo-ku, Tokyo, Japan

ABSTRACT

The relation between the energy absorption capacity and the maximum horizontal deformation of steel cylindrical structures after buckling is made clear based on the results of experiments and earthquake response analysises. The ultimate strength of steel cylindrical structures can be estimated by equating the energy absorption capacity of structures to the seismic energy input.

INTRODUCTION

The energy concept has been proved to be effective to develop the ultimate strength design method for structures against earthquakes[1]. The total energy input into a structure exerted by an earthquake is mainly governed by the total mass and the effective vibrational period of the structure. The effective vibrational period of the structure agrees with the fundamental natural period for the elastic system, and becomes a representative period of vibration for the inelastic system in which the softening in rigidity due to the plastification yields an elongation of the effective period.

The resistance of structures against earthquakes is the energy absorption capacity. Under the static loading, the occurrence of buckling implies the arrival of the ultimate collapse state, since no static equilibrium can exist beyond the buckling load. Contrary to the static loading, the occurrence of buckling does not mean the arrival of collapse state, since the energy absorption capacity is not limited by the buckling, and the structure can absorb further a considerable amount of energy due to the development of the local plastification along ridges of buckling patterns.

The energy input can be evaluated precisely for the individual earthquake by applying the elastic response analysis. On the other hand, to evaluate energy absorption capacity of structures in the post-buckling range, it is indispensable to gather experimental evidences. For the self-standing shell structures, the most important seismic loading is the horizontal forces. Therefore, the buckling strengths and the deformation characteristics of them under horizontal loadings must be sought after.

BUCKLING STRENGTH OF CYLINDRICAL SHELLS

Using more than one hundred specimens, a series of tests were performed on steel cylindrical shells subjected to changing bending and shearing forces under constant axial forces or internal pressures. The results of test are summerized as follows[2].

1) The lower limit of buckling load can be described as follows.

$$
\left.
\begin{array}{ll}
\text{When} \quad \sigma_h/\sigma_Y \leq 0.3 , & {}_b\sigma_m = {}_b\sigma_{m0} + (0.56 \, {}_0\sigma_{cr} - {}_b\sigma_{m0})(\sigma_h/\sigma_Y) / 0.3 \\
\text{When} \quad \sigma_h/\sigma_Y > 0.3 , & {}_b\sigma_m = 0.8 \, {}_0\sigma_{cr} \left(1 - \sigma_h/\sigma_Y\right)
\end{array}
\right\} \quad (1)
$$

where
${}_b\sigma_m$: the maximum extreme fibre stress due to bending at the bottom of shell
σ_h : the hoop stress due to internal pressure
σ_Y : the yield point stress
${}_b\sigma_{m0}$: the lower limit of bending buckling stress combined with axial compression without internal pressure
${}_0\sigma_{cr}$: the compressive buckling stress without internal pressure in the axisymmetric mode

The test results are compared with the predicted value in Fig.1.

RESTORING FORCE CHARACTERISYICS

In Fig.2 a cylindrical structure is shown. Fundamentally, it can be reduced to a single-mass system. The relationship between the lateral shear force Q and the inclination angle of of the structure θ under an arbitrary change of θ is identified to be the restoring force characteristics.
To predict the hysteresis rule, some definitions are introduced. The loading path and the unloading path are defined as follows.

$$Qd\theta > 0 \quad \text{(loading path), and} \quad Qd\theta < 0 \quad \text{(unloading path)}$$

The skeleton curve is identified by the $Q-\theta$ curve under the monotonic loading. The unloading point is defined by the point which rests on the skeleton curve and terminates the loading path. The initial unloading point is defined by the point on which buckling occurs. The intermediate unloading point is defined by the point which terminates the loading path but does not reach the skeleton curve.
Under the assumption that the initial unloading points under positive and negative loading domains have been already experienced, the hysteresis rule is described as follows.
The loading path points to the preceding unloading point in the same

loading domain. After the preceding loading point is reached, the loading path follows the skeleton curve. The unloading path from the unloading point points to the initial unloading point in the reverse loading domain. The unloading path from the intermediate unloading point has the same slope as that of the preceding unloading path in the same loading domain. Fig.3 illuatrates a typical pattern of restoring-force characteristics, according to the above-mentiond hysteresis rule.

In Fig.4, some test results are compared with the predicted curves based on the skeleton curves under the monotonic loading. In those figures, the lateral shear force Q is converted to the maximum bending stress ${}_b\sigma_m$ at the bottom of shell.

PRINCIPAL RESPONSES OF STRUCTURES TO EARTHQUAKES

Total Energy Input in a Structure[1]
The total energy input   is defined as

$$E = -\int_0^{t_0} m\ddot{y}_0\,\dot{y}\,dt \qquad (2)$$

where y_0: the ground motion
y : the relative diplacemet. of the mass
t_0 : the duration of ground motion

The energy spectrum is defined to be the relationship between the equivalent velocity V_E ($=\sqrt{2E/M}$ E: the total energy input, M: the total mass of structure) and the natural period T in the one-mass elastic vibrational system with $h = 0.1$, where h is the fraction of criti-cal damping.

When the structure undergoes plastic deformations, the substantial period of vibration becomes longer due to a softening effect of plastification. Therefore, it is important to find an essential period of vibration T_e in applying the energy spectrum for practical design purposes.

As an index of the plastification of system, the average plastic deformation ratio $\bar{\mu}$ is used. $\bar{\mu}$ is defined as

$$\bar{\mu} = \frac{\delta_{max}^+ + \delta_{max}^-}{2\delta_e} - 1 \qquad (3)$$

where $\delta_{max}^+, \delta_{max}^-$: the maximum story displacement in the positive and negative direction respectively.
δ_e : the elastic limit displacement.

The skeleton curve can be simplified as shown in Fig.5, where r_p is the nondimensionalyzed plastic deformation under the critical load Q_{cr} , k_d is the nondimentionalyzed slope in the post-buckling range, and q is the reserve of strength in the post-buckling range.

Applying the rostoring force characteristics shown in Fig.3 to the numerical response analysis, T_e was found to be expressed by

$$T_e = \sqrt{\frac{T_0^2 + T_0 T_m + T_m^2}{3}} \qquad (4)$$

where T_0: the natural period in the elastic range

$$T_m = \left(\frac{1}{\sqrt{a}} + \frac{1}{\sqrt{b}}\right)\frac{T_0}{2} \tag{5}$$

for $\quad \bar{\mu} \leq r_p$, $\qquad a = \dfrac{2}{2+\bar{\mu}}$, $\quad b = \dfrac{2}{2+3\bar{\mu}}$ (6)

for $\quad r_p + \dfrac{1-q}{-k_d} \geq \bar{\mu} > r_p$, $\qquad \left. \begin{aligned} a &= \frac{2 + k_d(\bar{\mu} - r_p)}{2+\bar{\mu}} \\[6pt] b &= \frac{[1 + k_d(\bar{\mu} - r_p)][2 + k_d(\bar{\mu} - r_p)]}{2 + 3\bar{\mu} + \bar{\mu}k_d(\bar{\mu} - r_p)} \end{aligned} \right\}$ (7)

for $\quad \bar{\mu} > r_p + \dfrac{1-q}{-k_d}$, $\qquad a = \dfrac{1+q}{\bar{\mu}+2}$, $\quad b = \dfrac{q(1+q)}{2+2\bar{\mu}+\bar{\mu}q}$ (8)

The energy spectrum for the Hachinohe record in the Tokachi-oki earthquake (1968) is shown in Fig.6. The energy responses of the system with the parameters of $r_p = 0$, $k_d = -1.0$, $q = 0.5$ are also shown in the figure, taking T_e in place of T_0 as the abscissa. It is seen that the energy responses are well represented by the energy spectrum with the use of $h = 0.1$.

Energy Absorption Capacity[3]
The equilibrium of energy under an earthquake is written as

$$W_e + W_p + W_h = E \tag{9}$$

wherw W_e: the elastic vibrational energy
$\qquad W_p$: the cumulative inelastic strain energy
$\qquad W_h$: the energy absorption due to damping

As for the energy input attributable to the strain energy $E - W_h$, the following rough estimate can be made[1].

$$E - W_h = \frac{E}{(1 + 3h + 1.2\sqrt{h})^2} \tag{10}$$

The elastic vibrational energy W_e corresponds to the area – A in Fig.5. The cumulative inelastic strain energy develops as $\bar{\mu}$ increases. Summarizing the results of the response analysis, the inelastic energy absorption W_p was found to be roughly estimated as

$$W_p = 10 \, A_p \, W_e \tag{11}$$

where $\quad A_p$: the nondimensionalized area of the area – B calculated by taking $\mu = \bar{\mu}$

A_p is given by

for $\quad \bar{\mu} \leq r_p$, $\qquad A_p = \bar{\mu}$ (12)

for $\quad r_p + \dfrac{1-q}{-k_d} \geq \bar{\mu} > r_p$, $\qquad A_p = r_p + [1 + 0.5 \, k_d(\bar{\mu} - r_p)](\bar{\mu} - r_p)$ (13)

for $\quad \bar{\mu} > r_\rho + \dfrac{1-q}{-k_d}$, $\quad A_\rho = r_\rho + \dfrac{q[k_d(r_\rho - \bar{\mu}) + q - 1] + 0.5(1 - q^2)}{-k_d}$ $\quad$ (14)

REQUIRED STRENGTH TO RESIST EARTHQUAKES

As is shown by broken lines in Fig.6, the design energy spectrum should be expressed by a bi-linear relationship described by

for $\qquad T_e \leqq T_G$, $\qquad V_E = \dfrac{V_0 T_e}{T_G}$ $\qquad\qquad$ (15)

for $\qquad T_e > T_G$, $\qquad V_E = V_0$ $\qquad\qquad\qquad$ (16)

where $\quad V_0$: the level of V_E in the longer period range
$\qquad T_G$: the period which divides the range of period

Introducing the yield shear coefficient α, the buckling strength of cylindrical structures is expressed as

$$\alpha = \frac{Q_{cr}}{Mg} \qquad\qquad (17)$$

where $\quad g$: the acceleration of gravity

Based on the equilibrium of energy, the required strength for the structure in which the allowable deformation is specified by $\bar{\mu}$ is obtained as follows.

$$\alpha = \frac{1}{\sqrt{1 + 10 A_\rho}} \cdot \frac{1}{1 + 3h + 1.2\sqrt{h}} \cdot \frac{2\pi V_E(T_e)}{T_0 g} \qquad (18)$$

The required strength for the elastic system α_e is expressed by

$$\alpha_e = \frac{2\pi V_E(T_0)}{T_0 g} \cdot \frac{1}{1 + 3h + 1.2\sqrt{h}} \qquad (19)$$

Therefore, the ratio of α to α_e implies the strength reduction factor which can be applied to the practical design of cylindrical structures in which the maximum allowable deformation is specified by $\bar{\mu}$. α / α_e is denoted by D_s. The D_s-values is obtained as follows.

for the shorter period structures with

$$T_e \leqq T_G , \qquad D_s = \frac{1}{\sqrt{1 + 10 A_\rho}} \cdot \frac{T_e}{T_0} \qquad (20)$$

for the longer period structures with

$$T_0 \geqq T_G , \qquad D_s = \frac{1}{\sqrt{1 + 10 A_\rho}} \qquad\qquad (21)$$

To estimate an upper-bound value of D_s, a lean skeleton curve with poor energy absorption capacity in assumed, taking parameters of $r_\rho = 0$, and $k_d = -1.0$.
The D_s-values calculated from Eqs(20) and (21) are shown in Fig.7.
The difference in Eqs(20) and (21) is only that Eq(20) has the term of

T_e/T_o. This difference corresponds to the increase of input energy in the shorter period structures and causes a marked increase of D_s-values as seen in Fig.7.

Since the most of cylindrical structures belong to the shorter period structure, the D_s-values for the shorter period structures are of practical importance.

Therefore, the D_s-values in Fig.7(a) can be referred for the design purposes.

It is seen in the figure that the D_s-value can descend at a level of 0.7 irrespective of the value of q, and can be reduced to 0.5 as far as the level of q is maintained to be greater than 0.4.

In Fig.8 the nondimensionalized skeleton curves which were obtained by experiments are shown. The skeleton curves assumed in deriving the D_s-values are also shown by broken lines with open circles which correspond to the values of $\bar{\mu}$ found in Fig.7(a) on the condition of $D_s = 0.5$.

The line which connects those open circles can be identified to be almost a lower bound of the skeleton curves. Therefore, the D_s-values which can be generally applied to cylindrical structures can be summarized as follows.

$$\left. \begin{array}{ll} \text{for} \quad q < 0.4, & D_s = 0.7 \\ \\ \text{for} \quad q \geqq 0.4, & D_s = 0.5 \end{array} \right\} \tag{22}$$

CONCLUSION

The buckling strength and the deformation characteristics of cylindrical shells under the seismic loading were made clear experimentally. The energy input exerted by an earthquake was found to be well predicted by the energy spectrum of the earthquake by taking account of the elongation of substantial period of vibration of the cylindrical structure due to the plastification in the post-buckling range.

The energy absorption capacity of the cylindrical structure in the post-buckling range was found to be expressed in terms of the maximum deformation.

By equating the energy absorption capacity to the energy input, the required strength of structure to resist to the earthquake was derived. It was found that the required strength can be reduced to one-half of the required strength of the elastically designed structure by taking account of the deformation capacity in the post-buckling range.

REFERENCES

1. Akiyama, H.: Earthquake - Resistant Limit - State Design for Buildings, University of Tokyo Press, 1985.
2. Akiyama, H., Takahashi, M. and Nomura, S.: Buckling Tests of Steel Cylindrical Shells Subjected to Combined Bending, Shear and Internal Pressure, Trans. of Architectural Institute of Japan, Vol. 400, 1989.
3. Akiyama, H.: Post - buckling Behavior of Steel Cylindrical Shells during Earthquakes, Trans. of SMiRT Conference, Lausanne, 1987.

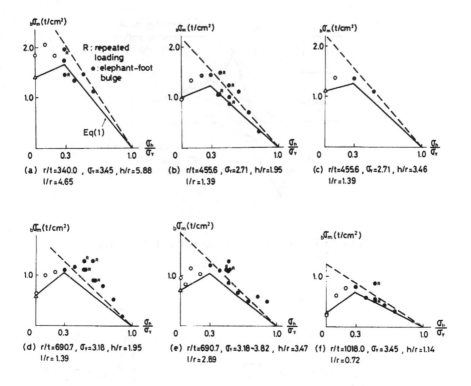

Figure 1. Maximum bending stresses

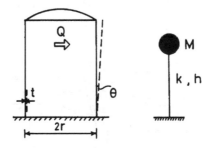

Figure 2. Cylindrical structure

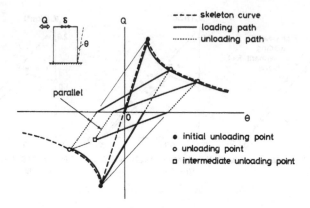

Figure 3. Hysteresis rule in $Q-\theta$ relationship

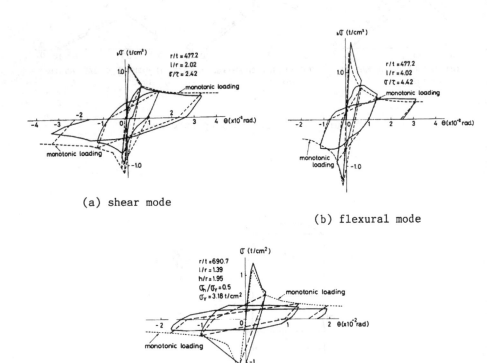

(a) shear mode

(b) flexural mode

(c) elephant-foot bulge mode (with internal pressure)

Figure 4. Load-deformation characteristics in post-buckling range

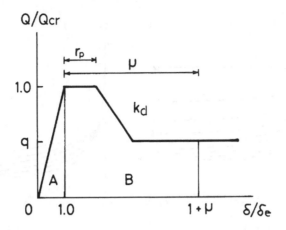

Figure 5. Simplified $Q-\delta$ relationship

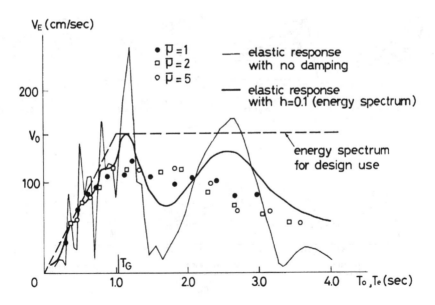

Figure 6. Energy spectrum of hachinohe record

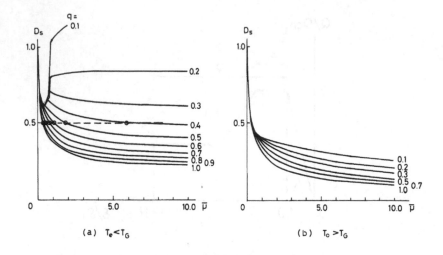

Figure 7. D_s-values for $\tau_p=0$, $k_d=-1.0$, and $a_p=2.5$

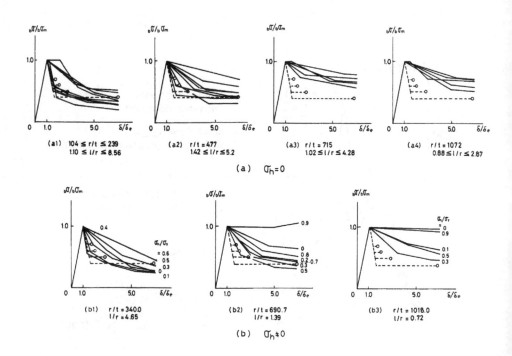

Figure 8. Comparision with test results

The ECCS *) Recommendations on Shell Stability,
Design Philosophy and Practical Applications

By Lars Å Samuelson, The Swedish Plant Inspectorate, Stockholm, Sweden
Chairman, ECCS TWG 8.4

Abstract

Rules for design of structures with respect to buckling have existed for a long time. The stability behavior of columns and plates is fairly well known among engineers and design methods are available for many of the problems arising in practical applications. Shell buckling has been studied intensively during the last decades and the behavior is now fairly well known. The tendency of shells to be imperfection sensitive has caused problems in defining a safe design procedure and the code cases developed up till the present time are fairly limited in scope. This fact has also resulted in a reluctance to address other practical problems such as the effect of local disturbances because "we do not even have sufficient knowledge on how to design shells subjected to uniform loads". Such special problems have so far been left to the designer to solve through for instance testing, rigorous nonlinear analyses or alternative design avoiding the problems.

The ECCS Recommendations on shell stability include design algorithms for a number of elements under various types of external loads. In the philosophy used the classical buckling load (or linear bifurcation if based on a numerical analysis) is taken as reference value for the design and reduction factors determined from tests are applied in order to arrive at a lower bound value for the carrying capacity. Although the design rules are primarily intended for use on steel structures, the methodology admits extension to other types of material. In different areas of application, however, such as aerospace stability problems, the rules would be too conservative and would lead to an uneconomic design. In such applications other levels of reduction (knock down) factors may be defined based on more strict tolerance levels for geometric imperfections.

Basic design philosophy

The main difficulty in the design of shell structures subjected to buckling stems from the fact that shells are imperfection sensitive and the (simple, classical) buckling theories often provide unsafe predictions of their carrying capacity. Comprehensive nonlinear analyses by use of numerical methods may result in a close estimate of the shell stability behavior but, since the initial imperfections and other disturbances need to be included, this method is not suitable for design purposes. Moreover, the effort, in terms of computer time and engineering judgement, still seldom warrants use of refined computation methods because of the higher costs involved.

---

*) European Convention for Constructional Steelwork

The ECCS design recommendations for shell stability problems are intended for use in steel constructions, eg silos, tanks, roofs, chimneys and similar structures. The computational methods, to some extent, reflect this background although there has been an effort to retain some generality. The design rules should provide results which are:

- Safe
- Easy to use
- Economic

The basic design procedure of the 1988 Edition of the Recommendations involves the following steps:

The carrying capacity σ_u of the shell is derived from the classical (linear) buckling limit σ_{cr} of the shell reduced by a reduction factor α:

For **moderately imperfection sensitive** shells :

$$\sigma_u = \alpha\sigma_{cr} \qquad \text{if} \qquad \alpha\sigma_{cr} \leq 0.5\, f_y \qquad (1)$$

$$\sigma_u = f_y[1 - 0.25\frac{f_y}{\alpha\sigma_{cr}}] \qquad \text{if} \qquad \alpha\sigma_{cr} > 0.5\, f_y \qquad (2)$$

For shells which are **extremely sensitive to imperfections**, (spherical segments under external pressure and cylinders and cones under axial compression):

$$\sigma_u = 0{,}75\alpha\sigma_{cr} \qquad \text{if} \qquad \alpha\sigma_{cr} \leq 0.5\, f_y \qquad (3)$$

$$\sigma_u = f_y[1 - 0.4123(\frac{f_y}{\alpha\sigma_{cr}})^{0.6}] \qquad \text{if} \qquad \alpha\sigma_{cr} > 0.5\, f_y \qquad (4)$$

This design philosopy is demonstrated in Fig. 1 for an imperfection sensitive shell.

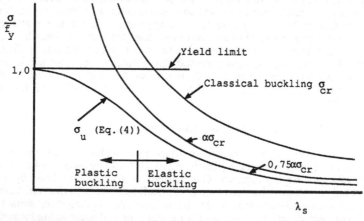

Figure 1. Definition of reduction factors for a shell subjected to buckling.

The reduction factor α is in general a function of shell geometry, loading conditions, initial imperfection amplitude and other factors and has been evaluated from comparison with experimental results. The α-value is selected such that 95 per cent of all experimental data should exceed the estimated carrying capacity σ_u. Recommended values of α are given in each specific case. In many cases the number of test results is inadequate and a conservative estimate of α has to be applied.

The present (1988) Edition of the Recommendations is based on this design philosophy although the specific rules for different shell/load cases may differ from the scheme given above. A harmonization of the various design formulas will be introduced in the next edition.

At the same time, design by use of alternative methods is not excluded provided they offer an equivalent margin of safety against collapse.

Scope of the Recommendations

A fairly large number of shell geometries and load cases is included in the present version of the Recommendations. These are:

Geometry	Loading	Special conditions
Circular cylinders and Cones: cylinders	Axial compression Bending	Unstiffened and ring and stringer stiffened
	External pressure (hydrostatic) Torsion and shear Combined loading	
Spherical shells	External pressure	
Torispherical shells	External / internal pressure	
Unstiffened cones	Liquid pressure	

Current research activities

The ECCS recommendations (and most other codes for that matter) include only shells of revolution with uniform thickness and subjected to uniform pressure distributions. In practical applications various problems arise which are not covered in any of the current design codes. Many of these problems have been noted by the members of the working group. At the present time a number of research activities are going on with the objective to provide simple rules for these cases. At the same time the existing recommendations will be recast in a more unified form.

Design by use of numerical methods (eg FEM). Development of recommendations for interpretation of the results.

Methods for analysis of shells subjected to local loads acting in the mid-surface of the shell, eg loads at supports.

Evaluation of the reduction in carrying capacity of a shell under uniform pressure loads when subjected to local forces perpendicular to the shell mid-surface.

The influence of circular or rectangular cutouts on the buckling resistance of shells. The influence of stiffening at the edge of the cutout will be included.

A powerful program for analysis of shells of revolution under nonsymmetric loading, Ref 2, has been developed and is being used for design method verification for wind loaded silos and tanks.

Rational rules for tanks on saddle supports are being developed.

Combination of shell elements. Definition of limits for application of the rules for individual elements in design against buckling.

Development of design rules for stiffened conical shells including experimental verification.

Examples

Application of the design rules for specific shell elements is fairly straight forward. In some cases, like for instance stringer stiffened cylinders, the recommended design formula is somewhat complex and efficient use of the formula requires a computer program. However, the design diagrams included may be used to calculate the carrying capacity of cylindrical shells with rectangular section stiffeners. An example is given in Fig. 2.

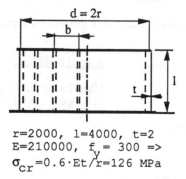

```
d = 2r
      b

r=2000, l=4000, t=2
E=210000, f_y = 300 =>
σ_cr=0.6·Et/r=126 MPa
```

Design diagrams B-D of Fig.
16a,b of Ref.[1] yield:

B) n=12 => $n_s \geq 40$ recommended
C) β=0.6 => $\bar{\sigma}_{cr,p}$=1.25
D) 1/r=2 => $\bar{\gamma}$=0.3 =>

$$\sigma_{cr,p}=1.25 \cdot 0.3 \cdot 126=47.25 \text{ MPa}$$

Figure 2 Design of an internally stiffened cylindrical shell.

Among the current research projects the development of rules for numerical analysis of the buckling capacity of shells is one of the most important. Some of the problems involved in the evaluation of the design value are demonstrated in Fig.3.

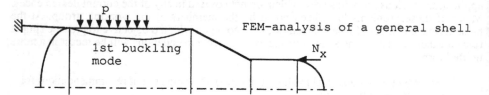

```
    p

1st buckling
   mode
```

FEM-analysis of a general shell

N_x

The buckling load must be evaluated at each station of
the shell even if bifurcation analysis indicates a specific
part of the shell is critical. Reduction factors relevant for
different shell elements are to be applied. Calculation of multiple eigenvalues may improve the accuracy of the prediction.

Fig. 3. General shell of revolution subjected to buckling.

References

1 European Convention for Constructional Steelwork, ECCS Report No.56, Brussels 1988.

2 Esslinger, M., Weiss, H. P.: Users Manual for Program F04B10 and F04B12. Non-linear, Non-axisymmetric Stress and Buckling Analysis of Shells of Revolution with or without Initial Imperfections, DLR Braunschweig, IB 131-89/36.

THE GERMAN CODE DIN 18800 PART 4:
STABILITY OF SHELL-TYPE STEEL STRUCTURES,
DESIGN PHILOSOPHY AND PRACTICAL APPLICATIONS

Herbert Schmidt
Chairman, Working Group DIN 18800 part 4
University of Essen, Dep. Civil Engineering, D-4300 Essen 1

ABSTRACT

The DASt Ri 013, published in 1980 by the German Constructional Steelwork Committee as Recommendations, was the first attempt in Germany to provide general shell stability design rules for civil engineering works in steel. After having proved for 10 years, DASt-Ri 013 has 1990 been developed into a DIN-code. This development included, besides some factual improvements and corrections based on practical experiences or new research results, a formal adaption to the design concept of Eurocode 3 "Design of Steel Structures". In the following notes the basic design philosophy of DIN 18800 Part 4 is described very shortly, and an example is given for its application to a design problem.

1 BASIC DESIGN PHILOSOPHY

DASt-Ri 013 used a design philosophy which was very similar to that of the ECCS Recommendations. The ultimate buckling stress σ_u of the shell was derived in a 4 step-procedure:

(a) Calculate the linear (classical) buckling stress σ_{cr} of the idealizedly geometrically perfect elastic shell by use of given (more or less approximate) formulas.

(b) Apply an imperfection knock-down factor α to arrive at a lower bound estimate of the ultimate buckling stress

$$\sigma_{u,el} = \alpha \cdot \sigma_{cr} \tag{1}$$

of the realistically imperfect, but purely elastic shell.

(c) If the latter is higher than 40% of the yield stress, apply a plasticity reduction factor

$$\sigma_u/f_y = f(f_y/\sigma_{u,el}) \tag{2}$$

to arrive at a lower bound estimate for the real steel shell.

d) Make sure that the shell is fabricated and/or manufactured and/or erected to the imperfection limits (tolerances) within which the imperfection factor is meant to be applicable.

Following the format of Eurocode 3, the two steps (b) and (c) have in DIN 18800/4 been combined into one stability reduction factor. As well known, it relates - generally speaking for all stability problems - the ultimate buckling resistance R_u of a structure to its plastic reference resistance R_{pl} and depends on the non-dimensional structural slenderness parameter

$$\bar{\lambda} = \sqrt{R_{pl}/R_{cr}} \ . \qquad (3)$$

For the shell stability cases covered by DIN 18800/4, the three buckling-relevant **membrane** stresses

σ_x = meridional compressive membrane stress,
σ_φ = circumferential compressive membrane stress,
τ = shear membrane stress

are used for defining the ultimate buckling resistance. Eqn.(3) then becomes

$$\bar{\lambda}_{Sx} = \sqrt{f_{y,k}/\sigma_{xSi}} \ , \qquad (4a)$$
$$\bar{\lambda}_{S\varphi} = \sqrt{f_{y,k}/\sigma_{\varphi Si}} \ , \qquad (4b)$$
$$\bar{\lambda}_{S\tau} = \sqrt{f_{y,k}/(\sqrt{3}\cdot\tau_{Si})} \ , \qquad (4c)$$

where the subscript S denotes "shell", the subscript k denotes "characteristic" and the subscript i denotes "idealized perfect elastic shell", i.e. the critical buckling stress as result of step (a). In contrary to DASt-Ri 013, it is explicitly allowed in DIN 18800/4 to calculate the critical buckling stresses σ_{xSi}, $\sigma_{\varphi Si}$ and τ_{Si} by numerical linear bifurcation analysis in case of configurations which are not covered by the given formulas.

As in the ECCS Recommendations, two categories of imperfection sensitivity are defined (figure 1).

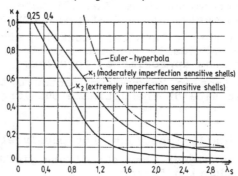

Figure 1. Reduction factors (buckling curves) for shell buckling acc. to DIN 18800/4

The reduction factors are determined by the following formulas:
For **moderately imperfection sensitive** shells:

$$\varkappa_1 = 1 \qquad\qquad\qquad \text{for} \qquad \bar{\lambda}_S \leq 0,4,$$
$$\varkappa_1 = 1,274 - 0,686\,\bar{\lambda}_S \qquad \text{for } 0,4 < \bar{\lambda}_S < 1,2, \qquad (5)$$
$$\varkappa_1 = 0,65/\bar{\lambda}_S^2 \qquad\qquad \text{for } 1,2 \leq \bar{\lambda}_S.$$

For **extremely imperfection sensitive** shells:

$$\varkappa_2 = 1 \qquad\qquad\qquad \text{for} \qquad \bar{\lambda}_S \leq 0,25,$$
$$\varkappa_2 = 1,233 - 0,933\,\bar{\lambda}_S \qquad \text{for } 0,25 < \bar{\lambda}_S \leq 1,00, \qquad (6)$$
$$\varkappa_2 = 0,3/\bar{\lambda}_S^3 \qquad\qquad \text{for } 1,00 < \bar{\lambda}_S < 1,50,$$
$$\varkappa_2 = 0,2/\bar{\lambda}_S^2 \qquad\qquad \text{for } 1,50 \leq \bar{\lambda}_S.$$

The design buckling (membrane) stresses are determined from

$$\sigma_{xS,R,d} = \varkappa_2 \cdot f_{y,k}/\gamma_{M2}, \qquad\qquad\qquad (7a)$$
$$\sigma_{\varphi S,R,d} = \varkappa_1 \cdot f_{y,k}/\gamma_{M1} \text{ (cylinders and cones)}, \qquad (7b)$$
$$\sigma_{\varphi S,R,d} = \varkappa_2 \cdot f_{y,k}/\gamma_{M2} \text{ (spheres)}, \qquad\qquad (7c)$$
$$\tau_{S,R,d} = \varkappa_1 \cdot f_{y,k}/(\sqrt{3} \cdot \gamma_{M1}), \qquad\qquad\qquad (7d)$$

where the partial safety factors for resistance are

$$\gamma_{M1} = 1,1\,, \qquad\qquad\qquad\qquad\qquad\qquad\qquad (8a)$$

$$\gamma_{M2} = 1,1 \qquad\qquad\qquad\qquad \text{for} \qquad \bar{\lambda}_S \leq 0,25,$$
$$\gamma_{M2} = 1,1\,(1+0,318\,\frac{s-0,25}{1,75}) \text{ for } 0,25 < \bar{\lambda}_S < 2,00, (8b)$$
$$\gamma_{M2} = 1,45 \qquad\qquad\qquad\qquad \text{for } 2,00 \leq \bar{\lambda}_S.$$

The partial safety factor γ_{M2} includes the same additional factor 0,75 as used in the ECCS-Recommendations (1,1/1,45 = 0,75).

If two or three buckling-relevant membrane stresses are present, they have to fulfill the following interaction condition:

$$\left(\frac{\sigma_x}{\sigma_{xS,R,d}}\right)^{1,25} + \left(\frac{\sigma_\varphi}{\sigma_{\varphi S,R,d}}\right)^{1,25} + \left(\frac{\tau}{\tau_{S,R,d}}\right)^2 \leq 1\,. \qquad (9)$$

2 COVERED SHELL BUCKLING CASES

DIN 18800/4 covers **unstiffened circular cylinders, cones and spheres** (including spherical caps) with **constant wall thickness** under all three buckling-relevant membrane stresses; furthermore **unstiffened circular cylinders** with **stepwise variable wall thickness** under compressive membrane stresses in both directions. Stiffened shells are not covered.

Concerning **load cases**, the basic idea of DIN 18800/4 is to provide a simple tool which may be conservatively applied to any action or loading resp. which creates one or more of the three buckling-relevant membrane stresses (except concentrated radial loads). If variable stress fields result from the action or loading resp., the maximum membrane stress values have to be introduced as numerators into eqn. (9). Of course, this may in some cases result in an unreasonably uneconomic design when using in eqn. (4) the given simple formulas for the critical buckling stresses. These formulas are, as in the ECCS Recommendations and in other shell stability design rules, based on basic load cases with constant membrane stress fields over the whole shell, e.g. a circular cylinder under constant torsional shear or a sperical cap under constant external pressure etc. In such cases, as has been mentioned, the problem may be overcome by a numerical bifurcation analysis yielding more realistic values for the critical buckling stresses in eqn. (4).

An example for a common shell buckling case in which the beforementioned problem shows up, is a circular cylinder under wind loading (figure 2a). For this case DIN 18800/4 gives a formula with which the maximum external pressure max q_w may be replaced by a fictitious axisymmetric external pressure (figure 2b)

$$q = \max q_w \cdot 0,46[1+0,1\sqrt{C_\varphi \cdot (r/l) \cdot \sqrt{r/t}}]. \tag{10}$$

The factor C_φ depends on the boundary conditions; it is 1 for simply supported cylinder edges.

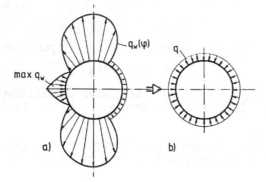

Figure 2. Wind loaded circular cylinder.

3 EXAMPLE

Of a high circular cylindrical vacuum vessel with additional axial loads (figure 3), the lowest cylinder portion between two ring stiffeners shall be checked.

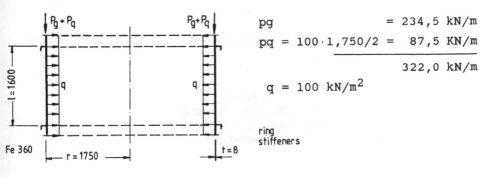

$$p_g = 234,5 \text{ kN/m}$$
$$p_q = 100 \cdot 1,750/2 = 87,5 \text{ KN/m}$$
$$\overline{\phantom{322,0 \text{ kN/m}}}$$
$$322,0 \text{ kN/m}$$
$$q = 100 \text{ kN/m}^2$$

Figure 3. Example for stability design with DIN 18800/4.

Design stresses under factored loads:

$$\sigma_x \quad = 1,35 \cdot 322,0/8 \quad\quad = 54,3 \text{ N/mm2}$$
$$\sigma_\varphi \quad = 1,35 \cdot 100 \cdot 1,750/8 \quad = 29,5 \text{ N/mm2}$$

Design meridional buckling stress:

$$\sigma_{xSi} \quad = 0,605 \cdot 1,0 \cdot 2,1 \cdot 10^6 \cdot 8/1750 = 580,8 \text{ N/mm}^2$$
$$\overline{\lambda}_{Sx} \quad = \sqrt{235/580,8} = 0,6361$$
$$\varkappa_2 \quad = 1,233 - 0,933 \cdot 0,6361 = 0,6395$$
$$\sigma_{xS,R,k} = 0,6395 \cdot 235 = 150,3 \text{ N/mm}^2$$
$$\gamma_{M2} \quad = 1,10 \cdot [1 + 0,318 \cdot (0,636 - 0,25)/1,75] = 1,177$$
$$\sigma_{xS,R,d} = 150,3/1,177 = 127,7 \text{ N/mm}^2$$

Design circumferential buckling stress:

$$c^*_\varphi \quad = 1,0 + 3,0/(0,91 \cdot \sqrt{1750/8})^{1,35} = 1,09$$
$$\sigma_{\varphi Si} = 0,92 \cdot 1,09 \cdot 2,1 \cdot 10^5 \cdot (1750/1600)(8/1750)^{1,5} = 71,2 \text{ N/mm}^2$$
$$\overline{\lambda}_{S\varphi} = \sqrt{235/71,2} = 1,817$$
$$\varkappa_1 \quad = 0,65/1,817^2 = 0,1969$$
$$\sigma_{\varphi S,R,k} = 0,1969 \cdot 235 = 46,3 \text{ N/mm}^2$$
$$\sigma_{\varphi S,R,d} = 46,3/1,1 = 42,0 \text{ N/mm}^2$$

Stability check:

$$\left(\frac{54,3}{127,7}\right)^{1,25} + \left(\frac{29,5}{42,0}\right)^{1,25} = 0,343 + 0,643 = 0,99 < 1.$$

TOWARDS AN IMPROVED DESIGN PROCEDURE FOR BUCKLING CRITICAL STRUCTURES

by

J. Arbocz
Delft University of Technology
Faculty of Aerospace Engineering
The Netherlands

I appreciate the opportunity and gladly share my thoughts with you, a very distinguished and knowledgeable audience, about how to introduce the analysis of uncertainties in buckling critical structures into the design process in a systematic manner. This new procedure could supplement existing design recommendations in the different branches like aerospace or off-shore applications.

I shall use initially the case of axially compressed isotropic shells to illustrate the main points, however, whatever I say can also be applied to shells with any type of wall construction and to shells carrying any type of loading. The main points are the same, however, the details of the steps that must be carried out may vary from case to case.

When Vic Weingarten and Paul Seide (both are at USC at the moment)[1] completed the initial shell design recommendations for NASA they proposed essentially the use of the following basic formula

$$P_a \leq \frac{\gamma}{F.\,S.} P_c$$

where

P_a	= allowable axial load
P_c	= critical theoretical axial load
γ	= "empirical knock-down factor" to account for the discrepancies observed between the critical theoretical axial loads and the experimentally observed critical loads
F.S. ≥ 1.0	= factor of safety

Figure 1 presents the way Weingarten and Seide arrived at their "empirical knock-down factor" for axially compressed isotropic shells of

$$\gamma = 1 - 0.902 \left(1 - e^{-\frac{1}{16}\sqrt{R/t}}\right) \tag{1}$$

That is, they curve fitted the available experimental data with the lower bound curve given by Eq. (1).

This approach has worked in the past satisfactorily. There have been incidents where shells have collapsed unexpectedly, but, by large, the buckling problem appears to be under control.

However, at the age of structural optimization there are cases where such overly conservative design approaches are unacceptable. Thus it has been hoped that with the large scale introduction of advanced computer codes, which incorporate the latest theoretical findings, an alternate design approach could be developed which would no longer penalize innovative shell design because of the poor experimental results obtained elsewhere.

Thus returning to Figure 1, I have been looking in the past 15 years or so for ways to define an Improved Lower Bound curve for a specific set of nominally identical shells. In Figure 1 the specific set is formed by the 7 seamless copper electroplated shells tested by me at Caltech in 1969. Please notice, that "Improved" means a "Higher" knock-down factor.

The turning point in my search came in 1978 when Isaac Elishakoff spent his Sabbatical with me in Delft and he introduced me, among other things, to the concept of reliability function $R(\lambda)$. Notice that, by definition, the reliability function is equal to the probability that the random buckling load Λ is greater or equal to a specified buckling load λ. Thus

$$R(\lambda) = \text{Prob}\{\Lambda \geq \lambda\} \tag{2}$$

If you consider now Figure 2, you can see that for a specified reliability of say 0.98, one can obtain from the plot of the reliability function the allowable buckling load of $\lambda_a = 0.5$, which corresponds to the knock-down factor γ in Figure 1.

Thus knowing the reliability function for certain type of shells makes it possible to arrive at "improved knock-down factors λ_a" in a straight forward manner. Notice that the shape of the reliability function $R(\lambda)$ depends on the particular shell geometry and external load combination, whereby the effect of boundary conditions and other features that might influence the buckling load can be included in the derivation of the reliability function $R(\lambda)$.

Initially we have used with Isaac the Monte Carlo Procedure to derive the reliability function $R(\lambda)$. What always disturbed me with this approach was the large number of deterministic buckling load calculations that had to be done in order to define the shape of the reliability function $R(\lambda)$ accurately.

Thus I was very pleased when in 1980 our attention was focused on the First Order, Second Moment Method by a young Dutch Civil Engineer S. van Manen[2]. This approach, in its initial implementation, requires only one more deterministic buckling load calculation than the number of random variables used. Notice that with the currently available supercomputer facilities we have the means to model the deterministic buckling behaviour with the accuracy required in order to make the use of statistical methods meaningful. However, in my opinion if one is unable to reproduce numerically the experimentally observed buckling process quite accurately, then it is of questionable value to use the same deterministic computational models repeatedly for statistical predictions.

What do we need then in order to make the introduction of a reliability based shell design procedure possible?

First, one needs detailed information about the type of characteristic imperfections that can be associated with a specific manufacturing process. Luckily, it appears that data on a relatively small sample of nominally identical shells (say, about 4) made by the same fabrication process is sufficient to get some statistically meaningful information about the expected mean vectors and variance-covariance matrices of the measured Fourier coefficients.

Second, one needs some information about the type of operational boundary conditions that the shells will be subjected to in the applications.

Third, one needs both a nonlinear program of sufficient generality in order to be able to model the measured initial imperfections and the operational boundary conditions accurately, and a computer big and fast enough to carry out the numerical calculations needed for the derivation of the reliability function $R(\lambda)$.

Considering the first requirement, Joe Singer and myself have been trying to convince the technical community that as a first step towards more accurate buckling predictions there must be a systematic effort trying to establish the characteristic imperfection distributions associated with the different manufacturing processes. Up to now we had only a limited success in getting contributions from different industries to our Initial Imperfection Data Bank.

In principal there are two reasons why companies do not want to get involved in this activity. At first, the people involved in producing the shells look at the suggested imperfection surveys as an additional quality control hurdle that can only cause trouble for them. Further no company wants to see a statement in the open literature stating that their fabrication process is producing more damaging initial imperfections than the process used by their competitors. Second, whereas it costs relatively little to carry out the complete imperfection surveys needed, most companies do not have the experience to carry out the necessary data reduction process in order to arrive at the Fourier coefficients of the measured data. Thus, besides pointing out to the people involved in the manufacturing of shells that the advantages of the proposed new procedure by far outweigh their possible reservations, there would have to be one central organization (say, NASA or Engineering Societies) where all the initial imperfection measurements from the different companies would be gathered and a catalog of the different initial imperfection distributions would be maintained and where the necessary data reduction would be done. In return, all participating companies would have access to all the data assembled.

Turning now to the second requirement, the situation is that we know very little about the precise nature of the different in-plane and out-of-plane boundary conditions that exist in the different shell applications. Especially wether $u = u_o$ (prescribed end-displacement) or $N_x = - N_o$ (prescribed end-load) should be used is a major question. As one can see from Table 1[3] for the stringer stiffened shell AS-2 there are bigger differences in the calculated buckling loads when we change from $N_x = - N_0$ to $u = u_0$, than when we change from simple support ($M_x = 0$) to clamped ($w_{,x} = 0$). Especially dramatic are the changes if one uses elastic end-rings for boundary conditions. As one can see from Figure 3 there is a critical stiffness of the symmetrically placed symmetric end-ring below which the collapse behaviour changes from general overall buckling to inextensional buckling whereby the buckling load decreases sharply[4].

Turning to the third requirement, severall of the currently available structural analysis computer codes (ex. STAGS, ABAQUS, MARC etc.) have the capability to calculate the collapse load of shells with general imperfections. However, it is well known that the success of such a deterministic collapse load analysis depends very heavily on the appropriate choice of the model used, which in turn requires considerable knowledge by the user as to the physical behaviour of the imperfect shell structure. Thus it is imperative that the deterministic buckling load calculations be carried out by persons of sufficiently high technical qualifications. What I am trying to say, it is not sufficient that the analyses be carried out by people, who are capable to use the proper INPUT but have insufficient theoretical back-ground to interpret the complex nonlinear behaviour that these structures will exhibit close to the limit point.

Finally I think that by your participation in an initiative to look at the introduction of uncertainty analysis in the design process of thin-walled structures you can make an important contribution to the development of improved shell design criteria. This is especially true due to the accumulated experience with shell buckling problems that the audience here present represents.

REFERENCES

1. Weingarten, V.I., Morgan, E.J. and Seide, P., "Elastic stability of thin-walled cylindrical and conical shells under axial compression", AIAA Journal, Vol. 3, 1956, pp. 500-505.

2. Elishakoff, I., Manen, S. van, Vermeulen, P.G. and Arbocz, J., "First-Order Second-Moment Analysis of the Buckling of Shells with Random Imperfections", AIAA Journal, Vol. 25, 1987, pp. 1113-1117.

3. Arbocz, J. and Sechler, F.E., "On the Buckling of Stiffened Imperfect Cylindrical Shells", AIAA Journal, Vol. 15, 1976, pp. 1611-1617.

4. Arbocz, J., Vermeulen, P.G. and van Geer, J., "The Buckling of Axially Compressed Imperfect Shells with Elastic Edge Supports", in: Buckling of Structures, Theory and Experiment, The Josef Singer Anniversary Volume, Edited by I. Elishakoff et al, Studies in Applied Mechanics No. 19, Elsevier Science Publishers B.V., Amsterdam, 1988, pp. 1-27.

	SS-3	SS-4	C-3	C-4
MEMBRANE PREBUCKLING (N/cm)	229.8 (10)	300.7 (14)	256.9 (10)	320.8 (14)
NONLINEAR PREBUCKLING (N/cm)	224.0 (10)	280.0 (14)	256.4 (10)	316.8 (14)

NUMBERS IN PARENTHESIS ARE THE NUMBER OF CIRCUMFERENTIAL WAVES , n

SS-3 $N_x = v = w = M_x = 0$ C-3 $N_x = v = w = w,_x = 0$

SS-4 $u = v = w = M_x = 0$ C-4 $u = v = w = w,_x = 0$

Table 1   Buckling loads of the perfect stringer stiffened shell AS-2[3]

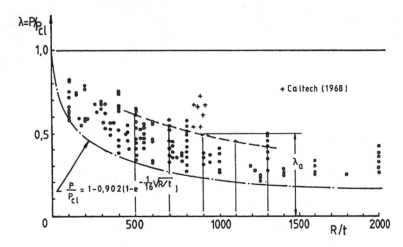

Fig. 1   Test data for isotropic cylinders under axial compression

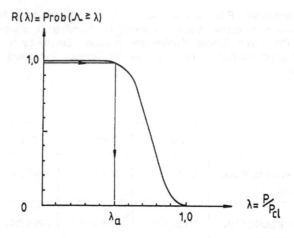

Fig. 2   Reliability function R(λ) for a given r/t ratio

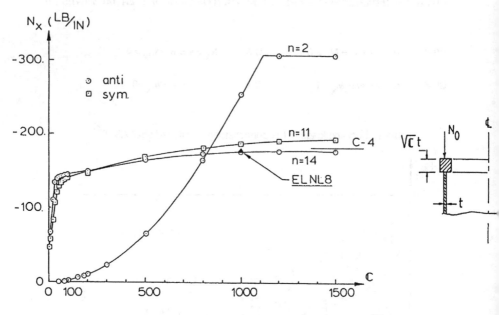

Fig. 3   Critical loads for shell AS-2 with elastic end rings[4]

DESIGN CODES FOR OFFSHORE STRUCTURES, BUCKLING OF CYLINDRICAL SHELLS

JONAS ODLAND, Dr. ing
Statoil, Norway, Member of ECCS TWG 8.4

ABSTRACT

Stiffened and unstiffened cylindrical shells are frequently used offshore as primary structural members. Typically such structures are exposed to external lateral pressure, axial compression and other loads that under extreme conditions may lead to collapse by buckling. The paper deals with pratical design criteria developed and implemented by the offshore industry. Emphasis is put on safety, consistency and simplicity. The effects of shape imperfections and residual stresses as observed in tests are incorporated, but well known classical cases appear as asymptotic solutions. In particular it is described how load combinations can be treated consistently both in the elastic and the elastic-plastic range. The method is valid not only for compression and shear, but also effects of tension can be included. The methodology is completely based on analytical formulations and therefore well suited for computerization.

1. INTRODUCTION

Cylindrical members are frequently used in offshore structures, see Fig. 1. The structures may be quite complex with longitudinal and transverse bulkheads and stiffness, see Fig. 2. The size may be considerable. The structure shown in Fig. 2 has a diameter of 25 m and a height of 63 m.

Buckling criteria are extremely important for many of the structures used in the offshore industry and have therefore been given great attention by designers as well as researchers, classification societies, national authorities, etc. The need for safe and efficient structures has initiated comprehensive research projects which have improved the understanding of buckling and produced a more reliable basis for design.

Design rules and recommendations issued by international classification societies like Det norske Veritas (DnV) or standardization institutions like the American Petroleum Institute (API) have been used as basis for design.

DnV issued their first rules for offshore structures in the early seventies. Shell buckling criteria were then to a great extent based on NASA CR912. The rules were completely rewritten in 1977 with the introduction of the limit state design method.

In 1990 the Norwegian Petroleum Directorate (NPD) issued a comprehensive set of design guidelines for steel structures in line with the latest DnV rules. These guidelines now constitue the basis for design of all structures for the Norwegian sector of the North Sea.

The following discussion of buckling criteria for cylindrical shells relates to this practice.

2. OBJECTIVE AND METHODOLOGY

In practical design work it is necessary to deal with different structures, different load cases and combinations of load cases. It is therefore a main objective to develop design methologies characterized by simplicity, consistency and generality.

The following features of the methodology are higlighted:
- the classical elastic buckling stress is maintained as a basic parameter
- the classical buckling stress is adjusted in a conservative manner to comply with experimental data
- linear interaction is adopted as a conservative assessment of elastic buckling under combined loading
- elasto-plastic buckling is considered as interaction between elastic buckling and yielding
- at low curvature, when shell buckling and plate buckling calculations produce similar results, it is advised to consider use of more sophisticated plate buckling anlysis also for shells

3. ELASTIC BUCKLING OF UNSTIFFENED CYLINDRICAL SHELLS

Exact and explicit solutions are available only for simple cases, but these solutions are very useful. For consistency it is useful to rearrange and write the classical solutions in the same way as plate buckling formulae, i.e.:

$$\sigma_{CL} = k_{CL} \frac{\pi^2 E}{12(1-\nu^2)} (\frac{t}{\ell})^2 \qquad (3.1)$$

where the buckling coefficient k_{CL} is the key parameter. The buckling coefficient can be approximated by the expression:

$$k_{CL} = \psi \sqrt{1 + (\frac{\zeta}{\psi})^2} \qquad (3.2)$$

As shown in Fig. 3, k_{CL} is a function of the curvature parameter, Z, which is also known as the Batdorf parameter. When the curvature approaches zero, k_{CL} approaches an asymptotic value, ψ, which actually represents a plate buckling solution. For increasing curvature k_{CL} approaches another asymptote, ζ, which represents the classical solution for a cylindrical shell of finite length. However, this solution is only valid up to a certain value of Z. Beyond this value special solutions apply.

While plate buckling formulae have been found to produce results in good agreement with experiments, it is well known that there is a clear disagreement between experimental data and classical shell buckling predictions. It has been accepted that the poor correlation between theory and experiments is mainly due to initial geometric imperfections.

For practical purposes it is necessary to define a modified elastic buckling stress. This may be done by application of a reduction factor, ρ. It should be emphasized that the empirical reduction factor shall only be applied to the term which represents shell buckling and not to the term representing plate behaviour. Hence the modified elastic buckling stress is:

$$\sigma_e = k \frac{\pi^2 E}{12(1 - \nu^2)} (\frac{t}{\ell})^2 \tag{3.3}$$

where

$$k = \psi\sqrt{1 + (\frac{\rho\zeta}{\psi})^2} \tag{3.4}$$

The effect of this modification is shown in Fig. 3. Values of ψ, ζ and ρ are given in Table 1 for the most important load cases.

TABLE 1
Buckling coefficients - cylindrical shells

	ψ	ζ	ρ
Axial compression	1	$0,702\ Z$	$0,5(1 + \frac{r}{150t})^{-0,5}$
Bending	1	$0,702\ Z$	$0,5(1 + \frac{r}{300t})^{-0,5}$
Torsion and shear	5,34	$0,856Z^{3/4}$	0,6
Local pressure	4	$1,04\sqrt{Z}$	0,6
Hydrostatic pressure	2	$1,04\sqrt{Z}$	0,6

Special solutions for long cylindrical shells:
As pointed out above the shell buckling solutions based on Eq. 3.1 are only valid up to a certain value of Z. Beyond this value special solutions apply:

Axial compression: Very long cylindrical shells buckle like columns with undeformed cross section. For the case with simply supported ends the elastic buckling stress is:

$$\sigma_{CL} = \frac{\pi^2}{2} E(\frac{r}{\ell})^2 \tag{3.5}$$

At the transition from shell buckling to column buckling, mode interaction has to be considered as described in Ch.6.

Bending: When a long cylindrical shell is subjected to bending the cross section starts to ovalize. When the bending moment reaches a limiting value, the shell will collapse. The critical moment is:

$$M_C = 0.3\pi E r t^2 \tag{3.6}$$

It should be noticed that the critical moment given by Eq. 3.6 is approximately equal to one half of the moment which would cause buckling according to linear classic theory. Local buckling may even occur before the limit moment is reached. This local instability occurs when the maximum axial stress according to non-linear theory reaches the critical stress of a cylinder with radius equal to the local radius of the ovalized cylinder. When the cylinder has a final length, the effect of end constraints will reduce the ovalization.

Torsion: For long cylindrical shells subjected to torsion the effect of boundary conditions becomes negligible. The buckling stress will thus be independent of length. The theoretical solution for this case is:

$$\tau_{CL} = 0,25 \ E \ (\frac{t}{r})^{3/2} \tag{3.7}$$

Another elastic buckling mode exists for extremely long cylinders. In that case the cylinder is deformed into a helix while the cross section remains circular.

Lateral pressure: A very long cylindrical shell subjected to external pressure behaves like a ring. The buckling stress is independent of length and the buckling mode is referred to as ovalization:

$$\sigma_{CL} = \frac{E}{4(1-\nu^2)} \ (\frac{t}{r})^2 \tag{3.8}$$

4. ELASTIC BUCKLING OF CURVED PANELS

A curved panel is a part of a circular cylindrical shell supported along all four edges. Buckling analyses of curved panels are very similar to buckling analyses of circular cylindrical shells. However, it is more difficult to find explicit solutions which satisfy all boundary conditions. Since the aspect ratio ℓ/s is normally greater than one, it is most convenient to write the elastic buckling stress formula as:

$$\sigma_e = k\frac{\pi^2 E}{12(1-\nu^2)}(\frac{t}{s})^2 \tag{4.1}$$

where k is the buckling coefficient which also for curved panels can be represented by the formula:

$$k = \psi\sqrt{1 + (\frac{\rho\zeta}{\psi})^2} \qquad\qquad (4.2)$$

k is a function of the curvature parameter $Z_s = Z(s/\ell)^2$. When the curvature approaches zero, k approaches the asymptotic value ψ, which represents a plate buckling solution, ref. Fig. 3. For increasing curvature k approaches another asymptote which actually represents the solution for a circular cylindrical shell. However, since s is the length parameter in Eq. 4.1 while ℓ is used in Eq. 3.3 the following relation exists:

$$\zeta_{(C)} = \zeta_{(P)}(\frac{\ell}{s})^2 \qquad\qquad (4.3)$$

where the subscripts C and P refer to "circular cylindrical shell" and "curved panel" respectively. Values of ψ, ζ and ρ are given in Table 2 for the most important load cases.

TABLE 2
Buckling coefficients - curved panels

	ψ	ζ	ρ
Longitudinal compression	4	$0,702Z_s$	$0,5(1 + \frac{r}{150t})^{-0,5}$
Shear	$5,34+4(\frac{s}{\ell})^2$	$0,856\sqrt{\frac{s}{\ell}}Z_s^{3/4}$	$0,6$
Transverse compression	$[1+(\frac{s}{\ell})^2]^2$	$1,04\frac{s}{\ell}\sqrt{Z_s}$	$0,6$

5. LONGITUDINALLY STIFFENED CYLINDRICAL SHELLS

Buckling analyses of longitudinally stiffened shells may be quite complicated, involving numerous buckling modes and possibilities for mode interaction. However, if the proportions are such that local buckling of stiffeners and panels may be excluded, conservative approximations can easily be established. The elastic buckling stress can be expressed by:

$$\sigma_e = k \frac{\pi^2 E}{12(1-\nu^2)} (\frac{t}{\ell})^2 \qquad\qquad (5.1)$$

where

$$k = \psi\sqrt{1 + (\frac{\rho\zeta}{\psi})^2} \qquad\qquad (5.2)$$

This formulation is identical to the formulation used for unstiffened cylindrical shells. However, the asymptote ψ is no longer a fixed value, but a function of the characteristic stiffener parameters, a and γ, and represents buckling of a stiffened flat plate. The other asymptote represents buckling of an unstiffened circular cylindrical shell. This means that longitudinal stiffeners become less effective as curvature increases.

It is generally assumed that stiffened shells are less imperfection sensitive than unstiffened shells. It is recommended to apply the same reduction factors as for unstiffened shells. For axial compression it appears reasonable to take the upper limit which is obtained for r/t = 0. Values for ψ, ζ and ρ are given in Table 3.

TABLE 3
Buckling coefficients-stiffened cylinders

	ψ	ζ	ρ
Axial compression and bending	$\dfrac{1 + \gamma}{1 + a}$	$0,702Z$	$0,5$
Torsion and shear	$5,34+1,82(\frac{\ell}{s})^{4/3}\gamma^{1/3}$	$0,856Z^{3/4}$	$0,6$
Lateral pressure	$2(1+\sqrt{1+\gamma})$	$1,04\sqrt{Z}$	$0,6$

There are many practical cases where the geometry is such that the effect of curvature is more or less negligible. It is then allowable to use design methods developed for stiffened plate structures. The advantage is that such methods are more sophisticated and accurate than the methods described above and that lower safety factor are permitted.

6. INTERACTION PROBLEMS

Interaction problems that are encountered in shell buckling analysis are essentially of the following categories:
- Interation between elastic buckling and yielding, i.e. elasto-plastic buckling
- Bukling under action of several load components, i.e. combined loading
- Interaction between different buckling modes when more than one buckling mode are associated with the same critical load, i.e. mode interaction

Elasto-plastic buckling:
It has been found, [1] that the following interaction formula leads to a reasonable prediction of the elasto-plastic buckling strength:

$$(\frac{\sigma_k}{\sigma_e})^2 + (\frac{\sigma_k}{\sigma_y})^2 = 1 \tag{6.1}$$

By rearrangement the following result is obtained:

$$\sigma_k = \frac{\sigma_y}{\sqrt{1 + \lambda^4}} \tag{6.2}$$

It has been a somewhat controversial question whether $\bar{\lambda}$ should be defined on the basis of a realistic/conservative elastic buckling stress estimate as suggested here, or the classical elastic buckling stress. There is no final answer to this, but it is this author's opinion that the suggested method has practical advantages over the alternative.

Combined loading:
Individual strength formulae are useless for most practical applicationa if they cannot be combined in a reasonable way in cases where several load components act simultaneously. The recommended method has also been described in [1] and is fully consistent with the approach for elasto-plastic buckling.

For a loading condition where the state of stress is defined by three stress components, σ_x, σ_θ and τ, the buckling stress may be determined from:

$$\left(\frac{\sigma_{x0}}{\sigma_{ex}} + \frac{\sigma_{\theta 0}}{\sigma_{e\theta}} + \frac{\tau}{\tau_e}\right)^2 + \left(\frac{\sigma_j}{\sigma_y}\right)^2 = 1 \tag{6.3}$$

where

$$\sigma_j = \sqrt{\sigma_x^2 + \sigma_\theta^2 - \sigma_x \sigma_\theta + 3\tau^2} \tag{6.4}$$

and

$$\sigma_{x0} = \begin{cases} 0 & \text{if } \sigma_x \geqslant 0 \quad \text{(tension)} \\ -\sigma_x & \text{if } \sigma_x < 0 \quad \text{(compression)} \end{cases}$$

$$\sigma_{\theta 0} = \begin{cases} 0 & \text{if } \sigma_\theta \geqslant 0 \\ -\sigma_\theta & \text{if } \sigma_\theta < 0 \end{cases} \tag{6.5}$$

By rearrangement the critical value of the equivalent stress is obtained:

$$\sigma_{kj} = \frac{\sigma_y}{\sqrt{1 + \bar{\lambda}_e^4}} \tag{6.6}$$

where

$$\bar{\lambda}_e^2 = \frac{\sigma_y}{\sigma_j}\left[\frac{\sigma_{x0}}{\sigma_{ex}} + \frac{\sigma_{\theta 0}}{\sigma_{e\theta}} + \frac{\tau}{\tau_e}\right] \tag{6.7}$$

This method is simple and straight-forward and it is also charac-terized by the following important features:
- For structures with high slenderness the failure criterion approaches elastic buckling with linear interaction
- For structures with low slenderness the failure criterion approaches the von Mises yield criterion

284

It should be emphasized that both shear stresses and tensile stresses can be consistently treated by use of this method. An evaluation of this and other methods may be found in [2].

Mode interaction:
Mode interaction is essentially a problem that has to be considered in connection with design of complex structures. A long, unstiffened cylindrical shell subjected to axial compression represents an important exception, however. If both local shell buckling and overall column buckling may occur, the effect of mode interaction has to be properly accounted for. One possibility is to address mode interaction in the column buckling analysis. This means that the long cylindrical shell is essentially considered as a column. The analysis is based on a "reduced yield stress" which is taken equal to the local shell buckling stress. For cases with combined loading the "reduced yield stress" has to be taken as:

$$\sigma_{yr} = \frac{\sigma_x}{\sigma_j}\sigma_{kj} \tag{6.8}$$

It could be added that it is normally considered as good practice to design structures such that mode interaction is avoided.

REFERENCES

[1] Odland, J.: "On the Strength of Welded Ring Stiffened Cylindrical Shells Primarily Subjected to Axial Compression", Department of Marine Technology, The Norwegian Institute of Technology, Report UR-81-15, 1981.

[2] Galletty, G.D., James, S., Kruzelecki, J., Pemsing, K.: "Interactive Buckling Tests on Cylinders Subjected to External Pressure and Axial Compression", Journal of Pressure Vessel Technology, Vol. 109 (1987) February.

NOMENCLATURE

A - cross sectional area of stiffener
E - modulus of elasticity
I - moment of inertia of stiffener with effective plate flange

$Z = \sqrt{1 - \nu^2}/\ell^2/rt$
$a = A/st$
ℓ - length in axial direction
r - radius
s - distance between stiffeners
t - shell thickness
$\gamma = 12(1-\nu^2)I/st^3$

$\bar{\lambda} = \sqrt{\sigma_y/\sigma_e}$
ν - Poisson's ratio
σ_a - axial stress

σ_b - bending stress
σ_e - elastic buckling stress

σ_j - effective stress (von Mises)
σ_k - critical stress
σ_x - axial stress
σ_θ - circumferential stress
σ_y - yield stress
τ - shear stress

285

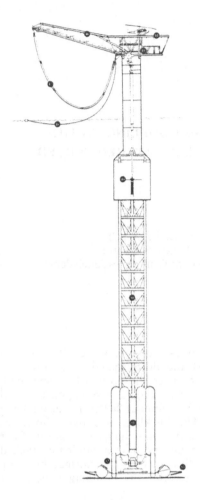

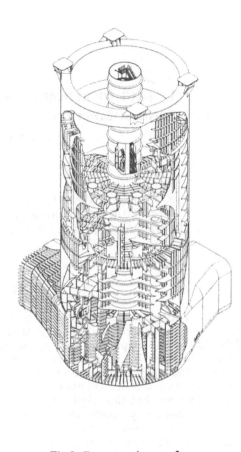

Fig.1 Offshore Loading Tower

Fig.2 Corner column of
Tension Leg Platform

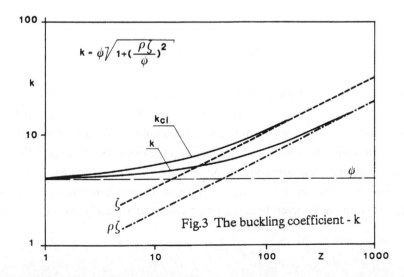

Fig.3 The buckling coefficient - k

INFLUENCE OF BENDING-TWISTING COUPLING ON THE BUCKLING LOADS OF SYMMETRICALLY LAYERED CURVED PANELS

K. ROHWER[1]), G. MALKI[2]), E. STECK[2])
[1]) DLR Braunschweig, Inst. für Strukturmechanik
[2]) TU Braunschweig, Inst. für Allgem. Mechanik und Festigkeitslehre

ABSTRACT

Bending-twisting coupling can hardly be avoided in layered composites. Its effect on the buckling loads of curved panels is studied with the aid of the computer program BEOS. As with flat plates, the coupling always reduces the normal buckling load, whereas the shear buckling load can be increased or reduced depending on the load direction. Increasing curvature results in a greater difference between the buckling loads with and without coupling, respectively. The buckling load itself, however, grows faster so that the relative difference decreases. That holds for various fiber directions, boundary conditions, and stacking sequences. For a laminate stacked $[(\pm\theta)_n]_s$ with constant thickness, the relative difference of the normal buckling loads was found to decline proportional to n^{-2} whereas the corresponding value under shear decreases with n^{-1} only.

INTRODUCTION

Curved panels are widely used elements in lightweight constructions. With improved experience and growing confidence in the reliability of advanced materials they are more and more often laminated from unidirectionally reinforced fiber composites. Symmetric stacking sequences are preferred; membrane-bending coupling, thus, results only from the shell curvature. Equilibrating the angle plies by arranging the same number of layers in $+\theta$ as in $-\theta$ direction assures orthotropic membrane behavior, but coupling between bending and twisting can usually not be avoided.

Under membrane compression and shear the panels are prone to buckle. In spite of the frequent application, buckling analysis of layered curved panels is seldom reported about in the open literature. For flat plates the influence of bending-twisting (b-t) coupling on the buckling loads is thoroughly investigated, as Leissa[1] has pointed out in his extensive review. But for curved panels the same effect has so far attracted very little attention. Buckling tests on axially loaded cylindrical panels made from laminated Carbon Fiber Reinforced Plastic (CFRP) are performed by Wilkins[2]. The results comprise b-t coupling but no comparison is drawn with corresponding orthotropic panels. Zhang and Matthews[3] analyzed cylindrically curved panels made from layered composite materials, but did not explicitly show the effect of b-t coupling. For the shear buckling of cylindrical panels this has been done by Whitney[4]. He stated that the coupling effect increases with increasing curvature. Arnold and Kedward[5] treated both, curved panels and the b-t coupling, but the mutual influence was not worked out.

The following study exclusively considers shallow curved panels over a quadratic ground view. The shell itself is set up from several homogeneous and transversely isotropic layers which are equilibrated and stacked symmetrically with respect to the middle surface. Based on the classical lamination theory, linear bifurcation buckling under normal load and shear load is considered. The b-t coupling influence is determined for a variety of curvatures, boundary conditions, stacking sequences and fiber angles. Inspecting the buckling mode turns out to be a useful aid for to obtain reasonable explanations if the analysis delivers unexpected buckling loads.

COMPUTATIONAL ASPECTS

As a special purpose computer program BEOS[6] was designed to calculate bifurcation buckling and natural vibrations of anisotropic panels. Based on Donnell's shallow shell theory bicubic Hermite interpolation polynomials are applied over subdivisions to approximate the natural modes. This leads to a procedure similar to the Finite Element method. For a known fundamental state the buckling loads are determined from a linear matrix eigenvalue problem which is solved using subspace iteration. Since a pure displacement method is used, the buckling loads always converge from above with mesh refinement.

All analyzed panels are built from CFRP, the properties of which are listed in Table 1. Application of the lamination theory determines the membrane, bending and coupling stiffness matrices **A**, **D** and **B**, respectively, which relate membrane, bending and torsional strains to the corresponding cross sectional forces and moments:

$$\left\{ \begin{matrix} N \\ M \end{matrix} \right\} = \begin{bmatrix} \mathbf{A} & \mathbf{B} \\ \mathbf{B}^T & \mathbf{D} \end{bmatrix} \cdot \left\{ \begin{matrix} \varepsilon \\ \kappa \end{matrix} \right\} .$$

Symmetric stacking is considered exlusively so that the matrix **B** vanishes. **A** and **D** are needed as input data for BEOS.

TABLE 1
Properties for CFRP Material

E_L =	125.0 kN/mm²
E_T =	8.8 kN/mm²
G_{LT} =	5.3 kN/mm²
v_{LT} =	0.35

The rate of convergence towards the actual buckling load depends on the waviness of the corresponding mode; the more waves there are, the finer the grid must be to assure a sufficient accuracy of the buckling load. For a cylindrical panel the number of waves increase with the curvature. In this sense, a panel with side lengths of 500×500 mm², a radius of curvature $R_y = 1000$ mm, and a thickness of 1 mm can be regarded a rather rigorous test case. A stacking sequence of $[+45, -45]_s$ renders high bending-twisting coupling effects. Such a panel, with all four edges clamped, and an equally distributed membrane force N_x applied, was analyzed using grids of increasing fineness. Buckling loads as given in Table 2 were obtained. They indicate, that a mesh of 10×10 subdivisions is fine enough to deliver reliable results. Therefore, such a mesh is used throughout in the following.

TABLE 2
Convergence with Mesh Refinement.

Grid	$N_{x_{cr}}$ [N/mm]
2×2	106.483
4×4	23.922
6×6	17.884
8×8	15.958
10×10	15.370

RESULTS

The great number of parameters involved in the buckling analysis of layered curved panels rules out to vary all of them simultaneously. A ground view of 500×500 mm² and a thickness of 1 mm is kept unchanged. The difference between simply supported and clamped boundary conditions is examined in a few examples only, and stacking sequences are limited to the simple cases of $[(\pm \theta)_n]_s$. Double curvature is not considered at all. Constant normal and shear stresses are treated separately; interaction effects must be left for further investigations.

Normal Load

In the first instance cylindrical CFRP panels with clamped boundary conditions and a stacking sequence of $[+45, -45]_s$, but with various radii of curvature are examined. The results are listed in Table 3. $N_{x_{cr}}$ (ortho) specifies the buckling load with neglected b-t coupling, whereas $N_{x_{cr}}$ (aniso) is obtained with the coupling considered.

TABLE 3
Curvature Influence on Normal Buckling Load.

R_y [mm]	$N_{x_{cr}}$ (ortho) [N/mm]	$N_{x_{cr}}$ (aniso) [N/mm]	Difference [N/mm]	Reduction [%]
∞	1.76180	1.37460	0.38720	22.0
10000	3.04670	2.44760	0.59910	19.7
5000	4.57099	3.80552	0.76547	16.7
4000	5.35510	4.48700	0.86810	16.2
3000	6.65351	5.65670	0.99681	15.0
2000	9.30424	8.01039	1.29385	13.9
1000	17.4732	15.3704	2.1028	12.0
500	34.2632	31.7427	2.5205	7.4

In accordance with Whitney's results[4] the differences between the two buckling loads increase with increasing curvature. This surprising fact can be explained by examining the corresponding buckling modes. A flat square plate would buckle with one half wave; nodal lines of that mode run along the boundaries, regardless whether or not the coupling is taken into account. Therefore, the single buckle has a fixed ground shape and that limits the freedom to develop a mode with a lower strain energy.

As the curvature increases, the number of half waves in axial direction increases too. The internal nodal lines of these buckling modes are not fixed. If coupling is neglected they run in circumferential direction. Due to coupling they change their directions so that each single buckle appears over a skew ground shape as depicted in Figure 1. Buckling modes can be longer in the direction of the outer layer which requires less strain energy.

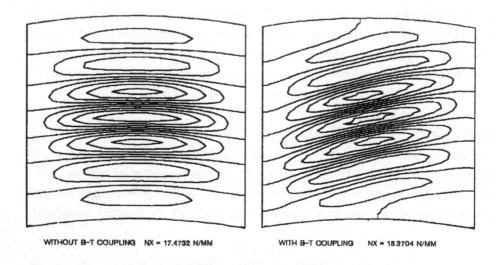

WITHOUT B–T COUPLING   NX = 17.4732 N/MM   WITH B–T COUPLING   NX = 15.3704 N/MM

Figure 1.   Buckling Mode of a CFRP Panel, $R_y = 1000$mm

Since the buckling load itself depends more on the curvature than on the b-t coupling the difference $[N_{cr}(\text{ortho}) - N_{cr}(\text{aniso})]$ will be related to $N_{cr}(\text{ortho})$. This relation, called reduction in the case of normal load, singles out the b-t coupling influence and therewith allows to specify the error due to its neglection more clearly. Table 3 shows that the buckling load reduction decreases with increasing curvature so that the increase in the difference values can be traced back to the curvature effect.

For flat plates with constant thickness and a stacking sequence of $[(\pm\theta)_n]_s$ the first author[7] has shown that the buckling load reduction decreases with n^{-2}. Figure 2 depicts corresponding values for cylindrical panels with $R_y = 5000$mm. The reduction rate, specified by the dotted line, is the same as for flat plates, independent of the boundary conditions. Only slight deviations from this result have been found in case of other radii of curvature in connection with the extreme fiber directions of $\theta = 15°$ or $75°$.

Shear Load

It is well known that the shear buckling load of a flat plates is increased compared to orthotropic material if the load results in compressive stress components in the outer layer fiber direction; otherwise it is decreased. This can be explained by a higher

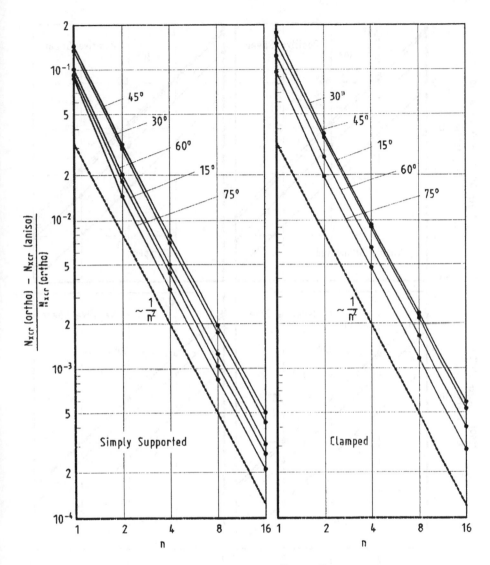

Figure 2. Reductions of Normal Buckling Load for $[(\pm\theta)_n]_s$ Panels, $R_y = 5000$mm

bending stiffness in the outer layer fiber direction and a lower one perpendicular to it. Considering a laminate with a stacking sequence of $[\pm\theta]_s$, for example, where the outer layer fiber direction is rotated $+\theta$ degrees from the x-axis towards the y-axis, a positive shear results in tensile stresses in the fiber direction, and therewith the b-t coupling reduces the buckling load. Under a negative shear the coupling causes an increase.

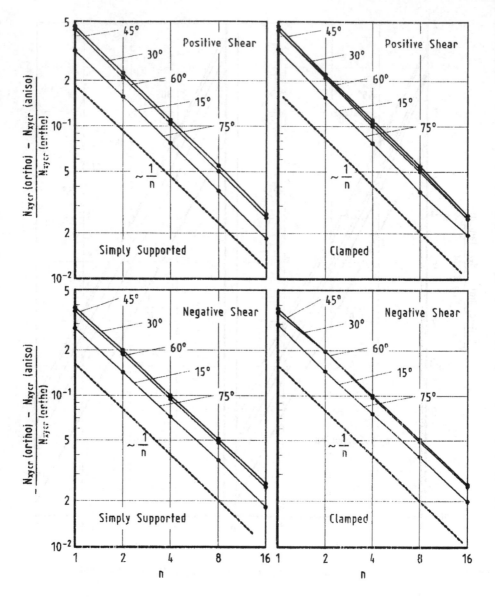

Figure 3. Reductions or Increase of Shear Buckling Load for $[(\pm\theta)_n]_s$ Panels, $R_y = 5000$mm

Since this effect is based on the stacking sequence alone it appears for curved panels likewise. Table 4 shows the b-t coupling effect for square panels with a stacking sequence of $[\pm 45]_s$ under positive and negative shear. In both cases the coupling effect is high compared to normal loads (see Table 3), and with increasing curvature the values decrease at a much slower rate. This holds for clamped (CC) as well as for simply supported (SS) boundary conditions; moreover, the influence of the boundary conditions on the coupling effect is rather small.

TABLE 4
B-T Coupling Effect for Shear Loaded Curved Panels.

R_y [mm]	Reduction [%] (Positive Shear)		Increase [%] (Negative Shear)	
	CC	SS	CC	SS
∞	52.7	55.1	45.7	44.5
10000	49.5	49.0	41.1	38.9
7500	48.5	48.5	39.7	38.0
5000	47.6	47.6	38.2	38.0
2500	45.9	45.8	37.4	36.8
1000	42.6	42.2	35.5	34.8
500	38.8	38.3	32.8	32.3

Differing from the behavior under normal loads, the b-t coupling influence on the shear buckling loads of $[(\pm\theta)_n]_s$ laminates with constant thickness varies with n^{-1} as can be seen in Figure 3. That holds for the buckling load reduction in case of positive shear as well as for the increase in case of negative shear. Furthermore, this dependency was found to be valid for every analyzed curvature. In order to condense

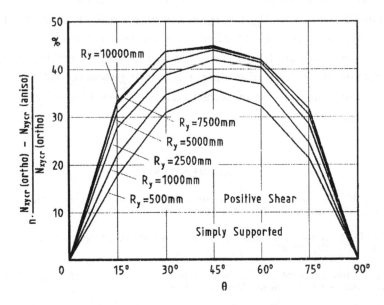

Figure 4. Reductions of Shear Buckling Load for $[(\pm\theta)_n]_s$ Panels, Depending on Fiber Angle and Curvature.

the results it is therefore admissible to multiply the buckling load reduction and increase, respectively, by n and therewith eliminate the influence of the ply number.

Such a procedure is applied in Figure 4, which shows how the buckling load reduction depends on the fiber angle θ. The biggest reduction appears for 45° which is not surprising since there the product of the two coupling coefficients, $D_{13} \cdot D_{23}$, reaches its maximum value. As under normal load, increasing curvature reduces the b-t coupling effect also under shear.

CONCLUSION

With the aid of the BEOS program the buckling loads for a great number of symmetrically layered curved panels were computed. Attention was focussed on the effect of bending-twisting coupling, which often may be small but can hardly be avoided even for equilibrated laminates. It has turned out that increasing curvature increases the difference between the buckling loads for considered and neglected coupling, respectively. In order to eliminate the predominating curvature influence this difference is related to the buckling load with neglected coupling. The resulting value which represents the coupling influence becomes smaller with increasing curvature.

The layer number directly affects the b-t coupling; the more layers the smaller are the coupling coefficients D_{13} and D_{23} as compared to the bending and torsional stiffnesses. Therefore, the coupling influence on the buckling load must fade if the layer number becomes high. It could be shown that the rate with which the influence declines depends on the type of loading. For stacking sequences of $[(\pm \theta)_n]_s$ the buckling load reduction in case of normal loads decreases with n^{-2} whereas under shear load the effect vanishes only with n^{-1}.

REFERENCES

[1] Leissa, A.W., *Buckling of laminated composite plates and shell panels.* Air Force Wright Aeronautical Laboratories, AFWAL-TR-85-3069, AD-A162723, 1985.

[2] Wilkins, D.J., *Compression buckling tests of laminated graphite-epoxy curved panels.* AIAA J., 1975, 13, 465-470.

[3] Zhang, Y. and Matthews, F.L., *Initial buckling of curved panels of generally layered composite materials.* Composite Structures, 1983, 1, 3-30.

[4] Whitney, J.M., *Buckling of anisotropic laminated cylindrical plates.* AIAA J., 1984, 22, 1641-1645.

[5] Arnold, R.R. and Kedward, K.T., *Stability critical stiffened panels*. In: Handbook of Composites (Ed.: Kelly, A. and Rabotnov, Y.N.), Vol. 2: Structures and Design (Ed.: Herakovich, C.T. and Tarnopol'skii, Y.M.) North-Holland, Amsterdam, New York, Oxford, Tokyo, 1989, Chapter 12, pp. 623-665.

[6] Rohwer, K., *BEOS, Buckling Loads and Natural Vibrations of Eccentrically Orthotropic Shells*. DFVLR-Mitt. 81-07 (1981).

[7] Rohwer, K., *Einfluß der Biege-Torsionskopplung auf die Beullasten symmetrisch geschichteter Platten*. DGLR-Bericht 87-02, pp. 325-347.

CYLINDER-CONE-CYLINDER INTERSECTIONS UNDER AXIAL COMPRESSION

PETER KNOEDEL
Versuchsanstalt fuer Stahl, Holz und Steine
University of Karlsruhe
Kaiserstrasse 12, W-7500 Karlsruhe 1 (FRG)

ABSTRACT

In the construction of chimneys and tubular masts adjacent cylindrical shells often have to be built with different radii according to the global bending moment or operational requirements. With a steep conical shell in between a cylinder-cone-cylinder intersection is built. This paper reports on a numerical study on the buckling stability of these structures.
Unstiffened and ring stiffened structures are considered, where the rings are attached to the edges of the cone or set off into the adjacent cylindrical strakes. It is shown, that the critical load of the entire structure remains under the critical load of a straight cylinder, even if heavy stiffeners are used and local buckling is relevant.

INTRODUCTION

The stability of the basic types of thin-walled shells, such as cylinders, cones, spheres, has become well known through theoretical and experimental research for most of the relevant load cases. Only little is known about shell structures, which are assembled by several of those basic types, e.g. silos, watertowers or tanks.

The present paper reports on the results of a numerical study on cylinder-cone-cylinder intersections. These are commonly used with steel stacks, when a conical strake is set between two cylindrical parts of different radii. Global bending due to wind action results in a $\cos\vartheta$ distribution of the membrane forces, and the stability of the shell wall will govern the dimensions of the structure in most cases.

Due to the radial deviating forces it is advantegeous to have ring stiffeners connected to the edges of the cone (discontinuities of the meridional slope). Due to reasons of economic manufacturing as well as reduction of the welding residual stresses, it is desireable to set the stiffeners off the edges into the adjacent cylindrical strakes.

According to the relevant stability codes such a structure would be divided into the basic shell types (i.e. cylinder, cone, cylinder) and the stability analysis would be performed under the assumption of 'classical boundary conditions'. This method is taken to give correct results if, due to sufficiently sized stiffeners, local buckling of the single shell parts of the structure is the relevant failure mode.

We did a numerical study to find out

-   which should be the minimum stiffener to prevent global buckling;
-   how much would be the loss of buckling load, if the required minimum stiffness is not available.

For reasons of simplification the strucutre has been investigated under pure axial compressive load [10]. The following results haven discussed in more detail in a recent publication of *Knoedel* and *Thiel* [8].

METHOD EMPLOYED

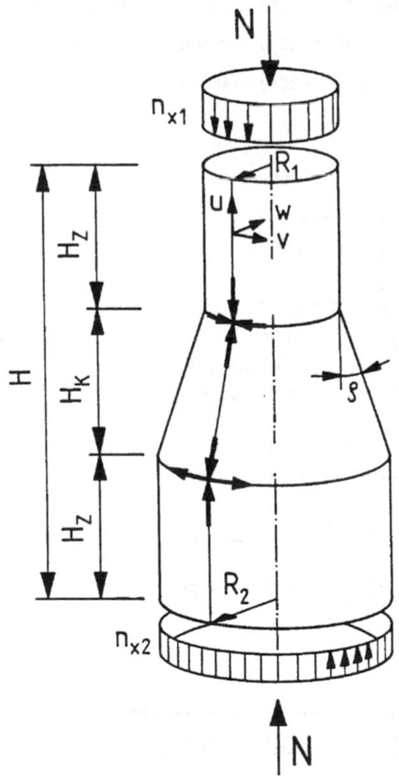

Geometry and Properties

Consider a structure, which is made up of three strakes, where the height H_Z of the cylinders and the height H_K of the cone is 2000 mm each (see fig. 1). The radius $R_1 = 1000$ mm of the top cylinder is fixed. Different slopes ρ of the cone (0°, 10°, 20°) give different radii R_2 of the bottom cylinder. The use of three wall thicknesses T = (1; 3; 10) mm gives R_1/T-ratios of 1000, 333 and 100. Although real chimneys have R/T-ratios well below 100, we choose those to investigate the purely elastic buckling of this type of structure in a first step.

For the same reason infinite linear material has been assumed for the shell wall (*Young*'s modulus being E = $2.1 \cdot 10^5$ N/mm^2 and *Poisson*'s ratio $\mu = 0.3$) and the initial shape of the structure was taken to be perfect.

The boundary conditions were 'classical', i.e. radial displacements w and bending moments m_x were set to zero along top and bottom edge. The support was at the bottom edge (u = 0), a line-load $n_{x,1}$ was applied to the top edge.

Figure 1.   Structure under investigation (taken from [8])

Numerical Procedure

For the solution of the geometrical nonlinear problem a known computer code was used, which has been described elsewhere [2], [6], [8]. It is based on mixed four-term ring elements, where a transfer matrix is calculated for each element. After determining the nonlinear load-deflection-path the lowest eigenvalue is sought by variation of the number N of the circumferential harmonics.

UNSTIFFENED STRUCTURE

At both edges of the cone, the discontinuity in the direction of the meridional membrane forces causes deviating forces, which act in radial direction. This results in inward or outward deflections of the intersection line, so that a certain area of both adjacent shells would act as compression- or tension-ring. In conventional strength design, this area would be referred to as effective width. As well, this action causes meridional bending moments along the edges of each of the adjacent shells. A recent paper reports these edge moments to lower the critical axial load significantly [6].

Figure 2 shows numerical results for the unstiffened structure. The ratios R_1/T plotted along the abscissa refer to the top cylinder, the bifurcation load n_{cr} on the ordinate is normalized by the critical load $n_{cr,z}$ of a straight cylinder. It can be seen, that compared to a cylinder ($\rho = 0°$) an angle ρ of about 10° is sufficient to reduce the buckling load by half.

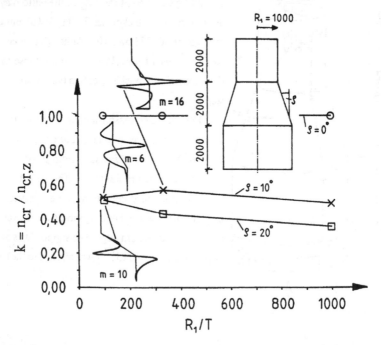

Figure 2.   Normalized bifurcation loads of the unstiffened structure with different slopes ρ of the cone (taken from [8]).

RINGSTIFFENED STRUCTURE

Geometry of Stiffeners

In the numerical studies annular plates of width b and thickness t were used, which had a fixed ratio b/t = 5. At both edges of the cone the rings were of same size. They were attached to the outside of the shell, so that the centers of their cross sections were r = R + b/2 apart from the axis of revolution.

For the wider edge of the cone the area A^* of an 'optimum ringstiffener' is defined such, that the radial deflection of the stiffener under the above deviating forces is equal to the expansion of the adjacent cylinder's shell wall under the applied axial load. For a given b^* the thickness of the optimum stiffener is given by

$$t^* = T \cdot \tan \rho \cdot (R \pm b^*/2)^2 / (R \cdot \mu \cdot b^*) , \qquad (1)$$

where '+' denotes a stiffener outside, and '-' a stiffener inside the shell wall. If the stiffener is centered with the shell wall, the term $b^*/2$ is zero.

In the following diagrams, the cross sectional area A of the actual stiffener is normalized by the area A^* of the optimum stiffener.

Variation of Stiffener Size

The influence of the actual stiffener area A on the critical load for different R_1/T and angles ρ is given in fig. 3.

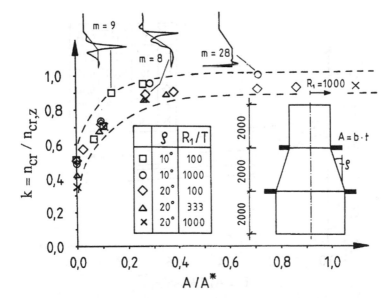

Figure 3.   Normalized critical load vs. normalized area A of stiffener, where A^* is the 'optimum stiffener' in strength design (taken from [8]).

Two important features can be drawn from fig. 3:

- no matter how stocky the ringstiffener is built, the critical load of the straight cylinder cannot be regained;

- a stiffener of 30 % A* is sufficient to provide some 90 % of the possible critical load.

Variation of Stiffener Position

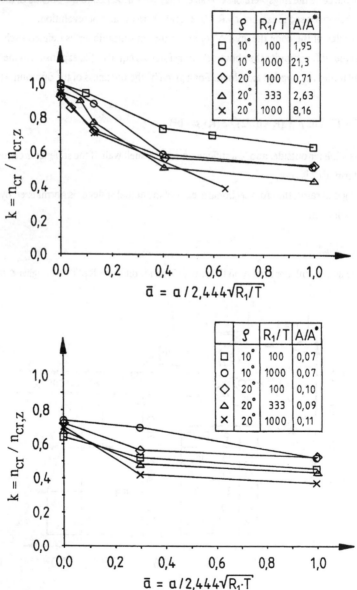

Figures 4. Influence of a distance 'a' between stiffener and edge (taken from [8])
a) stocky stiffeners, $21 \geq A/A^* \geq 0.7$
b) light stiffeners, $0.11 \geq A/A^* \geq 0.07$

The more the ring is shifted from the edge into the adjacent cylindrical strake, the more it will loose its effectivity, since the deviating forces first have to be transferred from the edge where they origin, through the weak shell, before they can be taken by the stiffener.

In order to normalize the distance 'a' between edge and center of the ring we use the bending half-wavelength of the top cylinder, so that

$$a = a \cdot \sqrt{\sqrt{(3-3\mu^2)}} / (\pi \sqrt{(R_1 T)}) \approx a / (2.444 \sqrt{(R_1 T)}). \qquad (2)$$

The following conclusions can be drawn from fig.s 4:

-   Even big stiffeners (over 100 % A^*) have lost their powers, if their distance from the cone's edge is of the order of the bending half-wavelength;
-   Even small stiffeners (about 10 % A^*) can increase the buckling load, if their distance is not wider than 50 % of the bending half-wavelength.

REMARKS ON IMPERFECTIONS

Unless the stability of a shell structure has not been investigated for different shapes and depths of initial imperfections (which we had not), it is hardly possible to exactly determine the imperfection sensitivity. But interpretation of our results might give a base for a good estimate.

From the point of view that an 'optimum structure' should transfer the applied loads directly to the support (compare [7]), a cylinder-cone-cylinder intersection is non-optimum, since it has global axisymmetric imperfections being $0 \leq w_0 \leq (R_2 - R_1)/2$ deep.

Some researchers worked on local axisymmetric imperfections (e.g. [3], [5], [9]), since these seem to be relevant for welded shell structures [4], [1].

In our study we varied the height H_K of the cone for a shell $R_1/T = 1000$. If the cone is very short, it is only a local imperfection in a straight cylinder. If it is long, it strongly affects the axial stiffness of the structure. In fig. 5 the normalized critical load is plotted against the height of the cone, which is normalized by the bending half-wavelength $2.444 \sqrt{(R_1 T)}$. The value $k = 0.82$ on the left hand side of the diagram is for $H_K = 1$ mm, the next one ($k = 0.53$) is for $H_K = 15$ mm.

Following the abscissa from the straight cylinder ($H_K = 0$) the first imperfection ($H_K = 1$ mm) has a depth of less than $w_0/T = 0.18$ and causes a drop of the critical load of 18 %. The next imperfection ($H_K = 15$ mm) has a depth of less than $w_0/T = 2.7$ and causes a loss of 'only' 47 %. So the effect of the imperfection decreases with increasing depth, and the bottom level $k \approx 0.4$ for $R1/T = 1000$ is reached if the height of the cone is in the magnitude of the bending half-wavelength.

Other researchers obtained even lower results for $w_0/T = 1$ (e.g. [9]). This might be contributed to the fact, that local axisymmetric imperfections are generally considered to be symmetrical to a plane perpendicular to the axis of revolution, while the structure in this paper is not. Thus, the radius is varied in two steps by 'general imperfections' - e.g enlarged to tip of the bulge, and reduced back to the original size. A cylinder-cone-cylinder intersection does the first step only, which might result in a smaller loss of global axial stiffness, and consecutively slightly higher bifurcation loads.

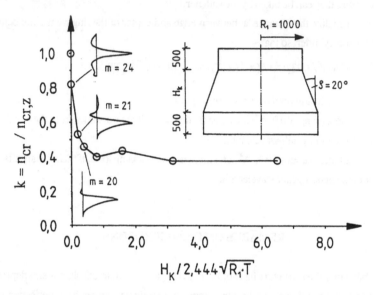

Figure 5.   From local to global imperfection - variation of the height of the cone (taken from [8]).

CONCLUSIONS

The structural behaviour of cylinder-cone-cylinder intersections is strongly affected by the deviating forces at the edges of the cone, and the meridional bending moments in the adjacent parts of the shell.

- Height and slope of the cone and the wall thickness of the shell decide, whether global or local buckling will be relevant. Hints can be taken from fig.s 2 and 5, although no general rule could be found with the present results.
- Compared to a straight cylinder of same radius, the bifurcation loads of the unstiffened structure drop to some 40 %, even if local buckling of the single parts of the shell will be relevant.
- Even if there are heavy ring stiffeners attached to the edges of the cone, the buckling strength of a straight cylinder cannot be regained.
- Stiffeners of only 30 % of the 'optimum size' are sufficient to supply some 90 % of the bifurcation load of a straight cylinder.

FUTURE WORK

The present results are considered to have been a first step towards the investigation of cylinder-cone-cylinder intersections. Further steps would be the investigation of ratios R/T down to 50, where plastic failure of the edges will interact with stability. If we found a sponsor, we would like to perform some large-scale experiments as well.

In general, we would like to work towards a better understanding of imperfections. Problems like collecting imperfection data by in-situ measurements on big welded structures, classification of the relevant types of imperfection, investigation of the effect of irregular imperfections, finding equivalent regular imperfections, are tasks, which seem to need the coordinated efforts of the community of shell stability.

REFERENCES

1. *Clarke, M.J., Rotter, J.M.*, A Technique for Measurement of Imperfections in Prototype Silos and Tanks. Research Report No. R565, School of Civil and Mining Engineering, University of Sydney, 1988.

2. *Esslinger, M., Geier, B., Wendt, U.*, Calculation of stresses and deformations of shells of revolution in the elasto-plastic range. Stahlbau, 1984, **53**, 17-25. (in German)

3. *Haefner, L.*, Influence of a circumferential weld on the stability and ultimate load of a cylinder under axial load. Ph.D., University of Stuttgart, 1982. (in German)

4. *Jetteur, P., Frey, F.*, Parametric study of a collapsed water tank. Proc., ECCS Colloquium on Stability of Plate and Shell Structures, Ghent University, 6-8 April, pp. 437-442.

5. *Juercke, R.K., Kraetzig, W.B., Wittek, U.*, Circular cylindrical shells with bulge imperfections. Stahlbau, 1983, **52**, 241-244. (in German)

6. *Knoedel, P., Maierhoefer, D.*, On the Stability of Cylindrical Shells Subjected to Axial Compressive Loads and Edge Moments. Stahlbau, 1989, **58**, 81-86. (in German)

7. *Knoedel, P.*, On Optimal Structures in Engineering. In *Eschenauer, H.A., Mattheck, C., Olhoff, N.* (Eds.), Engineering Optimization in Design Processes. Proc. of the Int. Conf., Karlsruhe Nuclear Research Center, Germany, September 3-4, 1990. Lecture Notes in Engineering 63, Springer-Verlag, Berlin, Heidelberg, 1991, pp. 317-323.

8. *Knoedel, P., Thiel, A.*, Stability of thin-walled cylinder cone intersections subjected to axial load. Stahlbau, 1991, **60**, 139-146. (in German)

9. *Rotter, J.M., Zhang, Q.*, The Strengthening Effect of Stored Solids on the Buckling of Cylindrical Steel Silos. Proc., 3rd Int. Conf. on Bulk Materials, Storage, Handling and Transportation. Newcastle 27-29 June 1989, Inst. of Engineers, Australia.

10. *Thiel, A.*, Investigations on the stability of ring-stiffened cylinder-cone-cylinder intersections under axial load. Diploma thesis with Prof. *F. Mang*, Versuchsanstalt fuer Stahl, Holz und Steine, University of Karlsruhe, 1989.

ELASTIC PLASTIC BUCKLING AT CONE-CYLINDER JUNCTIONS OF SILOS

RICHARD GREINER and ROBERT OFNER
Technical University of Graz,
Rechbauerstraße 12, A-8010 Graz, Austria

ABSTRACT

Elevated silos and tanks in steel usually consist of a cylindrical vessel and a conical hopper. One of the main areas of interest in design is the junction of the cylindrical and the conical structure where large circumferential compressive stresses may cause failure by instability. This paper describes the buckling behaviour of the cylindrical and the conical wall in the junction area and the overall stability of a stiffening ring located at the junction.

INTRODUCTION

The meridional tension forces of the hopper induce radial inward components in the cone-cylinder junction, which usually result in large circumferential compressive stresses in the adjoining shell wall and the stiffening ring, if any (Fig.1). Failures in the junction area may appear in various modes:
- plastic collapse,
- buckling of the shell wall,
- in-plane ring buckling,
- out-of-plane ring buckling.
Problems of this kind have been treated by Rotter in [1] for silos with ring stiffened cone-cylinder junctions presenting solutions for plastic collapse and out-of-plane ring buckling. This paper presents results [2] for the elastic and inelastic buckling of the shell walls in the case of unstiffened or ring-stiffened junctions as well as results for the overall, in-plane-stability of the stiffening ring (Fig.2).

ELASTIC BUCKLING OF THE SHELL WALLS

In the shell walls immediately adjacent to the cone-cylinder

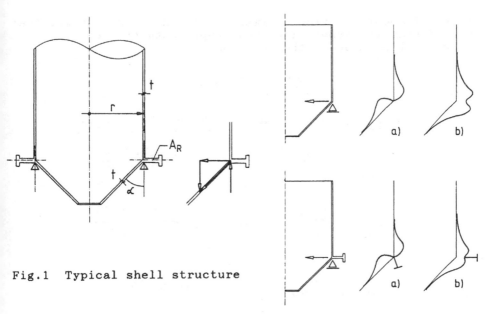

Fig.1 Typical shell structure

Fig.2 Buckling modes of stiffened
and unstiffened junction

junction buckling may occur according to the circumferential
compressive stresses induced there by the radial inward
components of the hopper forces. The elastic buckling strength
partly depends on the type of loading, partly on the presence
of a stiffening ring. Fig.3 gives four examples of such
structures of equal shell geometry with unstiffened and
stiffened junctions and different loading. The buckling
strength is defined in terms of the ring load pw, which is
substituting the radial inward component of the actual load
acting merely on the shell walls.

   According to the different load cases,depending on
whether the hopper induces meridional forces only or whether
internal pressure is applied to the vessel, the buckling
strength pw varies significantly. This is due to the variable
width of the compression zone adjacent to the junction caused
by the two different load cases.

   In the case of ring-stiffened junctions the buckling
stress of the shell walls is significantly influenced by the
restraining effect of the torsional rigidity of the ring.

   For reasons of easy and conservative standardisation of
the various stress distributions in the junction area the load
case pw was chosen to represent the compressive forces of the
hopper.

   Fig.4 illustrates elastic buckling loads pw and buckling
wave numbers m for structures with different radius to
thickness ratios 500 and 1000 as well as for varying cone
angles a. It may be observed that the buckling mode changes at

angles of about 20⁰ from a symmetrical mode to an asymmetrical one and that the symmetrical mode represents an overall in-plane-buckling mode of the junction.

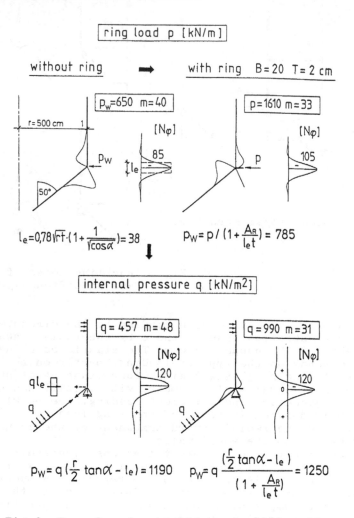

Fig.3   Examples for buckling of shell walls

The buckling loads given in Fig.4 are solutions of non-linear analyses by use of the BOSOR program [3]. The numerical results for the two r/t-ratios coincide in a large area when specific factors are chosen for the load p_w and the wave-number m.

The results in Fig.4 can also be used for ring stiffened shells, due to the fact that the two components p_w and p_R acting on shell wall and ring add up in simple form to the total load

$$p = p_w + p_R = p_w(1 + A_R/A_W).$$

A_R represents the area of the ring and $A_W = (l_e.t)$ is the effective area of the shell wall.

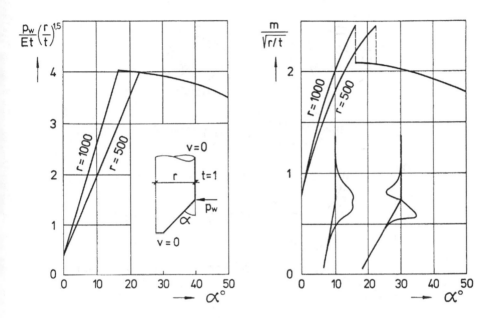

Fig.4   Elastic buckling loads and wave numbers of shell walls

In a current study elastic buckling loads of cone-cylinder junctions will be given more generally by way of a formula.

OVERALL STABILITY OF STIFFENING RINGS

In vessels of larger diameters rings are frequently placed at the cone-cylinder junctions in order to assist the shell structure in carrying the radial inward components of the hopper forces. Consequently the ring is subject to overall circumferential compression, which could possibly lead to failure by in-plane buckling. While authors in previous literature [4], [5] have treated the ring like a free ring with the buckling mode m = 2 or like a radially restrained ring, Rotter [1] indicates that in-plane ring buckling is prevented by the shell structure. The present paper reviews this overall failure mode in connection with the local buckling of the shell wall.
    In order to derive a simplified mechanical model of the in-plane-buckling behaviour the stiffening ring was treated as

a simple ring separated from the shell, supported by
tangential elastic restraint and subject to radial loads, the
restraint being considered as the membrane stiffness of the
shell structure in tangential direction. The buckling load
follows the classical solution for ring buckling and is given
by equation 1 in Fig. 5.

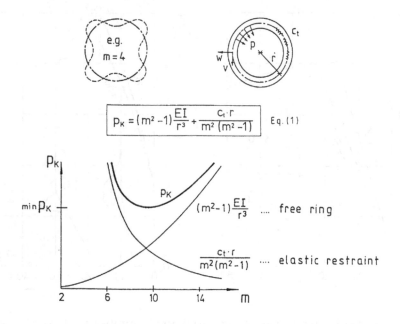

Fig. 5   Ring with tangential elastic restraint

The tangential stiffness moduli c_t of the shell structure have
been analysed for a ratio of $r/t = 500$, a cone angle of 45^0,
fixed boundary conditions of the shells and varying ring areas
(Fig.6). The classical buckling solution could be proved true
by comparison with ring buckling loads obtained by use of the
BOSOR program only in special cases, and then only
approximately. In these cases local buckling of the shell wall
had to be prevented in order to produce overall buckling modes
of the ring. Modeling overall ring buckling in terms of
equation 1 therefore is to be considered as a rough, but
conservative approximation.

In order to investigate whether a certain structure is likely
to fail by overall ring buckling or by local buckling of the
shell wall Fig.7 illustrates results of an example with a cone
angle of 45^0. Increasing ring areas bring about rising
buckling load p: in case of overall ring buckling this is due
to the increasing stiffness modulus c_t, in case of local shell
buckling the load increases as result of the reduction of the
compressive stresses in the wall. For practical values of
second moment of area of the ring section failure by overall

buckling should occur only with very high values of the ring area, which hardly fit to practical ring sections.

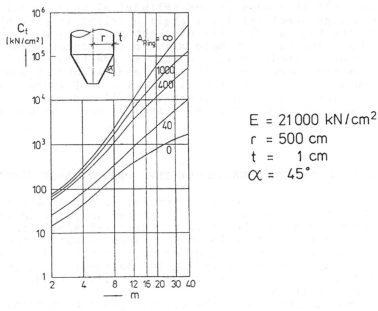

$$E = 21000 \text{ kN/cm}^2$$
$$r = 500 \text{ cm}$$
$$t = 1 \text{ cm}$$
$$\alpha = 45°$$

Fig. 6   Tangential membrane stiffness of shell structures

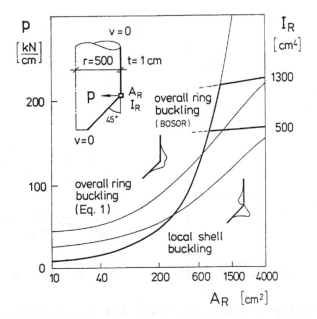

Fig. 7   Ring stiffened shell - local shell buckling and overall ring buckling

In case of smaller ring areas the buckling strength is
governed by the local shell buckling mode. This behaviour
corresponds to the assumption of the example in Fig.7, i.e.
constant wall thickness of the cylindrical and conical shell.
If, in design practice, local shell buckling is prevented by
reinforcing the wall locally adjacent to the ring, failure in
an overall ring buckling mode might become determining.

The above study indicates that in practical cases overall
ring buckling cannot be excluded in general and it should
conservatively be recommended to check this failure mode too.
In particular shell structures with small tangential restraint
- like open-topped cylindrical vessels - could be prone to
ring buckling.

ELASTIC-PLASTIC BUCKLING OF THE CONE-CYLINDER JUNCTION

The above unstiffened shell structures with ratios r/t = 500
and 1000 have been studied in the inelastic buckling range,
assuming material of the yield strength 240 N/mm². The
elastic-plastic buckling loads, obtained by use of the BOSOR 5
program, are given in Fig.8 a. as well as the plastic limit
loads (collapse loads) according to the study of Rotter [1].
The loads pw are related to the plastic limit load belonging
to the cone angle of 0⁰.

A smooth approach of the elastic-plastic buckling loads
towards the plastic limit loads can be seen when the cone

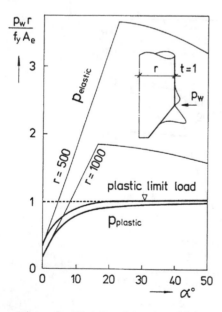 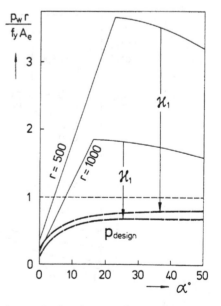

Fig. 8 Inelastic buckling loads and design values for wall
buckling

angle increases. Even in the case of very thin-walled structures, e.g. r/t = 1000, the plastic limit load is closely reached for larger cone angles commonly used in practice.

However, these results are based on the assumption of geometrically perfect shell structures. That means, that the values of buckling strength have to be reduced for design purposes taking into account the effect of imperfections. Fig. 8 b. illustrates this reduction when the "knock-down factor" $\varkappa_1$ according to the DIN 18800/4 [6] is applied to the elastic buckling stress caused by p_w. In the absence of test results relevant to this specific problem the reduced values may be considered as design loads predicting a lower bound of local buckling of the shell wall.

The load carrying capacity of the ring stiffened structure in the plastic range is demonstrated by Fig.9. The shell structure fails by plastic buckling and the ring by plastic collapse. According to the condition of equal strain in the junction the total strength of the structure may be obtained by adding up the two components of shell and ring. The same results are obtained by BOSOR 5 analyses too.

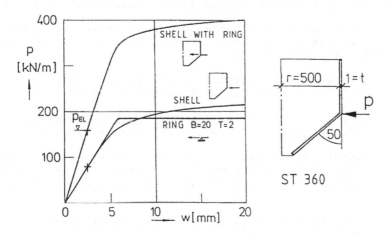

Fig. 9 Plastic buckling of ring stiffened structure

CONCLUSIONS

The shell wall adjacent to the cone-cylinder junction of elevated tanks and silos is subject to considerable compressive stresses, whether with unstiffened or ring stiffened junctions. These shell walls may fail locally by elastic or inelastic buckling depending on the ratio r/t and the cone angle. The results of this paper indicate that practical cone angles are likely to bring about plastic buckling which yields buckling loads of, or close to, the plastic collapse strength. Since these buckling loads correspond to perfect shells, design stresses should take into account imperfection effects by appropriate reduction factors.

In the absence of reduction factors based on proper test
results the knock-down-factor of the DIN 18800/4 may be
considered as a lower bound for design purposes.
In case of stiffening the cone-cylinder junctions by
rings the question of in-plane stability arises. It can be
demonstrated that such rings are efficiently restrained by the
stiffness of the shell structure. The results of this paper
indicate that in case of walls of constant thickness failure
by in-plane-ring buckling is unlikely to occur since local
buckling of the shell wall governs the problem. However, if
this local buckling can be prevented by purposeful design of
the shell wall, it is recommended to check these rings with
respect to overall in-plane buckling too.

REFERENCES

1. Rotter, I.M., The buckling and plastic collapse of ring
   stiffeners at cone/cylinder junctions,
   Proceed. of Internat.Colloquium 6.-8.April 1987, Ghent.

2. Ofner, R., Zum Tragverhalten des ringversteiften Zylinder-
   Konus Überganges von Behältern aus Stahl, Diplomarbeit
   TU Graz, November 1988.

3. BOSOR, Program for buckling of shells of revolution,
   D.Bushnell, Lockheed Palo Alto Research Laboratory

4. Schwaigerer, S., Festigkeitsberechnung im Dampfkessel-,
   Behälter- und Rohrleitungsbau, Springer Verlag, 1978

5. Kollbrunner, C.F., Haueter, O., Stabilität des Fußrings von
   Rippenkuppeldächern stehender Stahltanks, Mitteilung über
   Forschung und Konstruktion im Stahlbau, C.Zoschkke, 1949,
   H.7.

6. DIN 18800, Part 4: Stahlbauten, Stabilitätsfälle,
   Schalenbeulen, Nov.1990

BUCKLING OF CYLINDRICAL AND CONICAL SHELLS UNDER CONCENTRATED LOADING

EKKEHARD RAMM, NORBERT BÜCHTER
Institut für Baustatik, University of Stuttgart
Pfaffenwaldring 7, D-7000 Stuttgart, Germany

ABSTRACT

The bulk of all investigations with respect to buckling and ultimate load behavior of shells is devoted towards uniform loading conditions like constant axial load or lateral pressure. If non-uniform loads are present codes usually recommend to use an equivalent uniform loading condition instead. The present study describes a parametric study of cylindrical and conical shells under non-uniform and concentrated loading. In particular, cases are discussed where failure is caused by combined geometrical and material nonlinearity, like cylindrical shells under partial axial or external pressure load and under axisymmetric "ring" loads or liquid-filled conical shells. The analytical results are compared to experimental values, allowing to give an additional insight in the structural response and to verify the numerical model. Furthermore, codes may be checked with respect to reliability and efficiency.

FINITE ELEMENT FORMULATION

Isoparametric finite shell elements are applied, in which initial and deformed configurations are interpolated on the same polynomial basis. The displacement model is derived by the so-called "degeneration" introducing the shell assumptions in a discrete way at the nodes [1]. They are formulated for arbitrarily large displacements and rotations [2], [3], although the present problems only exhibit moderately large rotations.

The material nonlinearity is based on classical theory of plasticity using von Mises yield condition via a layered model. It is important to note that the introduction of the so-called consistent algorithmic material tangent renders a dramatic improvement of the convergence rate [4], [5]. A path-following algorithm utilizes the concept of arc-length or displacement control. For the parametric study the computer program CARAT [6] is applied. In order to avoid any kind of locking phenomena different versions of hybrid-mixed

The paper is dedicated to Professor Dr.-Ing. Walter Wunderlich, University of Munich, on the occasion of his 60th anniversary.

shell elements are applicable; but in this study the conventional 8–node displacement model with uniformly reduced integration is applied.

In a nonlinear analysis the ultimate load is defined whenever the first maximum in a load–deflection curve is obtained.

CIRCULAR CYLINDRICAL SHELL UNDER CONCENTRATED LOADING

Preliminary Remarks

<u>Objective:</u> Since codes usually are not detailed enough for localized loadings, e.g. [7], [8], the following parametric study gives a better insight into the structural response, eventually also allowing a reliable and more economic design for concentrated load conditions.

In general, codes recommend for partial loading to use a uniform load case with the same load intensity. This concept is analytically and experimentally verified for elastic structures but may influence the safety factors when material yielding occurs. The equivalent load case pretends an extended membrane stress state not present in reality.

<u>Scope:</u> Short to medium length steel cylinders, simply supported on both sides, have been investigated under partial axial and pressure load as well as ring loads. In particular, those radius to thickness ratios are considered which show a strong interaction between geometrical and material failure. In order to obtain realistic failure loads, geometrical imperfections, equivalent to the first buckling mode, are assumed with a maximum amplitude t_v as defined in [7], [8]. No hardening is introduced.

Cylindrical Shell under Concentrated Axial Load

<u>Data; structural system:</u> Load as well as geometrical and material data of the simply supported shell of medium length are given in figure 1: The relative slenderness is slightly larger than the limit value $\lambda_s = 1.58$ of the elastic region as defined in [7]. The nonlinear analyses are based on a 45º–sector.

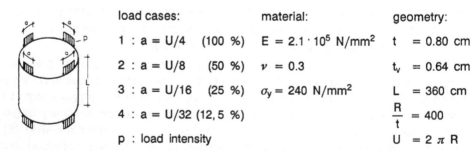

load cases:		material:	geometry:	
1 : a = U/4	(100 %)	$E = 2.1 \cdot 10^5$ N/mm²	t	= 0.80 cm
2 : a = U/8	(50 %)	$\nu = 0.3$	t_v	= 0.64 cm
3 : a = U/16	(25 %)	$\sigma_y = 240$ N/mm²	L	= 360 cm
4 : a = U/32	(12,5 %)		$\dfrac{R}{t}$	= 400
p : load intensity			U	= 2 π R

Figure 1: Geometrical, material and load data

<u>Results:</u> The cylinder buckles only in a region slightly wider than the loaded part, i.e. the deformations decay to the boundary of the 45º–sector (figure 2). Therefore, the results

can be transferred to shells stressed by less than 4 concentrated loads. The results are given in figure 3. The failure load normalized to the classical buckling load (uniform loading) is plotted versus the percentage of the loaded boundary.

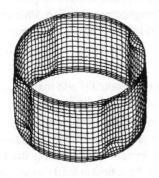

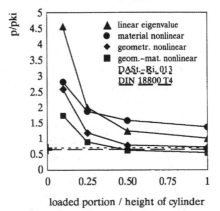

Figure 2: Displacement pattern at failure load (load case 3)

Figure 3: Failure loads for partial axial load

The following conclusions can be drawn:

• The concept of uniform equivalent axisymmetric loads as defined in [7], [8] is appropriate for load cases $a \geq U/16$.

• For smaller load regions the normalized ultimate load drastically increases; e.g. for $a = U/32$ the value is 1.67 of the normalized failure load for uniform loading.

• The bending stresses caused by the non-uniform loading play a minor role, also with respect to material failure.

Cylinder under Partial Axisymmetric Ring Load

Data, structural system: The simply supported shell falls into the category of a short cylinder; i.e. in [7] a factor $C = 1.07$ is needed for the classical buckling load. The material

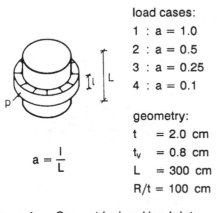

load cases:
1 : a = 1.0
2 : a = 0.5
3 : a = 0.25
4 : a = 0.1

geometry:
t   = 2.0 cm
t_v = 0.8 cm
L   = 300 cm
R/t = 100 cm

$$a = \frac{l}{L}$$

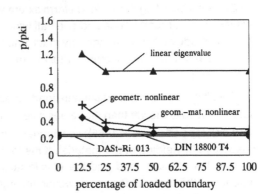

Figure 4:   Geometrical and load data

Figure 5:   Failure loads for partial ring load

data correspond to figure 1. The slenderness is $\bar{\lambda}_s = \sqrt{\sigma_y/a\,\sigma_{ki}} = 1.58$ which is the limit value of the elastic range. Again the first buckling mode with a maximum amplitude max $t_v = 0.4\,t$ is introduced as initial geometric imperfection.

Classical buckling theory for uniform external pressure results in a buckling mode with 7 waves in the circumference and one half wave in the longitudinal direction. Presumedly the failure mode does not change very much under partial loading. Therefore, the numerical analysis is restricted to a sector of 180/7 degree. A finite element mesh of 4 x 12 (hoop x axial) is used.

Results: The normalized ultimate load obtained by geometrically and materially non-linear analyses are lower than the values given in [7] ($\alpha = 0.7$) and [8] ($\lambda_{s\phi} = 1.32$, $\kappa_1 = 0.373$ corresponds to $\alpha = 0.65$); see figure 5. This reduction, not considered in [7], [8], is caused by the yielding of the material. If the load region is further reduced the difference between the ultimate load and the elastic failure load increases, indicating the major influence of the material failure. However, the intensity of the ultimate load increases; the uniform equivalent load drastically underestimates the load carrying capacity for narrow ring loads.

Cylinder under Partial External Pressure

Data, structural system: Two cylinders of medium length with two different radius-to-thickness ratios of 250 (cylinder A) and 100 (cylinder B) are investigated under four load cases (figure 6); for material data see figure 1. The shells are analysed with a sector of 180°.

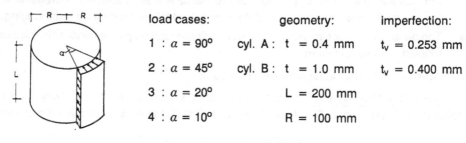

load cases:

1 : $a = 90°$

2 : $a = 45°$

3 : $a = 20°$

4 : $a = 10°$

geometry:

cyl. A : t = 0.4 mm

cyl. B : t = 1.0 mm

L = 200 mm

R = 100 mm

imperfection:

$t_v = 0.253$ mm

$t_v = 0.400$ mm

Figure 6: Geometrical and load data

Results: For both shells elastic and elastic-plastic geometrically nonlinear analyses have been performed with and without initial geometrical imperfections. The response is shown in figure 7 for load cases 1 and 4.

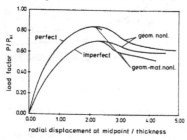

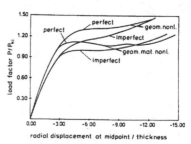

Figure 7 a: Cylinder A: load case 1

Cylinder A: load case 4

317

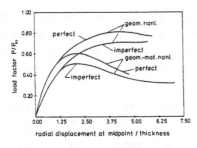

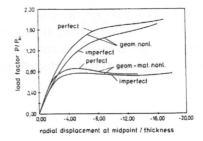

Figure 7 b: Cylinder B: load case 1   Cylinder B: load case 4

Failure loads are given in figure 8. Again the failure loads are normalized to the buckling load for uniform external pressure.

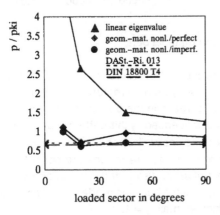

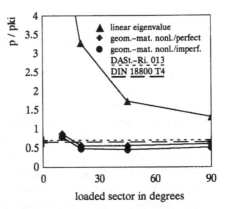

Figure 8:   Failure loads for partial external pressure (cylinder A: left, cylinder B: right)

Also here a distinct increase of the ultimate load can be recognized for small loaded sectors. The concept of a uniform equivalent load is well suited if the load sector is not small. However, again the design values given in [7], [8] are not conservative for the thick cylinder B (R/t = 100). The imperfection sensitivity is relatively small in this case and even reduces when the loaded part gets smaller. For cylinder A the ultimate load is only influenced by material failure in load cases 3 and 4 ($\alpha \leq 20°$). This is different for cylinder B. As expected the influence of material failure increases when the size of the loaded area is reduced.

LIQUID-FILLED CONICAL SHELLS

Parametric Study

Objective: Liquid-filled conical shells may buckle by localized failure due to compressive membrane stresses and material failure, mainly caused by tension hoop stresses. It is the purpose of this study to quantify this interaction for a selected example.

Data, structural system: In order to achieve the objective the level of the liquid – and therefore the ratio σ_ϕ/σ_x of hoop and meridional stresses – as well as the yield stress σ_y is varied (figure 9). All other data are kept constant. The specific weight γ of the liquid is used as load factor and increased within the nonlinear analysis.

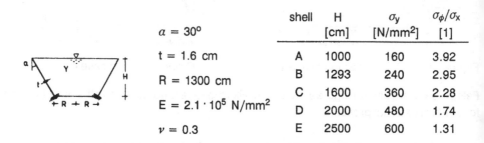

	shell	H [cm]	σ_y [N/mm²]	σ_ϕ/σ_x [1]
$a = 30°$				
$t = 1.6$ cm	A	1000	160	3.92
	B	1293	240	2.95
$R = 1300$ cm	C	1600	360	2.28
$E = 2.1 \cdot 10^5$ N/mm²	D	2000	480	1.74
$\nu = 0.3$	E	2500	600	1.31

Figure 9: Geometrical and material data

From cylindrical shells under axial load and internal pressure it is known that the axisymmetric deformation pattern changes into a nonsymmetric rhombic buckling mode with progressive loading, combined with a load decrease in the postcritical region [10]. However, elastic or elasto–plastic failure load is not influenced by this pattern change. With increasing ratio σ_ϕ/σ_x the effect even diminishes and finally disappears. Since this study concentrates on the prefailure range only axisymmetric failure modes are considered. 12 quadratic finite elements are introduced, non-uniformly distributed in meridional direction. Following the experience in [11] the first axisymmetric buckling mode is used as initial geometric imperfection with a maximum imperfection max $t_v = t$.

Results: In order to demonstrate the interaction of both nonlinearities the failure loads are presented in form of the well known $\kappa - \lambda -$ diagrams with the normalized load parameter $\kappa = \gamma/\gamma_y$ and the "slenderness parameter" $\lambda = \sqrt{\gamma_y/\gamma_e}$. γ is the failure load of the nonlinear analysis; γ_y and γ_e denote the plastic and elastic failure load. In this study γ_y is obtained using membrane stresses, however, γ_e is defined as the first maximum in a load deflection curve of a nonlinear elastic analysis. The results of the parametric study are given in figure 10. Related to a pure geometrically or pure materially nonlinear analysis the maximum reduction of the failure load obtained in a combined nonlinear analysis is 32 % for the perfect shell and 42 % for the imperfect shell. The results of these finite element analyses are also compared to the design values defined in [8] and [9], omitting the influence of the safety factors (figure 11).

The ultimate load capacity is best approximated by ECCS [9], in particular for shells A to C with high hoop tension forces; this is certainly due to the fact that the two–dimensional interaction is taken into account right from the beginning. In [8] the influence of the hoop tension forces is first of all neglected. In certain cases they may be considered later on by an increase of the factor κ, though not in these cases where the shell show on elephant-foot type of buckling. As a consequence, the ultimate load capacity of cones B to E is

319

underestimated, but of cone A overestimated. If the hoop tension is considered the ultimate load of cone D and E is obtained but it is exceeded for cones A to C.

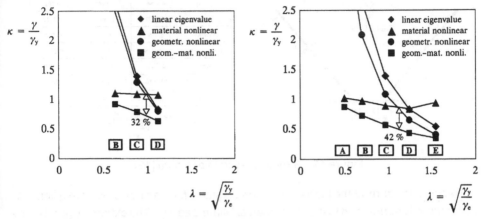

Figure 10: Failure loads for perfect (left) and imperfect (right) conical shell

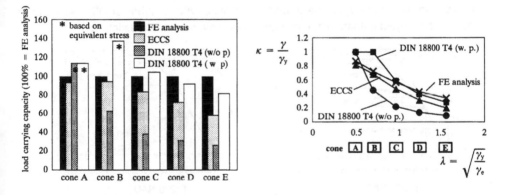

Figure 11: Comparison with codes (DIN 18800, T4 - ECCS)

Mercury-filled Conical Shells (Vandepitte [12])

Preliminary remarks: Utilizing the high specific weight of mercury, Vandepitte [12] succeeded to run experiments for steel model shells in the elasto-plastic region. Opposite to the previous study, this time the chosen material and geometry cause the shells to fail mainly by meridional compression. For three selected shells nonlinear analyses have been performed with and without imperfections.

Data, structural system: The structural system and the geometrical and material data are given in figure 12. Again, an axisymmetric model is used, although finally a nonsymmetric failure mode is likely, in particular in the post-critical region. The maximum geometrical imperfection was known from the experiment, however, its shape was not measured. Therefore, the imperfection was approximated by the first buckling mode of an initial eigenvalue analysis.

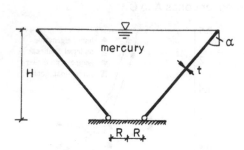

	F275SAD	F278SAD	F281SAD
R (mm)	100.60	100.00	99.92
t (mm)	0.5166	0.5843	0.6902
α (°)	39.9932	39.8981	39.8612
H (mm)	601.6	737.1	621.9
E (N/mm²)	174590.9	191245.3	199010.8
σ_F (N/mm²)	217.31	333.10	154.08
γ (N/mm³ * 10⁻⁶)	132.916	132.945	132.896
IMP (%) *	0.38	0.22	0.35

$$ * \quad \text{IMP (mm)} = \frac{\text{IMP(\%)} \cdot 3.6\sqrt{R \cdot t}/\cos \alpha}{100} $$

Figure 12: Structural system and data [12]

Results: In figure 13 the load – deflection diagrams for the three shells are given. The load factor γ is normalized to the experimental failure load γ_E. The extremely good correlation of the ultimate load in experiment and analysis (1 %, 1 % and 5 % difference) is certainly random; however, it is also an indication for the quality of tests and numerical simulation. In all three cases the imperfection sensitivity is rather small, see table in figure 13. The largest influence can be seen when both nonlinearities interact.

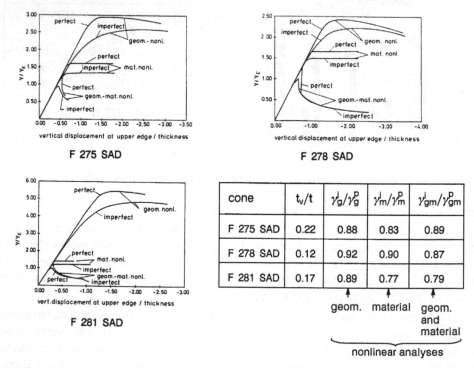

cone	t_v/t	γ_g^i/γ_g^p	γ_m^i/γ_m^p	$\gamma_{gm}^i/\gamma_{gm}^p$
F 275 SAD	0.22	0.88	0.83	0.89
F 278 SAD	0.12	0.92	0.90	0.87
F 281 SAD	0.17	0.89	0.77	0.79

geom.   material   geom. and material

nonlinear analyses

Figure 13: Load – deflection diagrams, failure loads (i = imperfect; p = perfect system)

In order to elucidate the interaction the results are added to the diagrams of the previous study, figure 14. Despite of different geometrical parameters and boundary conditions the

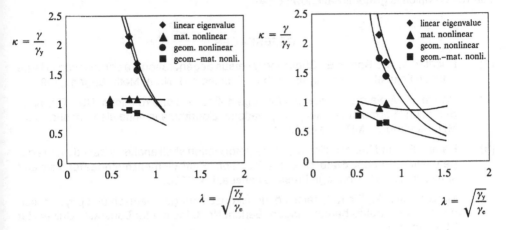

Figure 14: Failure loads for perfect (left) and imperfect (right) conical shell

results of the perfect shell coincide almost exactly with those values of the parametric study; the failure loads for the imperfect shell also show a good correlation. The maximum

cone	γ_{gm}^j/γ_m^j	$\gamma_{gm}^p/\gamma_{gm}^p$
F 275 SAD	0.71	0.83
F 278 SAD	0.67	0.78
F 281 SAD	0.82	0.90

interaction is obtained for cone F 278 SAD. The geometrical nonlinearity reduces the plastic limit load after all by 33 %.

CONCLUSIONS

The equivalent uniform load concept, often recommended in codes, is well suited as far as the load sectors are not too small; otherwise the approximation is too conservative. This also holds when the material failure is taken into account. For thick-walled cylindrical shells under external pressure (e.g. $R/t = 100$) the knock-down factor ought to be larger than 0.7 as already introduced by some codes.

The parametric study for liquid-filled conical shells again shows the important influence of the interaction of material and geometrical nonlinearities for structures of medium slenderness. The excellent correlation with experimental results also indicates that the quality of numerical simulation in ultimate load behavior of metal shells has reached a very good standard.

322

ACKNOWLEDGEMENT

The present study has been partially supported by a grant of the German Research Foundation (DFG) which is gratefully acknowledged.

REFERENCES

[1]   Büchter, N. and Ramm E., Shell theory versus degeneration – a comparison in large rotation finite element analysis. To be published in J. Num. Meth. Engng., 1991.

[2]   Ramm, E., A plate/shell element for large deflections and rotations. US – Germany Symp. on "Formulations and computational algorithms in finite element analysis", MIT, Cambridge, MIT-Press, 1977.

[3]   Ramm, E. and Matzenmiller, A., Large deformation shell analyses based on the degeneration concept. State-of-the-art texts on "Finite element methods for plate and shell structures", Pineridge Press, Swansea, UK, 1986.

[4]   Matzenmiller, A., Ein rationales Lösungskonzept für geometrisch und physikalisch nichtlineare Strukturberechnungen. Bericht Nr. 8, Institut für Baustatik, Universität Stuttgart, 1988.

[5]   Ramm, E. and Matzenmiller A., Computational aspects of elasto-plasticity in shell analysis. Int. Conf. on "Computational plasticity", Barcelona, Spain, Pineridge Press, Swansea, UK, 1987.

[6]   Stegmüller, H., Bletzinger, K.-U. and Kimmich, S., Eingabe- und Programmbeschreibung für das Programmsystem CARAT. Institut für Baustatik, Universität Stuttgart, 1988.

[7]   DASt-Richtlinie 013 "Beulsicherheitsnachweise für Schalen". 1980.

[8]   DIN 18800 Teil 4: Stahlbauten; Stabilitätsfälle, Schalenbeulen. November 1990.

[9]   ECCS-Recommendations R 4.6 "Buckling of shells", 4th edition. 1988.

[10]   Kollár, L. and Dulácska, E., Buckling of shells for engineers. Wiley & Sons, 1984.

[11]   Stegmüller, H., Grenzlastberechnungen flüssigkeitsgefüllter Schalen mit "degenerierten" Schalenelementen. Bericht Nr. 5, Institut für Baustatik, Universität Stuttgart, 1985.

[12]   Private communication Vandepitte-Schmidt-Bornscheuer on steel cones filled with mercury, October 1987.

BUCKLING OF CYLINDRICAL SHELLS UNDER LOCAL AXIAL LOADS

WERNER GUGGENBERGER
Technical University of Graz
Rechbauerstr. 12, A-8010 Graz, Austria

ABSTRACT

This paper deals with the nonlinear axial buckling behaviour of unstiffened and stiffened circular cylindrical shells with various local support constructions. Realistic buckling loads are calculated by Finite Element parametric studies. Geometric nonlinearity, elastoplastic material behaviour and different shapes of geometric imperfections are taken into account. For all structures considered classical buckling eigenvalues are computed, too. This should provide a starting point for simple design procedures using overall knock-down factors to allow for the nonlinear effects in an approximate manner.

INTRODUCTION

Large column-supported thin-walled steel silos are exposed to concentrated axial pressure above the column support regions. The cylindrical shell wall may be unstiffened or equipped with specific stiffening constructions, such as longitudinal stiffeners, additional ring stiffeners, local increase of the wall thickness or combinations of them (Fig. 1 and Fig. 3). If longitudinal stiffeners are provided the axial stress concentration moves upwards to the end of these stiffeners and may become even larger than in the case of the unstiffened shell (Fig. 2). Because of this unfavourable prebuckling behaviour a potential source of local buckling instability exists. An overview of possible buckling failure modes is shown in Fig. 1.

In literature only few recent studies of local loaded *unstiffened* cylindrical shells exist which take into account the nonlinear buckling behaviour, all of them using FE analyses [1–3]. Former authors treated the problem in a simplified manner by calculating classical buckling eigenvalues for axially constant prestress due to partial axial edge loading [4]. In design codes only the unstiffened cylindrical shell under uniform axial stress is adequately represented [7].

All numerical calculations of this study are performed with the FE program ABAQUS [6]. S4R5-shell elements with 7 thickness integration points are used in analyses including material nonlinearity. In all other cases biquadratic S8R5 elements were applied. The quasi-static nonlinear equation systems are solved by application of a path-following procedure.

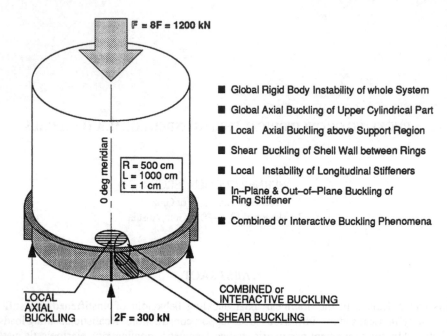

Figure 1. Overview of Local Buckling Instability Phenomena of Stiffened Cylindrical Shells under Concentrated Axial Support Loads

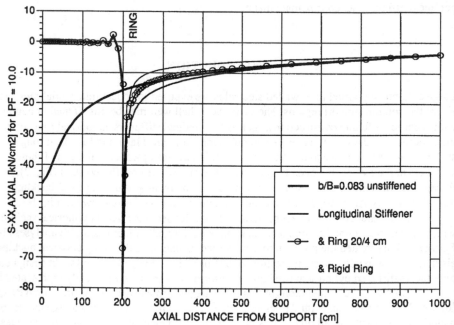

Figure 2. Axial Membrane Stress Distributions along Axial Line ´0 deg´ for various Support Constructions, LPF = 10.0, Linear Elastic

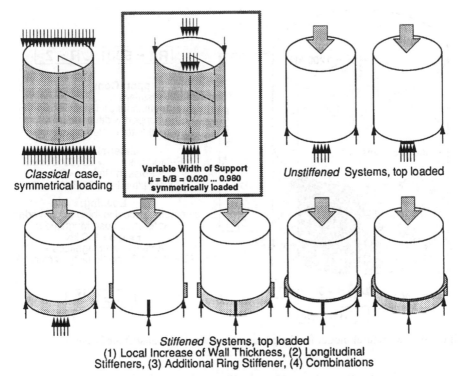

Figure 3.   Overview of Stiffened & Unstiffened Systems of axially loaded Cylindrical
Shells with Local Supports

UNSTIFFENED CYLINDRICAL SHELLS

Buckling analyses of the somewhat simpler model of the unstiffened shell with variable
support–width–ratios μ=b/B should assist the understanding of buckling phenomena of
stiffened cylindrical shells (Fig. 3). The following questions should be answered:
- How does buckling behaviour change when considering the variation of the support width:
  on the one end (μ=1) the case of the uniform axial load with its multiple bifurcations and
  high imperfection sensitivity, on the other end (μ=o) the limiting case of the point load ?
- Do *perfect systems* fail by snap–through or bifurcation buckling and what about the
  postbuckling behaviour ?
- How does *imperfection sensitivity* change with decreasing support–width–ratio, in the
  elastic as well as elastoplastic range ?
- What about the *real load bearing behaviour* when μ becomes small ? The interaction of
  plasticity with geometric imperfections and pure axial yielding are to be considered.

Modelling
Analysis parameters are summarized in Fig 4. This study was performed with R/t=500, L/R=2
and t=1 cm, considering symmetry in axial direction. The edges are classically supported. Ideal
Mises–elastoplasticity is assumed for nonlinear material behaviour.

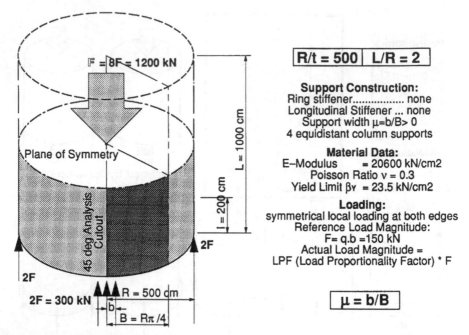

Figure 4.   Structural model for the parametric study of the unstiffened case

Nonlinear Buckling Behaviour of Perfect Elastic Systems

A summary of results can be found in Fig. 5, the critical stress ratio being the ratio of the actual critical stress $F_{cr}/(b*t)$ and the critical buckling stress $\sigma_{ci} = 0.605*E*t/R$. The limit stresses of the *perfect systems* are given by the curve GNL–MAX (maximum value of geometrically nonlinear analysis). Buckling failure occurs by snap–through but several distinct bifurcation points (3–4) exist slightly below the limit point. In any case, the situation of extreme multiple bifurcations of the uniformly loaded case vanishes completely.

The drastic drop of the perfect limit curve (GNL–MAX) in the right part of Fig. 5 may be interpreted as effect of load imperfection of the geometrically perfect structure. In the intermediate μ–range (μ = 0.25 to 0.75) the results are about 30–40% lower than for uniform loading. The shell behaves very flexibly. Particularly with μ≈0.5 large postbuckling deformations occur, accompanied by a large drop of load in the postbuckling range, while the GNL–MAX–value is somewhat higher than the average level of this range (Fig.6, curve A).

For narrow support widths – in the left part of Fig. 5 – the situation becomes clearer with respect to the failure mode: the shell fails by pure snap–through buckling, bifurcations vanish completely. The degree of instability (rank deficiency of the system stiffness matrix) is always 1 from the perfect limit point to the end of the observed postbuckling range. The perfect limit *stress* increases gradually and tends to infinity as μ approaches zero, the perfect limit *load* converging to a limiting value at the same time.

Buckling Eigenvalues

Buckling eigenvalues may be used as reference values for calculating design loads by applying overall knock–down factors according to design standards [5,7]. Such eigenvalues are given in Fig. 5 (curves BEV) due to partial edge loading: BEV–CLASS being based on the real

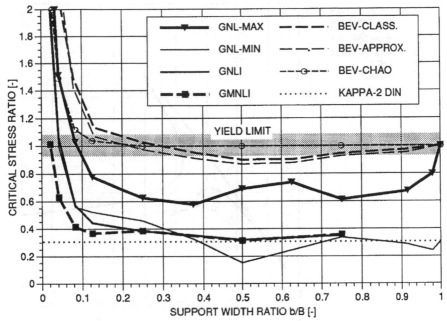

Figure 5.   Eigenvalue Buckling Stresses and Imperfect/Elastoplastic Buckling Stresses for variable support widths

prebuckling stresses and BEV–CHAO representing results for axially constant prestress [4].

Effect of Geometric Imperfections

A standard imperfection amplitude of $w_o/t = -1.0$ (inward directed prebuckle) has been chosen, approximately corresponding to [7]. Local imperfections of equal shape and amplitude (cubic spline–surfaces) have been applied in the range $\mu \leq 0.125$. For the larger μ–values (0.25; 0.5; 0.75) linear combinations of nonlinear eigenmodes have been used. Those were calculated by accompanying linear buckling eigenvalue analyses and refer to the above mentioned bifurcation and snap–through points. By application of this procedure all bifurcation points can be eliminated which is necessary to yield conservative buckling loads. Simple use of classical initial buckling eigenmodes would yield unfavourable (unsafe) results in this respect.

The limit stress of the imperfect shell is given in Fig. 5, curve GNLI (geometrically nonlinear analysis with imperfections). In the range $\mu > 0.125$ the buckling stresses are nearly constant and drop to a level close to the real buckling stress of the uniformly loaded shell (κ_2–value of DIN 18800/4). For narrow support widths ($\mu \leq 0.125$) the limit stresses increase and tend to infinity as μ approaches zero being considerably lower than in the perfect case. The load converges to a limiting value at the same time.

Generally it can be said that the overall decrease of the buckling stress is caused by two effects: the effect of the prebuckling deformations of the shell due to the partial edge loading on the one hand ('load imperfection') and the effect of the geometric imperfections on the other. As already mentioned the total drop is nearly the same as for the uniformly loaded case, about half of it caused by the effect of partial loading. In consequence the real *imperfection sensitivity* is only about half of that for uniform loading. If the support region

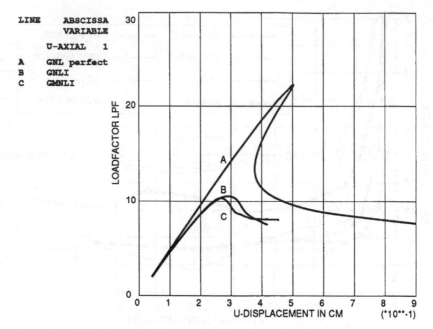

Figure 6.   Load–Displacement Diagram for *Unstiffened* System μ = b/B = 0.500, showing
Effects of Elastoplasticity and Imperfections

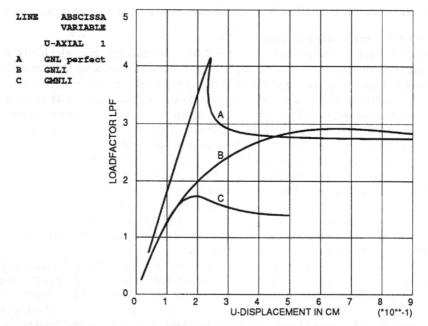

Figure 7.   Load–Displacement Diagram for *Unstiffened* System μ = b/B = 0.042, showing
Effects of Elastoplasticity and Imperfections

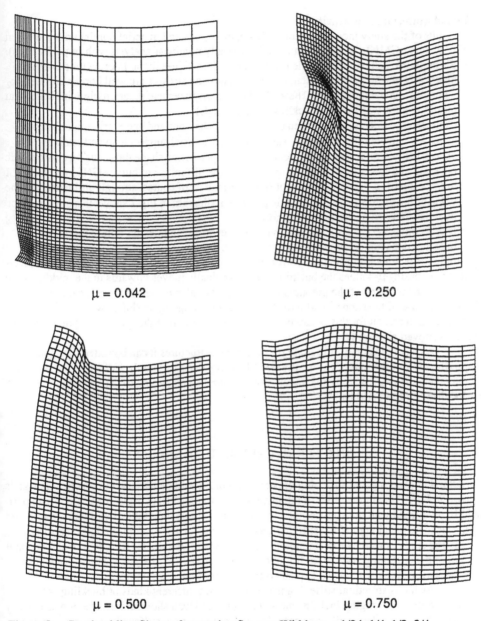

μ = 0.042

μ = 0.250

μ = 0.500

μ = 0.750

Figure 8.   Postbuckling Shapes for varying Support Widths μ = 1/24, 1/4, 1/2, 3/4
of Imperfect Elastoplastic Systems (GMNLI)

becomes increasingly localized ($\mu \to 0$) the imperfection sensitivity further decreases, which is
the result of high local bending in the prebuckling state.

Effect of Material Nonlinearity

The results of the study including geometric imperfections and material nonlinearity are given in Fig. 5, curve GMNLI (geometrically and materially nonlinear analysis with imperfections).

- Elastoplastic behaviour has little effect on buckling values in the intermediate μ–range; practically elastic buckling occurs. The real buckling stresses are well correlating with κ_2 – knock down factors of DIN 18800/4. The load displacement diagram Fig.6 indicates that plasticity hardly affects the buckling load (curve C)
- Considerable interaction between elastoplastic behaviour and geometric imperfections arises for narrow support widths, say $\mu \leq 0.125$. The axial support load acts radially excentrically with respect to the center of the imperfection shape thus causing elastoplastic buckling at a lower level. This is indicated by the difference of curves GNLI and GMNLI in the left part of Fig. 5. The drop of the limit load is, of course, strongly influenced by the value of the yield stress. The load displacement diagram in Fig.7 shows this strong effect by the difference of the curves B and C.
- If μ becomes very small pure axial yielding becomes the relevant failure mode and buckling does not occur any longer ($\mu \leq 0.021$).

Examples of postbuckling shapes for different μ–values are presented in Fig. 8. It can be seen that for increasing μ–values the buckling region gradually moves upwards to the middle of the shell since the maxima of the prebuckling stresses have also moved to these locations. With further increase of the support width ($\mu > 0.500$) the buckling region shifts away from the center line. The main reason for this behaviour is that the maximum of the prebuckling deformations has now moved sidewards.

As a conclusion of this study on unstiffened structures it can be stated that good approximation of the numerical results GMNLI is obtained by simply reducing the classical eigenvalues BEV–CLASS by the knock–down factors κ_2 according to [7] – even in the range of localized support widths.

STIFFENED CYLINDRICAL SHELLS

Structural reinforcing and stiffening of point–supported cylinders as used in practical design is shown in Fig. 3, i.e. in the order of the increasing stiffening effect: rigid constraint of support width, reinforced wall thickness of the bottom course, single longitudinal stiffeners, reinforced wall thickness plus longitudinal stiffeners, as well as the former cases plus additional stiffening rings. The study considers top loading, the shell geometry R=5m, L=10m, t=1cm and an axial length of the stiffening zone l=2m for all cases. The numerical analysis was performed in a similar way as has already been explained for the unstiffened structures.

As far as structural stiffening is concerned two different kinds of buckling behaviour could be observed: longitudinal stiffeners and reinforced wall show an imperfection sensitivity nearly as high as in the unstiffened case. On the contrary, ring–stiffened structures respond to imperfections rather moderately as result of the fact, that adequately sized stiffening members alone are able to carry the load without assistance of the shell structure above. This is demonstrated in Fig. 9 by the smooth postbuckling behaviour connected with large radial deformations, nearly independent of the assumed imperfections. A typical elastoplastic buckling pattern of the imperfect case is shown in Fig. 10.

The results of the numerical study are summarized in Fig. 11 and Fig. 12. The real buckling loads of ring–stiffened systems with *rigid* rings yield values as high as that for

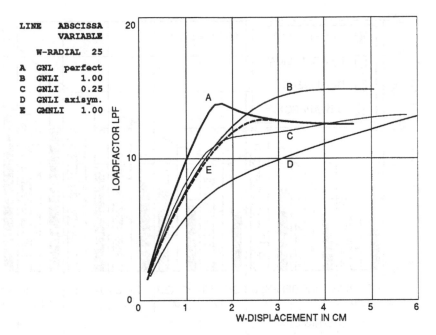

LINE	ABSCISSA VARIABLE	
	W-RADIAL	25
A	GNL	perfect
B	GNLI	1.00
C	GNLI	0.25
D	GNLI axisym.	
E	GMNLI	1.00

Figure 9.   Load–Displacement Diagram of *Stiffened* System (Elastic Ring 20/4 cm), showing Effects of Elastoplasticity and various Imperfections

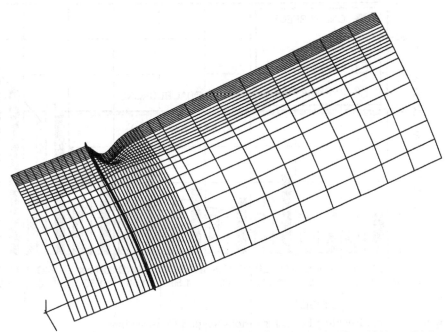

Figure 10.   Postbuckling Shape of Ring–Stiffened System (Elastic Ring 20/4 cm, Wall thickness ratio t_{bot}/t_{top} = 1.50), GMNLI

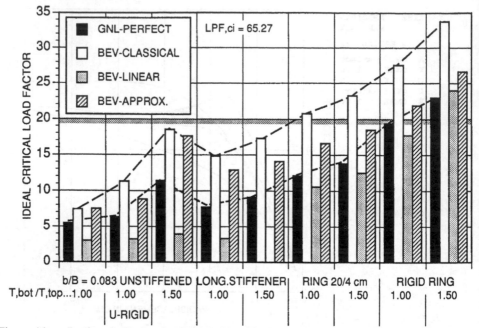

Figure 11. *Perfect & Eigenvalue* Buckling Loads for various Support Constructions

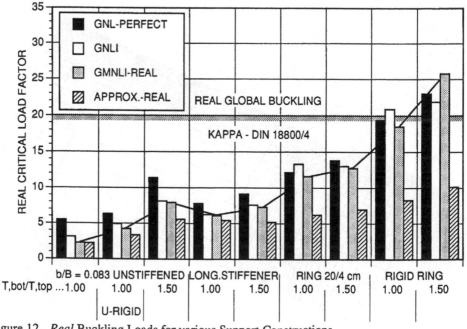

Figure 12. *Real* Buckling Loads for various Support Constructions

uniformly loaded unstiffened shells. That means that the point supports do not reduce the carrying capacity of the structure. In this case the upper unstiffened shell responds to the local stress concentration above the supports in a similar way as a shell subject to local foundation settlement. For elastic rings the load level is considerably lower since additional radial deformations of the ring cause unfavourable prebuckling deformations in the shell. In the case of longitudinal stiffeners and reinforced wall the load carrying capacity shows a further decrease. However, wall reinforcement alone can be very efficient and can raise the buckling load to a level similar or even higher than in longitudinally stiffened cases; this on the condition that the reinforcement is strong enough to shift the buckle above the reinforced course.

It could be shown (Fig. 12) that a fairly good approximation of real buckling loads can be obtained by reducing the classical eigenvalues (Fig. 11) by κ_2–factors in the cases of wall reinforcement and longitudinal stiffeners. However, applying this procedure to ring–stiffened cases results in conservative buckling loads of about 50% or less of the real values.

REFERENCES

1. Büchter, N., Ramm, E., Stabilitäts–und Traglastberechnungen von Zylinder– und Kegel-schalen bei konzentrierten Beanspruchungen. In Festschrift Richard Schardt, THD Schriftenreihe Wissenschaft und Technik, 1990, **51**, 277–97.

2. Rotter, J.M., Teng J.G., A Study of Buckling in Column–Supported Cylinders. Depart-ment of Civil Engineering, University of Edinburgh, U.K., 1990.

3. Samuelson, L.Å, Buckling of Cylindrical Shells under Axial Compression and subjected to Localized Loads. In Post–Buckling of Elastic Structures, ed. J. Szabó, Proceedings of the EuroMech Colloquium No. 200, Mátrafüred, Hungary, 5–7 Oct.1985, Elsevier, 1986.

4. Hoff, N.J., Chao, C.C., Madsen, W.A., Buckling of a Thin–Walled Circular Cylindrical Shell Heated Along an Axial Strip. J. Appl. Mech., June 1964, 253–58.

5. Schmidt, H., Krysik, R., Beulsicherheitsnachweis für baupraktische stählerne Rotations-schalen mit beliebiger Meridiangeometrie – mit oder ohne Versuche ? In Festschrift Richard Schardt, Inst. f. Statik, TU Braunschweig, Mai 1988, pp. 271–88.

6. ABAQUS Theory Manual and Users Manual, Version 4.8. Hibbit, Karlsson & Sorensen Inc., Providence, Rhode Island, 1989.

7. DIN 18800, Part 4: Steel structures, stability, buckling of shells. November 1990.

BUCKLING IN THIN ELASTIC CYLINDERS ON COLUMN SUPPORTS

J. MICHAEL ROTTER[1], JIN-GUANG TENG[2] and HONG-YU LI[3]

[1,3]Department of Civil Engineering and Building Science,
University of Edinburgh, Scotland, U.K.

[2]Department of Civil and Systems Engineering,
James Cook University of North Queensland,
Townsville, Q 4811, Australia.

ABSTRACT

Elevated silos and tanks are often supported on a number of columns. The discrete supports of the columns induce high stresses adjacent to the column terminations. In particular, very high meridional compressive stresses arise above the column termination and these can lead to buckling of the shell at a load much lower than that for a uniformly supported shell. This paper presents a brief summary of recent theoretical work on the elastic buckling strength of column-supported cylinders. Elastic calculations are relevant provided the shell is of typical practical thickness ($500 < R/t < 2000$).

No previous study is known of the buckling of column-supported cylinders. Very few studies have examined the buckling behaviour of shells under circumferentially varying axial loads in any pattern. Previous studies were almost all confined to linear bifurcation analyses of perfect elastic shells under harmonically varying loads, and it is difficult to draw design recommendations from them. The present investigation has examined many of the parameters which influence the buckling strength.

INTRODUCTION

Elevated silos and tanks are often supported on a number of columns (Fig. 1). The discrete supports of the columns induce high stresses near the column terminations. High local meridional compressive stresses arise directly above the column termination and these can lead to buckling of the shell at a load much lower than that for a uniformly supported shell.

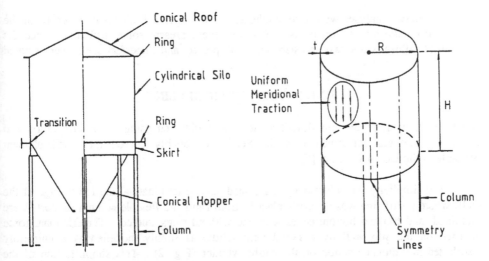

Figure 1. Typical column-supported storage structure

Figure 2. Column-supported cylinder

Most such shells are so thin that the buckling behaviour is entirely elastic.

The last few decades have seen an extensive research effort into shell buckling problems of many types, but only a few studies have examined the buckling behaviour of cylinders under circumferentially varying axial loads [e.g. 1,2,3]. These papers describe only linear bifurcation analyses of perfect shells, and are difficult to apply to practical imperfect shells.

Accurate predictions of the buckling or collapse loads of column-supported shells require a large deflection analysis which can take into account the possible geometric imperfections found in prototype structures [4,5], as well as the effect of local yielding on the buckling strength. However, before such accurate predictions can be understood, current knowledge of the linear bifurcation behaviour must be extended. A realistic model of a column-supported silo (Fig. 1) may need to account for interactions between the cylindrical shell, the transition ring and other shell segments.

This paper presents a brief summary of many calculations of elastic perfect and imperfect thin cylinders (Fig. 2) directly supported on columns. Finite element non-linear and bifurcation analyses were obtained using the LUSAS system [6]. Details of most of the calculations made to date can be found elsewhere [7,8,9,10].

The shell buckling strengths are defined in terms of mean stress above each column, related to the classical elastic critical stress for uniform axial compression. The variations of this mean buckling stress with column width, radius-to-thickness ratio, cylinder height, boundary conditions, and number of columns have all been examined [7,8,9]. Some of these effects are noted here.

Nonlinear analyses were also conducted, to try to find limit loads which might be different from the bifurcation loads. However, these present some difficulties since the criterion of failure is not easy to establish. Both perfect and imperfect shells were studied [10].

FINITE ELEMENT MODELLING

The finite element predictions described here were obtained using linear bifurcation and non-linear analyses performed with the Semi-Loof doubly curved thin shell element available in the LUSAS system [6].

For simplicity the columns were assumed here to terminate at the lower edge of the shell, where a stiff ring was normally placed. Both rigid and flexible column supports were examined, and top and bottom edges with and without rings studied. The frictional force imposed on the silo wall by a stored bulk solid was simply modelled as a uniformly distributed meridional traction on the entire cylinder (Fig. 2). This simplification of the real loading permits a conservative generalisation of the results.

Details of the finite element meshes, boundary conditions and exploitation of symmetry may be found elsewhere [8,9]. Where imperfect shells were studied [7,10], an artificial local axisymmetric imperfection was inserted at a critical point, to represent a weld depression. The assumed form of the depression, taken from earlier studies [11] was precisely modelled as a locally doubly curved shell.

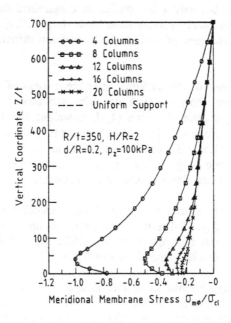

Figure 3. Meridional membrane stresses above column centreline (rigid support)

PREBUCKLING STRESS DISTRIBUTION

In perfect shells, the buckling behaviour is found to depend mostly on the meridional membrane stress distribution. The stress distribution above the centreline of the column is shown in Fig. 3 for rigid supports. A treatment of the support as flexible leads to higher local stresses, which later lead to lower buckling strengths. The region of high vertical compressive stress is quite limited, and comparable in size with typical buckles. The buckle forms in a zone of rapidly varying stress. Thus, any factors which change the rate of stress dispersal in the cylinder (radius-to-thickness ratio, ring stiffeners etc) lead to significantly changed buckling strengths.

In imperfect shells with deep imperfections, the circumferential stresses become locally large and compressive, leading to significantly reduced calculated buckling strengths.

CHARACTERISATION OF BUCKLING STRENGTH

To make the results immediately usable and comprehensible, the load at linear bifurcation was characterised by the mean membrane stress immediately above the column σ_m, divided by the classical elastic critical stress for uniform axial compression σ_{cl}. When the cylinder is subjected to a uniformly distributed traction p_z, the mean stress σ_m is given by

$$\sigma_m = \frac{2\pi R\, H\, p_z}{n\, d\, t} \tag{1}$$

where the cylinder is of radius R, height H and thickness t, and it is supported on n columns each of width d. The classical elastic critical stress is

$$\sigma_{cl} = 0.605\,\frac{Et}{R} \tag{2}$$

for a Poisson's ratio $\nu=0.3$. The ratio of these two stresses is then

$$\frac{\sigma_m}{\sigma_{cl}} = \frac{10.39}{n}\,\frac{H}{R}\,\frac{R}{d}\,\frac{R^2}{t^2}\,\frac{p_z}{E} \tag{3}$$

This relation also defines the dimensionless groups of the problem.

LINEAR BIFURCATION IN PERFECT CYLINDERS

The first results described here are those for linear bifurcation in perfect cylinders on rigid supports. The buckling deformations are found to be localised near the column (Fig. 4), so

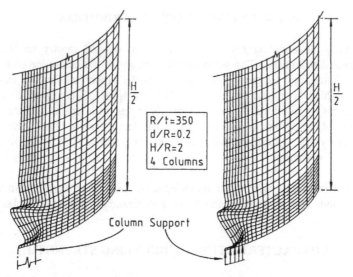

Figure 4. Buckling modes in perfect cylinders

the height of the shell is unimportant unless the cylinder is very short (Fig. 5). The variation of the calculated dimensionless mean buckling stress with the number of columns is shown in Fig. 6. If the number of columns is small and of practical width, the mean stress above the column at buckling is virtually independent of the number of columns. Thus, for most practical structures, the dimensionless mean buckling stress over each column may be considered to be independent of both the number of columns and the shell height.

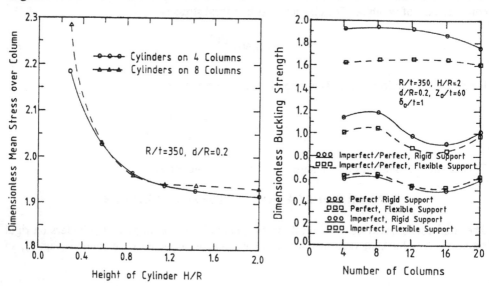

Figure 5. Effect of cylinder height (perfect cylinders)

Figure 6. Effect of number of columns

The dimensionless mean buckling stress is found to be more sensitive to the width of the column (Fig. 7). The reason is found in the changed size of the zone of high compressive stress. Most practical supports beneath a cylinder may be expected to induce a stress distribution comparable to that of the rigid support. However, if the support is flexible, the buckling strength is reduced. Calculations to investigate the reason showed that the greater part of the reduction was caused by the changed stress distribution, whilst the vertical restraint from the rigid column played a smaller role.

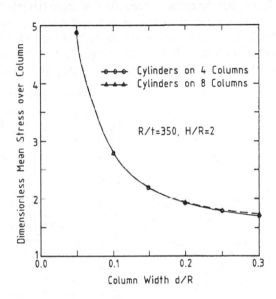

Figure 7. Effect of column width (perfect cylinders)

PERFECT AND IMPERFECT CYLINDERS

All previous studies of non-uniform axial compression buckling appear to have been confined to perfect shells. As the non-uniformity of the stress distribution is reduced, the predicted strength does not tend towards a recognised design strength, but towards a perfect shell buckling strength. Thus, it is difficult to exploit these studies in a practical design.

NONLINEAR ANALYSIS OF PERFECT AND IMPERFECT CYLINDERS

The calculations described above were all linear bifurcation analyses. Of course, linear bifurcation analyses may or may not give satisfactory predictions of buckling strength, depending on the geometry and the postbuckling path. Rigorous investigations must use a nonlinear analysis. Once a nonlinear analysis is being used, the question of imperfections in the shell can naturally be addressed. Nonlinear analyses are very much slower than linear bifurcation analyses, and the range of problems which can be explored in a given time is thus more restricted. Only a limited number of these results are therefore available to date.

Nonlinear analyses were performed on a number of different sample geometries to explore the effect of changing several parameters [7]. Typical results are shown in Fig. 8. The perfect cylinder always reached a limit load, but its value was sometimes a little imprecise using LUSAS. This limit load was generally around 0.6 of the linear bifurcation load for the same geometry.

The strength of the geometrically imperfect shell is more difficult to characterise. No limit load was found, but the cylinder becomes first more flexible and then stiffens, in a manner similar to the behaviour of very imperfect uniformly compressed cylinders [12]. The load which should be used to define the strength is then rather uncertain, and may depend on the onset of yielding. It has been suggested that the inflection point on the nonlinear curve should be used, but some geometries display very little inflection.

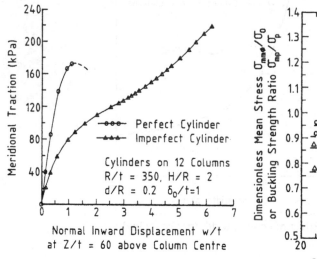

Figure 8. Load deflection curves from nonlinear analysis

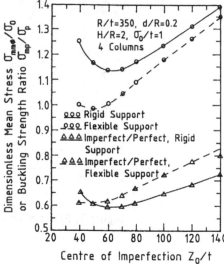

Figure 9. Effect of imperfection amplitude on bifurcation load

BIFURCATION IN IMPERFECT CYLINDERS

Because of the difficulty in characterising the strength of imperfect cylinders, linear bifurcation analyses were also performed on these. The reasons for believing that these can be useful analyses are set out elsewhere [10].

The linear bifurcation loads of the imperfect shells were found to be significantly lower than those of the corresponding perfect shells. A comparison is shown in Fig. 6 and the variation of bifurcation load with imperfection amplitude in Fig 9. These analyses identified the most serious location for an imperfection, and showed that the imperfection sensitivity does not depend very much on other shell parameters. One surprising result is that the bifurcation load of a cylinder with an imperfection amplitude of one wall thickness appears to be very close to the nonlinear limit load of a perfect cylinder. How general this finding might be has not been established.

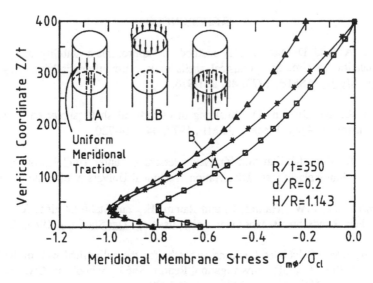

Figure 10. Stresses above column centreline for different load positions (rigid support)

EFFECT OF LOADING PATTERN

All the above calculations were performed on cylinders with a uniformly distributed traction on the whole wall. Tanks carry their fluid on the base, and stocky silos carry much of theirs on the hopper. Thus, the principal loading may sometimes arise as a uniform vertical line load around the lower edge of the cylinder. The effect of varying the loading position was investigated using bifurcation analyses on perfect cylinders. The position of the load was found to alter the stress distribution significantly (Fig. 10), especially in short cylinders on wide columns. Changes in both the buckling load and the buckling mode also occurred, with cylinder wall friction leading to lower strengths than hopper loading by typically 25%. Further work is needed to establish a more comprehensive picture of the sensitivity of the buckling strength to the position of the applied loading.

CONCLUSIONS

A brief summary has been given of many calculations of the buckling strength of column-supported cylinders. In general, the mean stress above the top of the column at either the nonlinear limit load or the linear bifurcation load is comparable with the classical elastic critical stress. The buckling mode and the strength are sensitive to the position of the applied loading and the boundary condition used at the bottom edge of the cylinder.

REFERENCES

1. Libai, A. and Durban, D., A method for approximate stability analysis and its application to circular cylindrical shells under circumferentially varying loads. J. Appl. Mech., ASME, 1973, **40**, 971-976.

2. Libai, A. and Durban, D., Buckling of cylindrical shells subjected to nonuniform axial loads. J. Appl. Mech., ASME, 1977, **44**, 714-720.

3. Peter, J., Zur Stabilitat von Kreiszylinderschalen unter Ungleich-mabig Verteilten Axialen Randbelastungen, Dissertation, Technical University of Hanover, 1974.

4. Bornscheuer, F.W., Hafner, L. and Ramm, E., Zur Stabilitat eines Kreinszylinders mit einer Rundschweissnaht unter Axialbelastung, Der Stahlbau, 1983, **52**, 11.

5. Rotter, J.M., Calculated buckling strengths for the cylindrical wall of 10 000 tonne silos at Port Kembla, Investigation Report S663, School of Civil and Mining Engineering, University of Sydney, June 1988.

6. LUSAS User's Manual, LUSAS theory manual, Version 9, Finite Element Analysis Ltd, Surrey, U.K., Dec. 1989.

7. Teng, J.G. and Rotter, J.M., A study of buckling in column-supported cylinders, Proc., International Union for Theoretical and Applied Mechanics, Symposium on Contact Loading and Local Effects in Thin-Walled Plated and Shell Structures, Prague, August 1990, Preliminary Report, pp. 39-48, Proceedings, Springer, Berlin, 1991.

8. Teng, J.G. and Rotter, J.M., Linear bifurcation of perfect cylinders on column supports, Research Report No. 91.02, Department of Civil Engineering and Building Science, University of Edinburgh, 1991.

9. Teng, J.G. and Rotter, J.M., Linear bifurcation of column-supported perfect cylinders: support modelling and boundary conditions, Research Report No. 91.03, Department of Civil Engineering and Building Science, University of Edinburgh, 1991.

10. Teng, J.G. and Rotter, J.M., Bifurcation Predictions for Imperfect Column-Supported Cylinders, Research Report No. 91.04, Department of Civil Engineering and Building Science, University of Edinburgh, 1991.

11. Rotter, J.M. and Teng, J.G., Elastic stability of cylindrical shells with weld depressions, J. Struct. Engg., ASCE, 1989, **115**, 5, 1244-1263.

12. Yamaki, N., Elastic Stability of Circular Cylindrical Shells, North Holland, Elsevier Applied Science Publishers, Amsterdam, 1984.

NOTATION

The following symbols are used in this paper:

d	Column width
E	Young's modulus
H	Height of cylinder
p_z	Uniform downward meridional traction
R	Radius of cylinder middle surface
t	Thickness of cylinder
X	Horizontal normal coordinate
Y	Horizontal tangential coordinate
Z	Vertical (meridional) coordinate
ν	Poisson's ratio
σ_m	Mean meridional membrane stress over each column
σ_{cl}	Classical elastic critical stress of a cylinder under uniform axial compression

AN EXPERIMENTAL AND NUMERICAL STUDY INTO THE COLLAPSE STRENGTH OF STEEL DOMES

J. Blachut, G.D. Galletly and D.G. Moffat
Department of Mechanical Engineering
University of Liverpool
P.O. Box 147
Liverpool L69 3BX

ABSTRACT

Experimental and numerical collapse pressures for four externally-pressurised 0.8m dia. steel torispheres are given in the paper. The numerical predictions were obtained using the BOSOR5 (axisymmetric 1-D analysis) and ABAQUS (2-D analysis) programs.

The 1-D analyses, based on the best-fit geometry and the meridian with the largest radial imperfections, gave ratios of p_{expt}/p_{calc} in the range 0.74-1.14. The 2-D finite element analyses used the imperfections measured at about 2000 points on the shell surface. The analyses were thus more complicated and time-consuming than the 1-D analyses. The 2-D FE analyses gave ratios of p_{expt}/p_{calc} in the range 0.89-1.14 and they also predicted the single lobe which characterized the shape of the dome at collapse.

A third numerical approach involved the extraction of a single localised flat patch from the measurements of initial shape of the domes. This imperfection was then assumed to be axisymmetric and located at the apex of the dome. Use of BOSOR5 to determine the lower-bound collapse pressures then gave p_{expt}/p_{calc} ratios in the range 0.97-1.13. These results are promising and they were also obtained at a fraction of the computational effort required for the FE analysis.

INTRODUCTION

This paper presents some selected results from a recent study into the elastic-plastic collapse strength of externally-pressurised steel torispherical domes [1, 2]. The domes were manufactured from high-strength steel, using standard industrial hot-pressing techniques. Geometrical imperfections, introduced by the manufacturing process, were present in all the domes. It is believed that the domes tested in this study had imperfections in them which were representative of industrial practice.

The domes were manufactured in pairs. Each dome of a pair had the same nominal geometry and the domes were made from the same material. The second dome in each pair was later cut into petals and then welded. This was done to mimic the real procedure of manufacturing full-scale components, where the dome is assembled from individual segments and then welded. The influence of welding on the collapse strength could also be determined, as each petalled/welded dome had a non-welded plain replica.

The diameter of the domes was about 0.8m and the nominal thickness was 6.4mm. The material was HY80 steel with a nominal 0.2% proof stress of about 550N/mm². Each dome had a short, 50mm, cylindrical flange. The shapes of the domes tested were all within the admissible region defined by the BS5500 code of practice [3]. An initial chord gauge assessment of shape was made in accordance with the BS5500 rules.

The shells investigated were expected to be sensitive to deviations from the perfect shape. It was, therefore, decided to monitor carefully a shell's shape prior to testing (i.e. after pressing, cutting and welding). The shape of each dome was measured along 72 meridians at 2.5° intervals within the spherical cap and at 5° intervals within the knuckle. The cylindrical portion was scanned at 5mm intervals. The shape measurements were made on the shell's inside surface using an LVDT transducer and a computer-based data acquisition system. On average, there were about 2000 measured points per dome. The same grid was used to measure the dome's thickness using an ultrasonic probe.

The above data was incorporated in several different ways into the numerical analyses which were carried out. The numerical calculations were based on the BOSOR5 [4] and ABAQUS [5] codes.

A one-dimensional, axisymmetric, discretization of each dome allowed the BOSOR5 program to be used. Results are given for the best-fit geometries of the domes and various types of axisymmetric imperfection. A recent paper [6] on the correlation of the DTMB experimental data on hemispherical shells [7] and BOSOR5 predictions suggests that it is possible to achieve reasonably good agreement between tests and theory, based on axisymmetric modelling only. This requires, however, a knowledge of a satisfactory substitute imperfection profile. An extension of this approach to the torispherical shells tested at Liverpool is discussed later in the present paper.

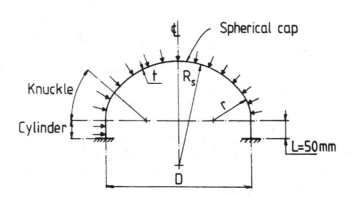

Fig. 1   Geometry of the torisphere.

The finite element modelling used in the 2-D analyses, and the finite element results, are also given in the paper. The pre- and post-processing of the data was done using the PATRAN package [8]. The conversion of the measured data into the finite element model (especially the shape measurements) is not a straight-forward operation. Therefore, the generation of the FE model is addressed in detail. Geometrical and material nonlinearities are taken into account in the ABAQUS calculations. The snap-through load is traced on the pressure-deflection path using the arc length method.

AXISYMMETRIC MODELLING OF THE DOMES WHICH WERE TESTED

References [1, 2] provide detailed information on tests of 16 torispheres and 2 hemispheres. The numerical results presented in Refs. [1, 2] were based on axisymmetric modelling of the domes and two types of modelling were used. The first was based on perfect best-fit shapes, where the measured shape of a single meridian was used to find the parameters describing that meridian (i.e. R_s, r and D - see Fig. 1). This procedure was repeated for all 72 measured meridians. The overall average values of R_s, r and D, together with the overall average wall thickness, t, were used as inputs in the axisymmetric modelling. The BOSOR5 program [4], with INDIC = 0 and a nonlinear stress-strain curve (given by six linear segments), was used to predict the axisymmetric collapse pressures. The ratios of $P_{expt}/P_{best-fit}$ varied from 0.756 to 1.132 [1, 2].

The second modelling incorporated axisymmetric imperfections and the worst meridian. For this case, the meridian which had the largest deviation from the best-fit shape was taken as the generator of an axisymmetric shell. The BOSOR5 program was used again and the ratios of P_{expt}/P_{worst} varied from 0.739 to 1.04.

The worst correlation between the experimental and the calculated collapse pressures (i.e. 0.756 and 0.739) was obtained for the non-petalled dome G/28P.

TABLE 1
Best-fit dimensions of torispheres analysed

Dome	R_S (mm)	D (mm)	r (mm)	R_S/D	r/D	D/t	$\bar{t}$ (mm)
G/28P	765.99	761.37	76.15	1.006	0.100	136.105	5.59
G/28W	767.13	780.85	85.44	0.982	0.109	136.919	5.70
G/23W1	610.94	791.03	145.73	0.772	0.184	138.728	5.70
G/23W2	611.29	791.12	145.91	0.773	0.184	139.281	5.68

Note: t = overall average thickness
       $\bar{t}$ = average thickness in spherical cap

As a result of the above calculations, it was decided to re-analyse torisphere G/28P using finite element, two-dimensional, modelling. The welded replica was also included in the finite element analyses, as was another pair of domes, i.e. G/23W1 and G/23W2, both of which had been petal-welded. All four domes were manufactured from the same material and their relevant geometric parameters are given in Table 1. The nonlinear stress-strain data and elastic constants are tabulated in Table 2.

TABLE 2
Piecewise-linear uniaxial stress-strain material data

σ (N/mm^2)	0	356.0	500	570	645	755	830
ϵ (×10^{-3})	0	1.8867	3.0	3.8	5.0	10.0	20.0

$E = 189$ kN/mm^2; $\nu = 0.30$

2-D FINITE ELEMENT MODELLING

In this section, the question of how a large number of measured points should be utilized when setting up a reliable finite element model is addressed. As it is known that the collapse strength of the torispherical shells being analyzed can be sensitive to deviations in shape, one has to be careful about an adequate generation of doubly-curved shell elements.

The procedures adopted for dome G/28P (i.e. the dome for which the axisymmetric modelling yielded the worst correlation with experiment) will now be illustrated.

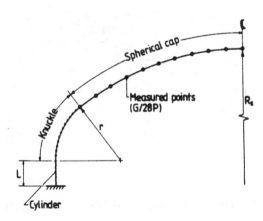

Fig. 2 Measured points along a single meridian in the G/28P dome.

Fig. 2 shows the locations of the 32 measured points for a single meridian (11 in the spherical cap, 13 in the knuckle and 8 in the cylindrical flange). In the first instance, the measured points were simply connected together to form 8-noded doubly-curved shell elements - as sketched in Fig. 3.

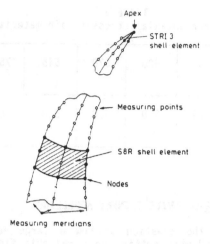

Fig. 3   Eight-noded and triangular shell elements vs. measured points.

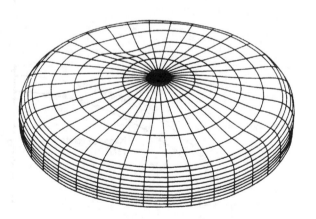

Fig. 4   Deformed shape of G/28P at the collapse pressure.

The resulting collapse pressure, based on the ABAQUS [5] calculations, is given in Table 3.   Although the aspect ratio for some shell elements was greater than 2:1, the calculated p_{expt}/p_{ABAQUS} was 0.88 (versus 0.76 for the axisymmetric BOSOR5 calculations).   Figure 4 depicts the dome's deformed shape at collapse, as calculated by the ABAQUS program.

TABLE 3
BOSOR5 and ABAQUS calculated collapse pressures

Dome	BOSOR5 (N/mm^2)		ABAQUS (N/mm^2)	Pexpt (N/mm^2)	Pexpt/PABAQUS
	Best-fit	Worst Meridian			
G/28P	3.56	3.64	3.07	2.69	0.876

3.1 Patran Generated Model

There are several options in PATRAN which can facilitate the generation of a line or lines through the experimentally-measured points (e.g. polynomial best-fit, C-spline, Bézier). The best-fit option in PATRAN was chosen herein to generate 10 line segments from the 32 measured points. Each line passes through the chosen end points but the interior points usually do not lie on the generated curve.

Also, adjacent line segments do not meet with continuous first derivatives. This may result in visible cusps between adjacent line segments. The cusps can be reduced in two ways: (i) by requiring continuity of the first derivatives - this can result in an inappropriate representation of the measured points or (ii) by moving the junction between two line segments to a different point. The second option, with no smoothing, was adopted in this paper.

It was checked that the junctions were at points where there were no visible cusps. This was repeated for every 4th meridian (out of 72 measured). At this stage, the measured points around the circumference were approximated and the technique used in the meridional direction was also applied here.

The above process led to the approximation of a shell's surface via a number of patches. The corner points of every patch were measured points, while the curved edges were best-fit to the experimental data. Some measured points, however, were left-out (i.e. points lying inside the domain bounded by the curved edges). On average, two data points per patch were not utilized in the above approximation.

The resulting surface had then to be divided into a number of shell elements. A convergence study in [9] showed that 10 elements along the meridian and 18 circumferentially were sufficient for a dome having a geometry similar to dome G/28P. This results, in the present case, in 162 eight-noded shell elements and 36 three-noded triangles used at the apex. The ABAQUS S8R and STRI3 elements were used in the calculations. The collapse pressure, based on geometry/material nonlinearities, was calculated using arc length-controlled loading - see column (2) in Table 4. The overall average thickness was used (see Table 1).

TABLE 4
Calculated collapse pressures for different Finite Element models

Dome	ABAQUS-based collapse pressure (N/mm^2)					Experiment (N/mm^2)
	(1)	(2)	(3)	(4)	(5)	
G/28P	3.07	3.15	3.01	3.01	2.91	2.69

Note: The FE models are described: (1) in section 3, (2) in section 3.1, (3), (4) and (5) in section 3.2.

3.2   Cubic Spline Modelling

It was mentioned earlier that the PATRAN approximation did not generate smooth curves passing through all the measured points. The aim in this section is to generate such a surface which contains all the measured points in it. This is achieved by spanning cubic splines between every two measured points. Further imposition of continuity and continuity of slope at the measured points, together with known boundary conditions at the apex and at the clamped end, results in a standard cubic spline approximation scheme [9].

A sample of the results is given in Table 4. Column (3) corresponds to a 12 meridional x 36 circumferential grid. The identical collapse pressure in column (4) was obtained for a 25 x 36 grid. In both cases, the overall average wall thickness was used.

The inclusion of variable dome thickness into the FE model was considered next. The dome's thickness was measured using the same grid as that used for the shape measurements. For the G/28P dome, there are 2304 thickness values available. The local minimum thickness was 5.03mm and the maximum 6.31mm. Cubic splines were used to approximate the thickness at the nodes of the elements. The current version of ABAQUS only accepts a constant thickness for an element. This average thickness for an element is obtained from its nodal values. Fig. 5 shows the meridional variation of thickness along three meridians and the resulting piecewise constant thickness profile for a meridional column of ABAQUS shell elements.

The collapse pressure corresponding to the variable wall thickness dome is given in Table 4 - column (5). It is evident from Table 4 that the best correlation is obtained with the cubic spline approximation of shape and piecewise constant thickness (column 5). However, the lack of continuity in the thickness distribution is artificial and does not resemble the real shell. In all subsequent calculations, an overall average thickness was used.

3.3   Comparison of Results

A comparison of the calculated and the experimental collapse pressures is given in Table 5. The finite element model had 12 meridional and 36 circumferential elements. The shape was approximated by cubic splines.

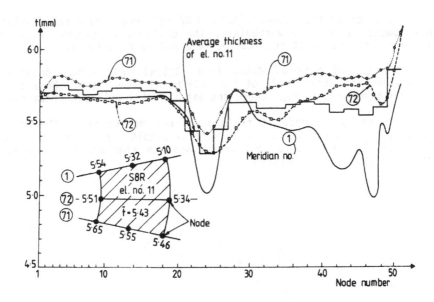

Fig. 5   Meridional thickness profile along three adjacent meridians and
resulting piecewise constant thickness used in the FE analysis. Also nodal
values of t in S8R shell element.

The shape of all domes, prior to testing, was assessed according to
the BS 5500 code, using the chord gauge method [3].   In the case of
torispheres, this method only applies to the spherical cap.   One of the
requirements states that the radius of local flattening, R_{imp}, should not
be greater than 1.3 times the nominal radius and it should be measured over
an arc length $s_{imp} = 2.4\sqrt{R_{imp}t}$. The ratios R_{imp}/R_s, for the four domes
discussed in this paper, are given in Table 5.

TABLE 5

Comparison of calculated and experimental collapse pressures (the overall
average thickness was used in the calculations)

Dome	BS 5500 R_{imp}/R_s	BOSOR5 (N/mm^2)		ABAQUS (N/mm^2)	Experiment (N/mm^2)	$\dfrac{P_{expt}}{P_{ABAQUS}}$
		Best-fit	Worst Meridian			
G/28P	1.67	3.56	3.60	3.01	2.69	0.89
G/28W1	1.59	3.82	3.72	2.97	3.38	1.14
G/23W1	1.29	6.99	7.14	5.85	5.79	0.99
G/23W2	1.32	7.01	7.15	6.31	6.48	1.03

It is clear that domes G/28P and G/28W1 were considerably outside the code specification. Nevertheless, the collapse pressures, based on the finite element calculations, can be regarded as acceptable.

The second pair of domes (i.e. G/23W1 and G/23W2) were slightly below/above the upper limit of the code requirement. For these cases, the FE results accurately reproduced the experimental collapse pressures.

The incipient mode of failure was not monitored in the experiments and only the post-collapse shapes were available. Figs. 6 and 7 show that the post-collapse shape of all four domes was similar, i.e. a single lobe near the cap/knuckle junction.

Fig. 6   Experimental post-collapse shape of G/28P (right) and G/28W1.

The FE simulated post-collapse shape is given in Figs. 8 and 9. It resembles the experimental single dimple at the junction but the circumferential location sometimes does not correspond to the location found experimentally.

FLAT PATCH APPROACH

In this section, the use of a localised imperfection in the spherical cap of a torispherical shell to predict the dome's collapse strength is discussed. The idea is to extract this shape from the available geometry measurements and model it as if it were an axisymmetric imperfection located at the apex. This would facilitate axisymmetric analysis of a dome.

Previous calculations showed that the collapse strength of externally-pressurised spherical cap/hemispheres or torispheres can be affected by a localised flat patch positioned at the pole. Different ways of representing that patch in the numerical modelling were tried in the past (i.e. localised smooth dimple, truncated Legendre polynomial dimple, local increased-radius segment, etc.) [10, 11]. The numerical predictions of the collapse strength were, by and large, not dependent on the actual form of the flattened area and local increased-radius flattening was found to be convenient for numerical purposes. The collapse strength depends, in this case, on the following three parameters:

- the amplitude of flattening at the apex δ_0,
- the arc length of the flattening s_{imp} and
- the radius of the imperfect shell R_{imp}.

Fig. 7   Experimental post-collapse shape of G/23W1 and G/23W2.

Out of these three parameters (δ_0, s_{imp}, R_{imp}) only two are independent, since the following relation holds

$$\frac{\delta_0}{t} = \frac{s_{imp}^2}{8} (1 - R_s/R_{imp}) \, 1/(R_s t) \tag{1}$$

The above on-axis imperfection approach has been verified experimentally at Liverpool. Several machined steel torispheres, with purposely-introduced increased-radius flattening at the apex, have been tested [12]. The ratios of the experimental collapse pressures to those numerically-predicted were in the range 0.94 - 1.0.

Some unusual numerical results have also been published recently which suggest that, in a hemispherical shell, an off-axis, local imperfection can reduce the collapse strength more than the equivalent on-axis imperfection [13]. A project currently being carried out at Liverpool is examining this topic both numerically (finite element calculations) and experimentally (elastic buckling/collapse of a torisphere with an off-axis localised flattening) [14].

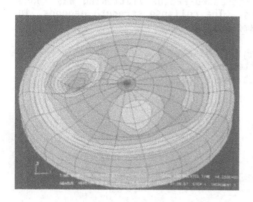 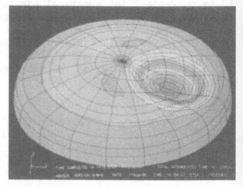

Fig. 8   Numerically simulated post-collapse mode of G/28P and G/28W1.

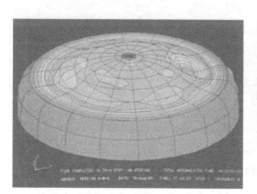 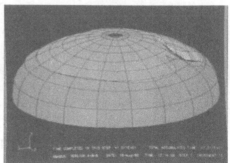

Fig. 9   Numerically simulated post-collapse mode of G/23W1 and G/23W2.

On the other hand, some 28 DTMB (U.S. Navy) hemispheres were scanned for shape prior to testing and a number of flat patches, usually not at the pole, were found in them [7].   The available shape parameters of these patches were utilised in building-up axisymmetric, imperfect numerical models [6].   In many cases, the calculated collapse pressures agreed well with the DTMB experiments (the ratios p_{expt}/p_{calc} were in the range 0.92-1.20).

Following on from the study on hemispherical shells, localised imperfect areas in the Liverpool torispherical domes were singled out as follows:

-       only the spherical cap was considered,
-       only the grid used for the shape measurements was employed.

Next, the best-fit radius was found for each meridian. Then each meridian was approximated by cubic splines. The cubics passed through all the measured points. The local radius of curvature was calculated from:

$$1/R_{imp} = - y''/\sqrt{[1 + (y')^2]^3} \qquad (2)$$

The arc length s_{imp}, for which $R_{imp}/R_s \geq 1.0$, was then established. The largest s_{imp} for every meridian was recorded, together with radial deviations from the best-fit circle. Then the same procedure was repeated, at a given latitude, along the measured points in the circumferential direction. Table 6 contains the largest s_{imp} and the corresponding $(\delta_o/t)_{max}$ within this imperfect arc length. The corresponding size in the perpendicular direction is also added. The maximum values of R_{imp}/R_s encountered are also included. They vary from 2.69 to 51.5 (this latter corresponds to a very small, localised, flattening).

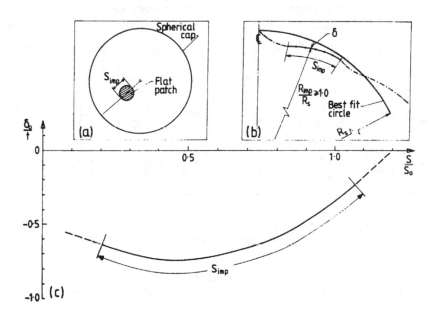

Fig. 10   Sketch of a localised flattening in spherical cap (a); parameters of the flattening (b) and the inward deviations from the best-fit circle in G/28P dome (c).

Fig. 10a shows the position of a local patch in dome G/28P. The meridional profile of the deviation from the best-fit is shown in Figs. 10b, c.

The following BOSOR5 calculations were performed using the localised patch imperfections given in Table 6:

(i)   the larger s_{imp} and the corresponding $(\delta_o/t)_{max}$ were taken as the parameters characterizing the flattening at the apex. The calculated collapse pressures are given in the last column of Table 6, and

(ii) for a given dome, the larger (δ_o/t) of the two shown in Table 6 was taken. A lower-bound collapse pressure was found by varying s_{imp} - see [10]. The s_{imp} corresponding to the lower bound is different from the s_{imp} approximated from the experimental data.

The results are given in Table 7 and it may be seen that there is good agreement between the experimental results and the calculations.

TABLE 6
Size of local imperfection in the spherical cap. Maximum values of δ_o and R_{imp} encountered are also given

Dome	Meridional direction			Circumferential direction			BOSOR5 Collapse Pressure (N/mm^2)
	$\dfrac{s_{imp}}{s_o}$	$\left[\dfrac{\delta_o}{t}\right]_{max}$	$\left[\dfrac{R_{imp}}{R_s}\right]_{max}$	$\dfrac{s_{imp}}{s_o}$	$\left[\dfrac{\delta_o}{t}\right]_{max}$	$\left[\dfrac{R_{imp}}{R_s}\right]_{max}$	
G/28P	0.916	0.749	4.03	0.794	0.668	4.20	3.66
G/28W1	0.911	0.586	51.29	1.202	0.340	2.14	3.86
G/23W1	0.952	0.271	2.69	0.956	0.330	4.01	6.18
G/23W2	1.053	0.244	6.17	1.229	0.300	3.10	6.22

Note: $s_o = 2.4 \sqrt{R_s \bar{t}}$; R_s = best-fit value
$\bar{t}$ = overall average thickness

TABLE 7
Collapse pressures for a singled-out flat patch positioned at the apex

Dome	$\left[\dfrac{\delta_o}{t}\right]_{max}$	$\dfrac{s_{imp}}{s_o}$	BOSOR5 $P_{COLLAPSE}$ (N/mm^2)	$\dfrac{P_{expt}}{P_{BOSOR5}}$
G/28P	0.749	1.375	2.58	1.04
G/28W	0.586	1.380	3.00	1.13
G/23W1	0.330	1.227	5.98	0.97
G/23W2	0.300	1.229	6.22	1.04

CONCLUSIONS

As expected, the two-dimensional FE analyses produced rather better agreement between the experimental and the calculated pressures than the axisymmetric analyses based on BOSOR5. The range of the ratio p_{expt}/p_{calc} improved from (0.76-1.13) to (0.89-1.14). These results were obtained using an overall average thickness for the domes. The influence of variable thickness on collapse pressures can only be assessed when a variable wall thickness option is introduced into the ABAQUS code.

The localised flat patch approach, with subsequent BOSOR5 analysis, gave good agreement with the experimental results, at a fraction of the computational effort. The ratios p_{expt}/p_{calc} were in the range 0.97-1.13.

In view of the above, it might be worthwhile developing a procedure which could identify critical localised imperfections, i.e. their sizes, shapes and relative locations. This would enable one to establish which type of imperfection plays the dominant role in weakening the collapse strength of a dome and whether there are interactions between neighbouring imperfections.

REFERENCES

1. D.G. Moffat, J. Blachut, S. James and G.D. Galletly, "Collapse of externally-pressurised torispherical and hemispherical domes", University of Liverpool, Department of Mechanical Engineering, Rep. No. A/151/89, 1989.

2. D.G. Moffat, J. Blachut, S. James and G.D. Galletly, "Collapse of externally-pressurised petal-welded torispherical and hemispherical end-closures", Proc. of Seventh International Conference on Pressure Vessel Technology - ICPVT7, Düsseldorf, May/June, 1992.

3. BS 5500: Specification of unfired fusion welded pressure vessels, 1988, British Standards Institution.

4. D. Bushnell, "BOSOR5 - A program for buckling of elastic-plastic complex shells of revolution including large deflections and creep", Comp. and Struct., Vol. 6, 221-229, 1976.

5. ABAQUS, User's Manual, Version 4.8, 1989, Hibbit, Karlsson and Sorensen Inc., Providence, R.I., U.S.A.

6. G.D. Galletly and J. Blachut, "Buckling design of imperfect welded hemispherical shells subjected to external pressure", Proc. IMechE, 1991, (in press).

7. T.J. Kiernan and K. Nishida, "The buckling strength of fabricated HY 80 steel spherical shells", U.S. Navy DTMB Rpt. 172, Washington D.C., July 1966.

8. PATRAN, Manuals-releases 2.1-2.4, PDA Engineering, Costa Mesa, CA 92626, U.S.A., 1990.

9. J. Blachut and G.D. Galletly, "Non-axisymmetric collapse analysis of steel dome ends using ABAQUS", - in preparation.

10. G.D. Galletly, J. Blachut and J. Kruzelecki, "Plastic buckling of imperfect hemispherical shells subjected to external pressure", Proc. Instn. Mech. Engrs., C3, 201, 153-170, 1987.

11. J. Blachut and G.D. Galletly, "Buckling strength of imperfect spherical caps - some remarks", AIAA J., 7, 28, 1317-1319, 1990.

12. G.D. Galletly, D.N. Moreton, and A. Muc, "Buckling of slightly flattened domed ends reinforced locally with fibre-reinforced plastic", Proc. Instn. Mech. Engrs., 204, 15-24, 1990.

13. W.J. Shao and P.A. Frieze, "Static and dynamic numerical analysis studies of hemispherical and spherical caps", Thin-Walled Structures, 8, 99-118 (Part I) and 183-201 (Part II), 1989.

14. J. Blachut and G.D. Galletly, "Influence of local imperfections on the collapse strength of a torisphere". To be presented at the 1st European Solid Mech. Conf., Munich, 9-13 Sept., 1991.

BUCKLING OF THIN-WALLED CYLINDER UNDER AXIAL COMPRESSION AND INTERNAL PRESSURE

by A. LIMAM, J.F. JULLIEN
Concretes and Structures Laboratory INSA Lyon (France)
E. GRECO
AEROSPATIALE Les Mureaux (France)
and D. LESTRAT
CNES Evry (France)

The results of investigations on the critical buckling load of shells subjected to combined internal pressure and axial load show that the effect of the internal pressure is to stabilise the shell. Nevertheless the specifications given in the note "NASA Space Vehicle Design" which are adopted by the E.C.C.S. design code, recommend that the reduction factors are higher than those given for axial compression only. This leads to very conservative design. One of the reason for the high coefficient of security is that the tests on which the reduction coefficents were based were carried out on Mylar specimens rather than on metal ones. Since the effect of local plasticity on the geometrical imperfections is to precipitate instability, it is questioned whether this very conservative aspect of the recommendations is justified.

The object of the study was to investigate the buckling behavior of metal cylindrical shells under combined internal pressure and axial load, and to determine the effect of this load on the critical buckling load. The radius/thickness ratio was approximately 700 and none of the shells were stiffened.

METHODOLOGY

Different types of tests were carried out. A series of loading tests executed on the same specimen gave similar buckling loads. This is because the factors which most influence the critical load are the geometrical imperfections and the support conditions which, of course, were constant in this case. The use of one specimen made it necessary that the buckling load was purely elastic and that the axial displacement could be stopped immediately after bifurcation to prevent plastic deformations. The same specimen was subject to a series of different internal pressures to obtain the corresponding buckling loads. However at one internal pressure two specimens were tested to establish the scatter of the results.

A second series of tests was made to determine the effect of geometric imperfections which were corresponding to the critical mode of the internal pressurised shell (axisymmetric mode) and another on the non axisymmetric mode (shell not internally pressurised). These tests were made on the same specimens that had previously been subjected to a plastic deformation subsequent to the elastic bifurcation. A series of tests on these specimens at different internal pressures allowed the consideration of the most unfavourable case, that is the presence of critical imperfections, local plasticity and residual stresses ; these results are compared with the curves given in the E.C.C.S. report.

All these experimental results are systematically compared with the numerical values obtained by using the program INCA of the system CASTEM. This latter is based on finite elements and is particularly well suited for instability calculations of shells of revolution. If these calculations are carried out using non linear geometry, elasto-plastic material, and the initial shape imperfections then they permit an interpretation of the experiment results.

SPECIMENS, APPARATUS AND TEST PROCEDURES

Two support conditions were used for the specimens. One perfectly encastre, the other created by thickening the shell near the support so that the fixity at the point of the shell where it obtains its original thickness is indeterminate.

The shells were built up using electrodeposition of copper ; this method offers many advantages : there is no joint in the shell and there are no residual stresses. The manufacture of the specimens is obtained by deposition on a duraluminium base. The separation of the base from the shell is achievied by dissolving the duraluminium in a bath of sodium hydroxide. The material properties for each shell were determined from a specimen manufactured at the same time as the shell. The values of Youngs Modulus, the limit of linearity and the 0,2% proof stress are given in Table 1.

N° of Specimen	$t(\mu)$	R / t	E(MPa)	σL(MPa)	$\sigma 0,2$ % (MPa)
A 1	175	770	82 680	234	357
A 2	195	686	98 180	204	368
A 3	150	867	70 220	247	401
A 4	215	620	95 740	192	354

Table 1 : Geometrical and mechanical caracteristics of specimens.

The test arrangement for the compression tests is shown in the photograph, Fig.1.

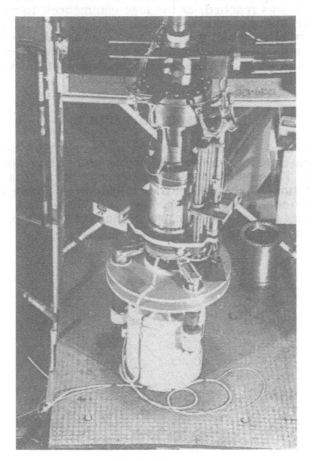

Fig. 1 : Test setup

Pressurization of the cylinders was accomplished with compressed air through ports in the bottom cap. Four displacement transducers equally spaced along the inside of the circumference of the cylinder permitted measurement of the relative movement between the top and bottom of the shell.

The same procedure was followed for all the tests. The first stage consisted of a study of the initial imperfections. The position of 50 000 points on the surface of the shell was obtained by means of an capacity transducer which was not in contact with the specimen. These readings enabled the geometry of the surface to be represented by a Fourier series and the maximum amplitude of an imperfection to be calculated.

The shell was then pressurised to the desired value. A new measurement of the surface geometry enabled the reduction in the amplitude of the imperfections to be established. The axial compression was then introduced by the application of an axial displacement and was increased until bifurcation was reached, or the load commenced to decrease slowly, at which point loading was immediately stopped. After recording the critical geometry of the shell it was totally unloaded. A new reading of the geometry then allowed a new quantification of the imperfections and also a new estimate of their magnitude and the degree of deterioration of the specimen.

EXPERIMENTAL RESULTS

Fig. 2 gives the load-shortening curve for specimen A1. The occurrence of diamond shape buckles was unmistakable since they appeared with a snapping noise. The precritical curve is perfectly linear, the buckling then leads to bifurcation. The values of the critical stress (Table 2) obtained for the two specimens presented (A1 and A2) show that the theory of buckling predicts the critical load, given that the test apparatus and the specimens are of high quality.

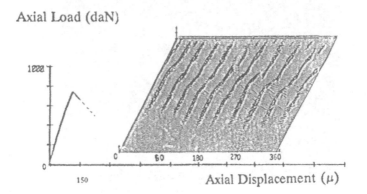

N° of Specimen	σ_{ex} (MPa)	σ_{ex} / Tcl
A 1	51	74 %
A 2	73	79 %

Table 2 : Critical axial stresses (P = 0)

Fig.3 shows a typical load-deflection curves for cylinders subjected to internal pressure and axial load. The buckling load obtained by bifurcation without internal pressure increases with increasing pressure towards a limit point, this latter follows from the loss of axial rigidity caused by non linear behavior (both geometric and material). For a low intensity of internal pressure (P = 0.2 bar) the critical load attains 93 % of the classical buckling load.

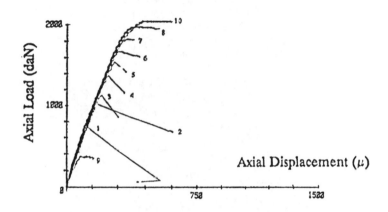

Fig. 3 : Load-Deflection curves for pressurised cylinders in axial compression

This increase in the carrying capacity shown in Fig.3 is perfectly compatible with the reduction of 50 % in the initial imperfection amplitudes which were initially approximately 20 % of the thickness. This reduction applies for all the harmonics greater than the fifth (Fig.4).

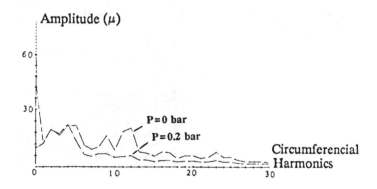

Fig. 4 : Geometric imperfections- Pressurised and unpressurised shell

On increasing the internal pressure the longitudinal axial and circumferential wave lengths diminish, the amplitude of the buckles are smaller (Fig.5) and attain the axisymmetric mode of higher pressure.

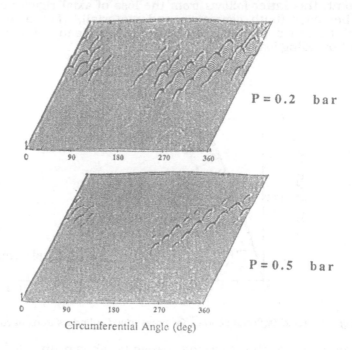

Fig. 5 : Post-Critical modes (Specimen A1)

The results of specimens A1 and A3 are presented in (Fig.6) in non dimensional form :

$$\lambda = \sqrt{3\left(1 - v^2\right)} \left(\frac{R}{Et}\right)\left[\sigma - \frac{PR}{2t}\right]$$

$$\tilde{p} = \left(\frac{PR^2}{Et^2}\right). \sqrt{3\left(1 - v^2\right)}$$

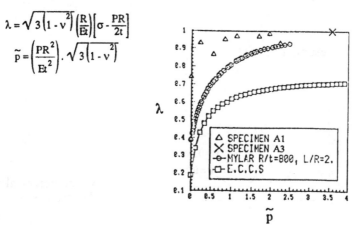

Fig. 6 : Variation of axial-stress coefficient with internal pressure parameter

For specimen A1 at higher values of P (2 bars) a ring shaped deformation appeared near the end caps, this corresponds to the onset of plasticity resulting from clamped end bending moments. The wave only grows at the boundaries of the shell, leads to instability, and is a localised plastic collapse phenomenon which causes a drop in the critical load (Fig.7).

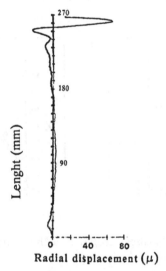

Fig. 7 : Buckling Deflection w- (Specimen A1 : P = 2 bars)

The test on specimen A3 confirmed this, the increase in the thickness of the shell close to the supports prevented local plasticity and allowed elastic buckling to occur by a limit point with a generalised axisymmetric mode. In this case the load and the critical mode are exactly predicted by classical theory (Fig.8).

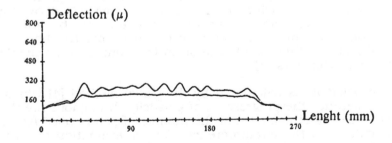

Fig. 8 : Axial Distributions of the prebuckling and buckling deflections w

The critical loads for specimens A2 and A4 are lower since the imperfections correspond to the critical mode and because of local plasticity, these make the enhancement of the load by internal pressure more difficult. It was noted that the multimodal imperfections are more dangerous than the axisymmetric ones (Fig. 9). This has since been confirmed by studies at the laboratory at INSA [2].

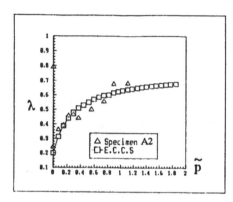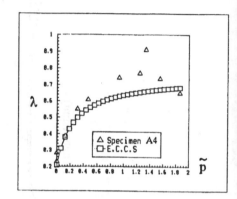

Fig. 9 : Variation of axial-stress coefficient with internal pressure parameter

In the cases presented the internal pressure usually has a stabilising effect but in the case of very high pressure, which is a function of the amplitude of the imperfections, the onset of plasticity causes a reduction of the critical load.

COMPARISON OF EXPERIMENTAL AND CALCULATED RESULTS

Among the different elements that exist in the program INCA we have made use of shell element, "COQUE", which is well adapted for the calculation of axisymmetric structures. It is an element with 2 nodes and 4 degrees of freedom per node. The element "COEP" is a variation of this element which allows the effect of the development of plasticity in thickness to be included. The "COMU" element is derived from the above two elements and represents an extension because it allows the consideration of structures with non axisymmetric imperfections without limitation on their amplitude or on the number of harmonics representing the imperfections [3].

The plasticity is calculated by the global method [4] which does not take into account the progression of plasticity throughout the thickness.
The user has the choice of the number of points used for the evaluation of plasticity on the circumference. For the evaluation of plastic bifurcation, the tangent modulus theory was used. A comparative calculation using

deformation theory was made on the perfect shell. The tangent modulus approach is more conservative, the difference between the two approaches is more important for high internal pressures (Fig.10). For the range of internal pressures used in the tests the difference between the two theories is of the order of 10%.

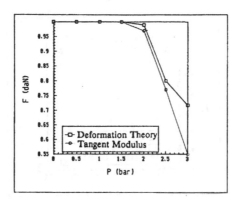

Fig. 10 : Comparison of deformation theory with tangent modulus approach

For studies of the instability of imperfect shells the methodology proposed by RCCMR[5]was used, this is characterised by the technique of modal imperfections which considers imperfections corresponding to the Euler critical mode with the amplitude equal to the maximum amplitude of the tolerances permitted during manufacture.

In the case under consideration, the precise knowledge of the spectrum of the geometric imperfections allows two calculations, that considering the mean amplitude on the surface, and that with the maximum observed local amplitude.

For the specimens having only their initial imperfections and not having plastic deformations, the calculations with the average amplitude of imperfection give a critical load less than 5% different from the experimental one [Fig. 11].

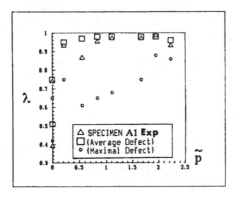

Fig. 11 : Comparison of theoritical and experimental results

For the other tests the maximum amplitude is more representative because the imperfections are generalised on the whole surface and correspond to the critical mode [Fig. 12].

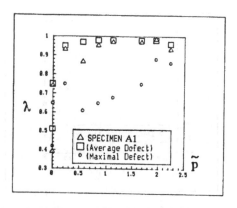

Fig. 12: Comparison of theoritical and experimental results

CONCLUSIONS

This study showed the importance of the influence of internal pressure on the geometrical imperfections and consequently on the elastic critical load. A similar conclusion is reached in the case shells having geometric imperfections and plastic deformations. The internal pressure normally has a benefical effect and allows an increase in the critical load but in the case of very high pressure the onset of plasticity causes a reduction in this load.

In the case of initial imperfections which do not correspond to the critical mode, the E.C.C.S. design recommendations are very conservative, but in the case where they correspond to this mode the recommendations are a very good approximation.

If the buckling load is to be predicted with precision by calculation it is important to consider the effect of every parameter on the numerical model chosen.

Acknowledgements

This research was part of an overall project on cryogenic tanks carried out in the laboratorty of concrete and structures at INSA in LYON at the request of the CNES and AEROSPATIALE

REFERENCES

[1] E.C.C.S, European Recommendations for steel construction.

[2] N. WAECKEL, J.F. JULLIEN, "Experimental study on the instability of cylindrical shells with initial geometrical imperfections", Session ASME (P.V.P), San Antonio, Juin 84

[3] A. COMBESCURE "Static And Dynamic Buckling of large thin shells", Nuclear Engineering and Design 92 (1986) 339-354, North-Holland, Amsterdam.

[4] A. HOFFMANN, M. LIVOLANT and R. ROCHE, "Plastic analysis of shells by finite element method and global Plasticity", Paper L6/2, 2nd SMIRT, Berlin.

[5] RCC-MR Règle de conception et de construction des matériels mécaniques des ilots nucléaires Tome I, Volume 2, Annexe A7, Edition AFCEN, Juin 85.

ON THE EFFECT OF LOCAL IMPERFECTIONS ON THE STABILITY OF ELASTIC-PLASTIC SHELLS IN BENDING

G. T. JU and S. KYRIAKIDES
Engineering Mechanics Research Laboratory
Department of Aerospace Engineering and Engineering Mechanics
The University of Texas at Austin
Austin, Texas 78712, USA

ABSTRACT

The plastic flexure of long circular cylindrical shells is limited by short wavelength wrinkles which appear on the compressed side of the shell. The ripples were found [1,2] to localize and to induce catastrophic collapse of the shell. In this paper the effect of local, small amplitude imperfections is investigated though a combination of experiment and analysis. Local imperfections (dents) of various amplitudes were introduced in aluminum shells which were then tested to collapse in curvature controlled pure bending experiments. The critical curvature was found to be very sensitive to the imperfection amplitude. A special purpose nonlinear shell analysis was used to reproduce the experimental results. The collapse curvature was found to correlate well with a bifurcation mode detected in the dented region.

INTRODUCTION

The buckling and subsequent collapse of long, circular, cylindrical shells, plastically deformed by pure bending, has been shown in [1,2] to be strongly influenced by the shell diameter-to-thickness ratio (D/t). In the case of thinner shells (for aluminum, $D/t > 40$), the following sequence of events was observed experimentally and reproduced numerically: Bending induces ovalization to the cross section of the shell [3]. This geometric nonlinearity, together with the reduction in modulus as the material proceeds into the plastic range, result in an inherent or *natural limit load* instability, as shown in the calculated moment-curvature response (uniform ovalization case) in Fig. 1. For this thinner class of shells, the natural limit load is preceded by a bifurcation type instability in the form of axial, short wavelength wrinkles on the compressed side of the shell (similar to elastic behavior [4,5]).

In practice, due to initial imperfections present in the shells, the wrinkles appear in pockets of a few wrinkles, as shown in Fig. 2a. A number of such pockets are usually detectable along the length of the shell. As the curvature of the shell increases, the amplitude of the wrinkles grows. In the process, the shell becomes more compliant. This reduction in rigidity induces a limit moment in the response of the shell. The *induced* limit moment occurs at a curvature which is significantly smaller than that corresponding to the natural limit load.

Following the limit load, the wrinkles, rather than continue growing uniformly, localize and the moment drops more precipitously (see also [6,7]). A second bifurcation, characterized by a number of circumferential waves, was detected in the trough of the wrinkle with the most severe deformation [2]. Such a buckling mode (see Fig. 2b) was found experimentally to lead to local catastrophic collapse of the shell (shell loses 80 – 90% of moment capacity). In practice, rather than through a bifurcation, the collapse is induced by amplification of geometric imperfections in the neighborhood of the calculated bifurcation points.

Figure 1 shows a comparison of an experimental moment-curvature response for an aluminum 6061-T6 shell with $D/t = 44.0$ with responses calculated by three levels of analysis having the following characteristics:

I.   The cross section is assumed to ovalize uniformly along the shell length.
II. Axial wrinkles are modeled but are assumed to grow uniformly along the length.
III. Axial wrinkles are allowed to localize.

In the case of the uniform wrinkling analysis, a small, initial axisymmetric imperfection defined as follows was used:

$$\overline{w} = -a_{oi} R \cos \frac{\pi s}{\lambda_D} , \qquad (1)$$

where λ_D is the critical wavelength of axial wrinkles obtained from a bifurcation analysis [2,8].

In the case of analysis (III) the initial imperfection used was as follows:

$$\overline{w} = -R \left[a_{oi} + a_i \cos \left(\frac{\pi s}{5\lambda_D} \right) \right] \cos \left(\frac{\pi s}{\lambda_D} \right) . \qquad (2)$$

This imperfection provides a small bias in the amplitude of one of the wrinkles and facilitates localization [6]. The following symbols are used in Fig. 1 to identify critical points on the responses shown:

$\wedge \equiv$ calculated natural limit load,
$\blacktriangle \equiv$ calculated induced limit load,
$\downarrow \equiv$ first observation of ripples,
$\uparrow \equiv$ calculated first bifurcation,
$\blacktriangle \equiv$ calculated second bifurcation,
$\downarrow \equiv$ catastrophic failure of shell.

The results shown in Fig. 1 clearly demonstrate that the calculated first bifurcation is in good agreement with the values at which the axial wrinkles were first observed to occur in the shell tested. In addition, by including representative initial imperfections the analysis which allows the wrinkles to localize yields a limit load instability and a second bifurcation which compare very well with the experimental onset of catastrophic collapse of the shell. This type of agreement between experiment and analysis was found to occur for a series of shells with different D/t values. More details on both the experiments and the analysis performed can be found in [1,2].

The experimental and numerical results presented in [1,2] clearly demonstrate that localization of axial wrinkles plays a crucial role in the ultimate curvature of the bent shell. Motivated by this observation, in this study we evaluate the effect of localized imperfections on the collapse curvature of such shells.

EXPERIMENTS

The experiments were conducted on Al-6061-T6 drawn tubes with diameter of 1.248 in (31.7 mm) and $D/t = 44.0$. The length of the test specimens (2L) was 24 diameters. A uniaxial test was conducted for each tube tested using axial specimens cut adjacent to the bending specimens. The stress-strain curves obtained were fitted with a Ramberg-Osgood fit given by

$$\varepsilon = \frac{\sigma}{E}\left[1 + \frac{3}{7}\left(\frac{\sigma}{\sigma_y}\right)^{n-1}\right]. \tag{3}$$

Variations in the diameter and thickness of the test specimens along the length were recorded prior to each experiment. The geometric and material parameters of the test specimens are listed in Table 1. A local dent was introduced at mid-span to a number of test specimens, as follows. A smooth, circular, steel solid rod, oriented at 90 degrees to the axis of the shell, was pressed into the shell, as shown in Fig. 3a. The shell was supported on a flat, rigid base and the indention was conducted in a quasi-static fashion in a "screw" type machine. This produced a local dent on one side of the shell.

In the cases discussed here, the diameter of the indenter, d, was 0.250 in (6.35 mm). The axial variation of an initial imperfection introduced with this method is shown in Fig. 3b. The amplitude of the induced imperfections were in the range of $0 < a_L < 9 \times 10^{-3}$. In this range of amplitude, it was found that the wavelength of the imperfections remained approximately constant. The wavelength is seen in Fig. 3b to be of the order of the calculated critical half wavelength λ_D.

The following expression was used to fit the induced initial imperfections

$$\overline{w}(s,\theta) = -a_L \frac{R}{4} [\cos\theta - 2\cos 2\theta + \cos 3\theta] e^{-\beta\left(\frac{s}{l_o}\right)^2}. \tag{4}$$

This representation is seen in Fig. 3b to reproduce the axial variation of the imperfection quite adequately ($l_o = 0.18$ and $\beta = 4.6$). The same degree of success was found to occur in the circumferential direction.

The bending experiments were conducted in a stiff, four-point bending facility under curvature control loading. The moment and curvature were continuously monitored and recorded through a computer-based data acquisition system. In addition, two transducers, mounted on the test specimen, were used to record the change in diameter in the plane of bending (ΔD) during the experiment. One was located in the trough of the initial dent and the other 6 diameters away from it. Details about the test facility, the transducers, the data acquisition system and the experimental procedure used can be found in [1].

Experimental Results

A set of experimental results for a shell with an initial imperfection with amplitude $a_L = 6 \times 10^{-3}$ is shown in Fig. 4. The moment-curvature response recorded (Fig. 4a) was found to be virtually unaffected by the imperfection. The shell buckled suddenly and catastrophically by developing one local buckle in the trough of the imperfection. The mode of buckling is similar to the one shown in Fig. 2b.

Figure 4b shows a plot of the measured change in diameter in the plane of bending at the two points mentioned above ($\Delta D/D_o$ will be called ovalization in the remainder of this paper). As the critical curvature is approached, the ovalization at point A is seen to grow faster than at point B, indicating local acceleration in the growth of deformation. Figure 5 shows a set of axial ovalization scans recorded at different values of curvature during a second experiment on a similar shell. Away from the boundaries and the region of the local imperfection, the ovalization is seen to remain essentially uniform along the length for all values of curvature shown. The ovalization in the region of the local imperfection is seen to grow in the scan corresponding to the last curvature recorded. A more detailed scan of ovalization of the region of the localized imperfection, recorded with a higher resolution instrument, is shown in Fig. 6. The wavelength and amplitude of the dent change gradually at small values of curvature. As the critical curvature is approached the dent experiences accelerated growth which quickly results in catastrophic collapse of the shell.

Similar experiments, in which the amplitude of the induced local imperfection was varied, were conducted. The behavior observed and the results recorded follow the same trend as the results discussed above. The moment-curvature responses measured from four such shells with imperfection amplitudes of $a_L = 0$, 4, 6 and 8×10^{-3} are shown in Fig. 7. The initial parts of the four responses are almost indistinguishable. The vertical arrows show the points at which each shell collapsed catastrophically. Clearly, the presence of such local imperfections can drastically reduce the maximum curvature to which the shell can be bent. For example, the critical curvature of the shell with imperfection amplitude of $a_L = 8 \times 10^{-3}$ was only 59% of the critical curvature of the shell with no dent. Increase in the amplitude of the dent beyond this value further reduces the critical curvature and the corresponding moment.

ANALYSIS AND PREDICTIONS

The imperfect shell was analyzed using the nonlinear, elastic-plastic shell analysis presented by the authors in [2]. The formulation adopted uses a special deformation composition scheme in which the shell is first deformed into a circular torus with a radius corresponding to the radius of curvature of the bent shell, k. Additional deformation is then applied in which the displacements $\{u, v, w\}$ are measured from the circular torus. This deformation field is represented through Sanders [10] nonlinear shell kinematic relationships.

For simplicity, the initial imperfection given in eq. (4) was assumed to be stress free. The presence of the imperfection altered the metric and bending tensors of the initial cylinder which was then deformed using the deformation composition scheme mentioned above.

The inelastic material behavior was modeled through the J_2 flow theory of plasticity with isotropic hardening. As is customary, bifurcation checks were conducted using the J_2 deformation theory of plasticity and the stress state from the prebuckling analysis.

The problem was solved incrementally using the principle of virtual work. The overall shell curvature was the prescribed variable. The structure was discretized by approximating the displacements through the following series:

$$
\begin{aligned}
u &= R \sum_{i=1}^{I_u} \sum_{j=0}^{J_u} c_{ij} \sin ips \cos j\theta , \qquad p = \frac{\pi}{L_a} \\
v &= R \sum_{n=2}^{n_v} b_n \sin n\theta + R \sum_{i=1}^{I_v} \sum_{j=1}^{J_v} d_{ij} \cos ips \sin j\theta , \\
w &= R \sum_{n=0}^{N_w} a_n \cos n\theta + R \sum_{i=1}^{I_w} \sum_{j=0}^{J_w} e_{ij} \cos ips \cos j\theta .
\end{aligned}
\tag{5}
$$

The induced deformation was assumed to be symmetric about the mid-span of the shell. Due to the local nature of the imperfection and of the deformation induced by bending, it was found that a shell (half) length of $L_a = 10l_o$ provided a more than sufficient domain. In eq. (5) $N_v = N_w = J_u = 6$ and $I_u = 8$ were found from convergence tests to provide sufficient numerical accuracy. Typically 12 integration points in the circumference, 45 along the length and 7 in the thickness direction were used. The calculations were conducted on a Cray YMP computer.

Numerical Results

A set of results obtained from the analysis for an imperfection amplitude of $a_L = 6 \times 10^{-3}$ are shown in Fig. 4 together with the corresponding experimental results. The calculated moment-curvature response is seen in Fig. 4a to be in excellent agreement with the one measured. The calculated change in ovalization in the trough of the imperfection and in the undisturbed region away from the imperfection are shown in Fig. 4b. The predicted results are in good agreement with the measurements up to the onset of failure. A full field view of the ovalization along the length of the shell analyzed is shown in Fig. 8. These results also compare well with the corresponding experimental results shown in Fig. 6.

The presence of the imperfection induces a limit load instability. The limit load occurs at a curvature which is somewhat larger than the one at which the shell failed catastrophically in the experiment. However, a bifurcation instability was detected prior to the limit load. This bifurcation mode is one characterized by a number of circumferential waves which develop in the trough of the dent. (Note: The length of the most deformed region can be seen in Fig. 8 to increase during bending. Thus the domain of the bifurcation check was varied appropriately to always be $2l$ as defined in Fig. 8.). The bifurcation instability is considered to be the most probable cause of catastrophic failure of the shell. As mentioned earlier in the experiment, rather than through a process of bifurcation, the shell buckles through a process involving amplification of imperfections which correspond to the buckling mode calculated above. This explains the upturn observed in the ovalization measured in the trough of the dent as the critical curvature is approached (see Fig. 4b).

Figure 9 shows a graphical reproduction of a calculated equilibrium configuration corresponding to a curvature just larger than the value corresponding to the limit moment. The nature of the localized deformation is quite well illustrated.

The same type of behavior was found to occur for other values of imperfection amplitudes analyzed. As in the experiments, it was found that the early parts of the calculated moment-curvature responses are relatively unaffected by the presence of the imperfections. The main effect of the imperfections was to precipitate the instability. Expanded views of the critical sections of moment-curvature responses calculated for a set of imperfection amplitudes are shown in Fig. 10. The induced limit moments and the predicted bifurcation points are identified on each response. As the amplitude of the imperfection, a_L, is increased both instabilities are seen to occur at smaller values of curvature.

Similar results to the ones shown in Fig. 10 were obtained for other imperfection amplitudes. The curvatures corresponding to the induced limit moment and to the bifurcation are plotted against the imperfection amplitude in Fig. 11. Included in the figure are the collapse curvatures of the four imperfect shells tested. From the results it can be concluded that the special bifurcation check developed yields critical curvatures which are very representative of the critical collapse curvatures of the imperfect structures.

The following conclusions can be drawn from the study:

(a) For the class of plastically deformed shells studied, bending deformation is limited by a sequence of instabilities involving the development (bifurcation) and localization of short wavelength axial wrinkles on the compressed side of the shell. Bending is terminated by catastrophic local collapse of the shell.

(b) It has been found that small amplitude local imperfections can significantly reduce the collapse curvature of the shell.

NOMENCLATURE

a_i, a_{oi}, a_L	amplitude of imperfections	s, θ	coordinates
D	shell outside diameter	t	shell wall thickness
D_o	shell mean diameter	w	initial imperfection
E	Young's modulus	$\Delta D/D_o$	ovalization
$2L, 2L_a$	shell length	κ	curvature
$2l_o$	imperfection length	κ_1	t/D_0^2
M	moment	κ_C	collapse curvature
M_o	$\sigma_o D_0^2 t$	λ_D	half-wave length of wrinkle
n	hardening parameter	σ_y	Ramberg-Osgood yield parameter
R	shell radius	σ_o	yield stress

ACKNOWLEDGEMENT

The work presented was conducted with financial support from the U.S. Office of Naval Research under contract N00014-91J-1103.

Table 1: Geometric and material parameters of shells tested

D, in (mm)	D/t	$2L/D$	E, ksi (GPa)	σ_o, ksi (MPa)	σ_y, ksi (MPa)	n
1.248 (31.70)	44.0	24	9.72×10^3 (67.03)	41.1 (283)	41.0 (283)	28

REFERENCES

1. Kyriakides, S. and Ju, G.T., Bifurcation and Localization Instabilities in Cylindrical Shells Under Bending: Part I Experiments, *Int'l J. of Solids and Structures* (to appear), also EMRL Rep. No. 90/9, 1990, Univ. Texas at Austin.

2. Ju, G.T. and Kyriakides, S., Bifurcation and Localization Instabilities in Cylindrical Shells Under Bending: Part II Predictions, *Int'l J. of Solids and Structures* (to appear), also EMRL Rep. No. 90/10, 1990, Univ. Texas at Austin.

3. Brazier, L. G., On the Flexure of Thin Cylindrical Shells and Other Thin Sections, *Proc. Royal Society of London*, Series A, **116**, 104-114, 1927.

4. Axelrad, E. L., Flexible Shells, *Proc. of the 15th IUTAM Congress*, Toronto, Canada, 45-56, 1980.

5. Fabian, O., Collapse of Cylindrical, Elastic Tubes Under Combined Bending, Pressure and Axial Loads, *Int'l J. of Solids and Structures*, **13**, 1257-1270, 1977.

6. Tvergaard, V., On the Transition from a Diamond Mode to an Axisymmetric Mode of Collapse in Cylindrical Shells, *Int'l J. of Solids and Structures*, **19**, 845-856, 1983.

7. Tvergaard, V. and Needleman, A., On the Localization of Buckling Patterns, *ASME J. of Applied Mechanics*, **47**, 613-619, 1980.

8. Bushnell, D., Bifurcation Buckling of Shells of Revolution Including Large Deflections, Plasticity and Creep, *Int'l J. of Solids and Structures*, **10**, 1287-1305, 1981.

9. Ju, G. T., and Kyriakides, S., Bifurcation Buckling vs. Limit Load Instabilities of Elastic-Plastic Tubes Under Bending and External Pressure, *ASME J. of Offshore Mechanics and Arctic Engineering*, **113**, 43-52, 1991.

10. Sanders, J. L., Nonlinear Theories of Thin Shells, *Quart. of Applied Mathematics*, **21**, 21-36, 1963.

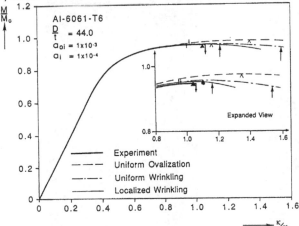

Fig. 1   Moment-curvature of Al-6061-T6 shell: Comparison between experiment and three types of analysis (from [2])

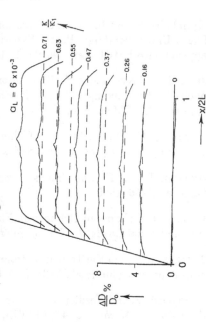

Fig. 3 (a) Crosswise indention of cylindrical shell
(b) Local change of shell diameter as a result of indention

Fig. 5 Measured ovalization along length of shell at different values of curvature

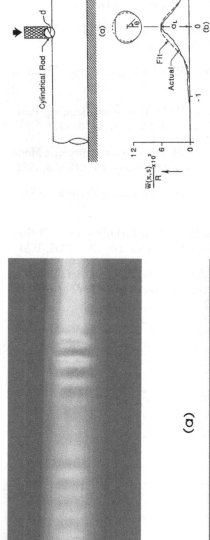

Fig. 2 (a) Axial wrinkles developed on compression side of shell in pure bending
(b) Local buckle which occurs after wrinkling

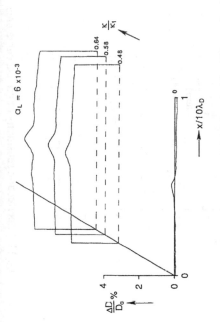

Fig. 6 Measured ovalization in neighborhood of local imperfection as a function of curvature

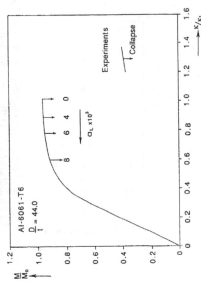

Fig. 7 Moment-curvature responses measured for different local imperfections

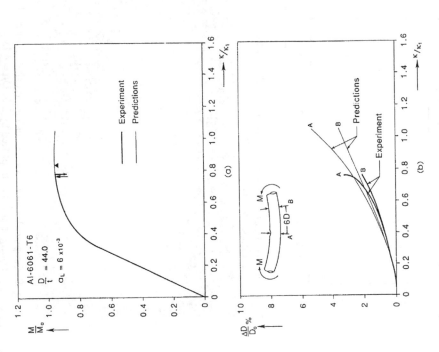

Fig. 4 Comparison of measured and predicted shell response. (a) Moment-curvature, (b) ovalization-curvature

380

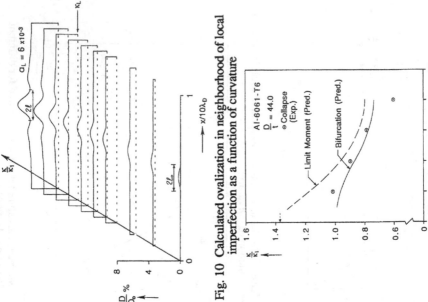

Fig. 10 Calculated ovalization in neighborhood of local imperfection as a function of curvature

Fig. 11 Imperfection sensitivity of critical curvatures

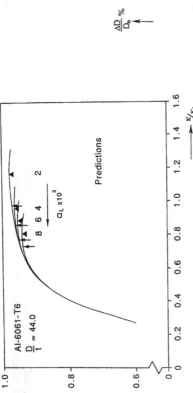

Fig. 8 Moment-curvature responses calculated for different local imperfection amplitudes

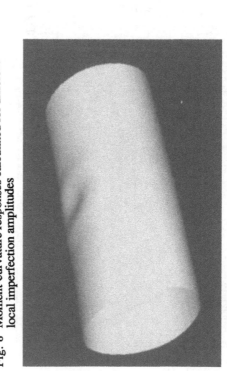

Fig. 9 Graphical reproduction of calculated shell deformed configuration (w amplified by 2.5. D/t = 44.0)

RECENT ADVANCES ON THERMAL BUCKLING
NEW RESULTS OBTAINED AT C.E.A

A COMBESCURE-J BROCHARD CEA DMT Saclay (FRANCE)

ABSTRACT

This paper shows the most recent results obtained at CEA DMT Saclay on thermal buckling.The buckling of a cylinder whose thickness to radius ratio is 500 is studied under the combination of a severe axial gradient of temperature and of an axial traction.The axial gradient is either fixed or axially moving.An analytical prediction of buckling is also performed to analyse the experimental results and to derive some general properties of these types of buckling.

1 INTRODUCTION

The buckling under these type of loads has been studied since six years at Saclay and the results start to be now at a mature state for the case of fixed axial gradient of temperature combined with an axial traction.Some more recent results are given in the case of a moving axial gradient of temperature .The observed mechanism is not,as one would see from the litterature a ratchetting,but a buckling which increases at each cycle of the thermal load;this result is obtained even without any axial load.

A rather detailed analytical work is also presented in the case of a fixed axial gradient of temperature combined with an axial tension.In this case a full interaction diagram is presented between the two loads.The influence of the theory of plastic buckling is shown as well as the influence of the load history.

The following notations shall be used throughout the paper as well as in the figures:

σ_y^{20} is the conventionnal yield stress at 20°c,

σ_n is the axial tensile stress,

σ^{th} is the maximum von mises equivalent stress due to the thermal load : this stress is computed with the elastic properties at 20°c.

2 EXPERIMENTAL RESULTS.

2.1 Geometry and loadings (experimental device)

The cylinder studied is made of 304 stainless steel.Its radius is .5 m and its thickness is 1. mm.It is loaded by an axial load imposed by an hydraulic jack of 50 tons.The axial temperature gradient is produced by an inductive coil.The experimental set up is given on figure 1.A typical axial gradient of temperature is given on figure 2.This experimental set up has already been described in a previous paper(ref 1).

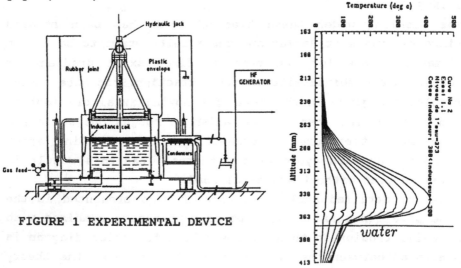

FIGURE 1 EXPERIMENTAL DEVICE

FIGURE 2 TYPICAL AXIAL GRADIENT
OF TEMPERATURE(FIXED LEVEL)

2.2 Fixed axial gradient of temperature.

A lot of experiments have been performed showing mainly the following results.The external pressure has a very small influence on the buckling under thermal load,so that the two buckling can be considered in this case as independant.The main interaction is between the axial tension and the axial gradient of temperature.Figure 3 shows a typical observed buckling.The amplitude of the buckles is of about three or four times the thickness,when the cylinder is hot,but the residual buckle is of about one tenth of the thickness.All experimental results are given on figure 4.One sees on this figure that no buckling could be observed when there is no sufficient axial tension.Different load histories were also studied and one sees that the worse load history is to apply first the traction and afterwards the axial gradient of temperature.

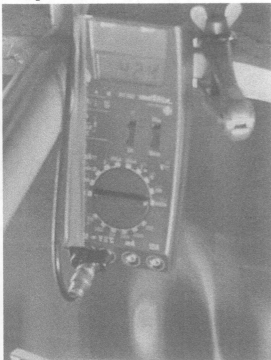

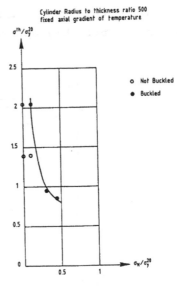

Cylinder Radius to thickness ratio 500
fixed axial gradient of temperature

o   Not Buckled
●   Buckled

EXPERIMENTAL BUCKLING
UNDER COMBINED LOADS
FIGURE 4

FIGURE 3 TYPICAL EXPERIMENTAL BUCKLING
MODE (FIXED LEVEL)

2.3 Moving axial gradient of temperature

The cylinder studied here had the geometry given on figure 5.One can observe that the moving level is very close to the free edge of the cylinder.Figure 6 defines the moving gradient of temperature obtained.With this load one has observed a buckling having the shape given on figure 7 at the first loading.Figure 8 gives the elastic buckling mode which is very similar to the observed experimental one When one continues to cycle the thermal gradient the buckles continue to develop and their amplitude reaches twenty times the thickness after twenty cycles.These results are rather interesting because they show that the problem of ratchetting is not,in this case,an axisymmetric ratchetting but more a non symetric ratchetting.This point is not usually adressed in the litterature.For instance the work performed by Ponter and his coauthors is not adressed to solve this non symmetric behavior and the results he gives are probably optimistic in this case.

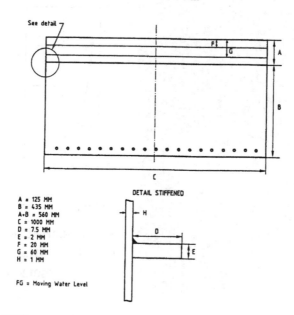

GEOMETRY STIFFENED SHELL

A = 125 MM
B = 435 MM
A+B = 560 MM
C = 1000 MM
D = 7.5 MM
E = 2 MM
F = 20 MM
G = 60 MM
H = 1 MM

FG = Moving Water Level

FIGURE 5 GEOMETRY OF THE MOVING GRADIENT CASE

385

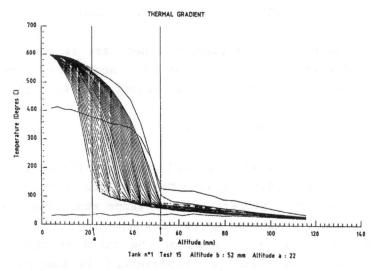

FIGURE 6 MOVING THERMAL GRADIENT OBTAINED

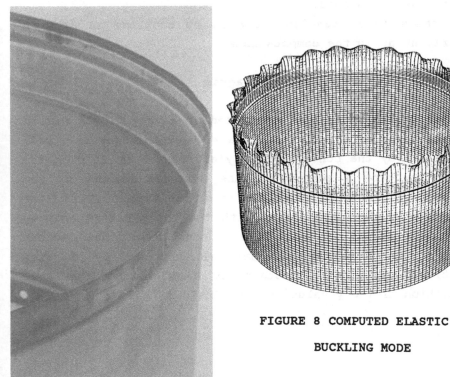

FIGURE 8 COMPUTED ELASTIC

BUCKLING MODE

FIGURE 7 EXPERIMENTAL BUCKLING MODE   OBSERVED

3 ANALYTICAL PREDICTIONS (interaction diagrams)

The analytical predictions presented here are relative only to the fixed gradient of temperature.The analytical work on the moving level case shall not be presented in this paper.

3.1 Interpretation of experiments

The experiments have been simulated using the finite element program INCA of CASTEM system,taking into account the variation of material properties with temperature.The theory of tangeant modulus is used to predict buckling.
In the case of an axial tension of one third of the yield stress at 20°C ,the buckling temperature is computed as 290°C for an experimental value of 280°C:the agreement can be considered as good.
When there is no tension there is no buckling as well in experiment as in the computations.

3.2 Basis of the interaction diagram

The two loadings can be represented in a very simple way on an interaction diagram.In abcissa we plot the tensile stress σ_n divided by the conventionnal yield stress at 20°C.On the y axis we plot a measure of the stresses induced by the gradient:this measure is the ratio of the maximum equivalent Von Mises stress ,associated with the thermal load only,to the yield stress at 20°C.It has been shown that we can replace the observed experimental gradient by a step function and that we obtain by this method the same prediction if the y value is identical.

3.3 Buckling prediction with different theories of plastic buckling

The effect of theory of plastic buckling has been studied in the case of the following history of loading:one applies first an axial load characterised by its x values and afterwards the temperature is increased until buckling is obtained.Figure 9 shows the computed buckling temperatures with the tangeant modulus theory compared with the finite deformation theory and with flow theory.All the results are then compared with experimental values.One sees that the flow theory is not conservative and then dangerous.The tangent modulus theory is pessimistic but not too much.The deformation theory is a bit unconservative.A typical buckling mode is shown on figure 10.

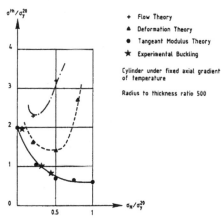

EFFECT OF DIFFERENT THEORIES
OF PLASTIC BUCKLING ON
INTERACTION DIAGRAM

FIGURE 9

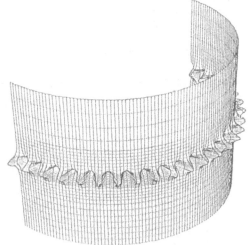

FIGURE 10 TYPICAL COMPUTED
PLASTIC BUCKLING

3.4 Effect of load histories on the buckling load.

The three load histories were studied with the tangeant modulus plastic buckling.The first history is to increase first the axial tension up to x (0. .25 .50 .75 1.) and then to look for the y value to obtain buckling.The second history consists of fixing first the temperature load value (y) and to increase the axial tension until buckling.The last one consists of increasing simultaneously x and y.

Figure 11 shows the diagram obtained for a thin shell (radius to thickness ratio of 1000):one can remark on this figure that the first load history is the most conservative.This load history can lead to a buckling for a load 20% lower than load history 3 which is the load history given by the codes(ASME or RCC-MR).

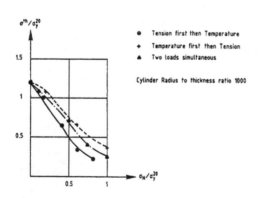

FIGURE 11 EFFECT OF LOAD HISTORY
ON INTERACTION DIAGRAM

4 DESIGN DIAGRAM

A design diagram has then been given for these type of cylinders under this type of loadings.It is shown on figure 12.This diagram is obtained from the minimum envelope of all computed load histories.To use this diagram one has to quantify the x and y values of the loadings as defined in paragraph 3.2.If one is under the curve no buckling is possible,if one is over the structure may buckle.The diagram given here is valid for cylinders having an aspect ratio between 100. and 1000.

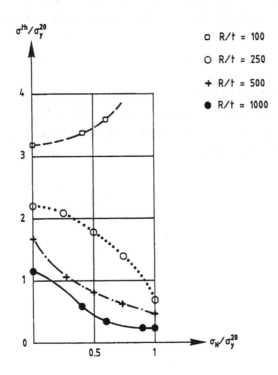

Fig. 12 - Design diagram for different aspect ratios Buckling of cylinders under combined tension and axial gradient of temperature (steel)

5 CONCLUSIONS

Some very interesting effects are shown in this work.The first new point is the existence of buckling and non symmetric progressive deformation in the case of thin cylinders under moving axial gradient of temperature.The second main result is the design diagram given in figure 10 that allows to avoid buckling in the case of combined axial tension and axial temperature gradient.The work is now continued in three directions.The first one is to extend the diagrams to structures having different aspect ratios. The moving axial temperature gradient is now being fully studied with two points of view:the first one is the stability aspect and the second is the ratchetting behavior.

REFERENCES

(1) J BROCHARD A COMBESCURE
Thermal load influence on the reduction of critical buckling loads.transactions of the 9th SMIRT 1989 volume E p209-218
A A BALKEMA/ROTTERDAM/BOSTON

STATIC AND DYNAMIC BUCKLING CHARACTERISTICS OF IMPERFECT CYLINDRICAL SHELLS UNDER TRANSVERSE SHEARING LOADS

T.MURAKAMI, H.YOGUCHI
Heavy Apparatus Engineering Lab., Toshiba Corporation, Yokohama, Japan
H.HIRAYAMA
Nuclear Energy Group, Toshiba Corporation, Yokohama, Japan
H.NAKAMURA, S.MATSUURA
Central Research Institute of Electric Power Industry, Chiba, Japan

ABSTRACT

Cylindrical shells containing vertical wrinkles made by press-forming were tested under static and pseudo-dynamic transverse loads to investigate the effect of geometrical imperfections on shear buckling characteristics. Buckling loads for imperfect cylinders were compared with those for nearly perfect cylinders and with estimated loads based on the theoretical torsional buckling stress and on the interaction relation with the yield stress. It was observed that shear buckling was less imperfection sensitive compared with axial one and quite stable even for imperfect cylinders in the post buckling region. It was clarified that there was no substantial difference in the buckling characteristics with imperfection for static and seismic loads. From these test results, the correction factor of the shear buckling load with imperfection was proposed in the interim buckling design guide.

INTRODUCTION

To establish the buckling design methodology and buckling design guide of FBR under seismic loads, the Demonstration Test and Research Program of Buckling sponsored by the Ministry of International Trade and Industry (MITI) has been made (1987-1993) by CRIEPI (Central Research Institute of Electric Power Industry) and nuclear power plant fabricators in Japan [1]-[7]. For short cylinders under transverse loads, wherein the shear buckling might occur, experimental studies on the elastic-plastic shear buckling were made in this project and some other authors [8][9] for nearly perfect cylinders. As to the geometrical imperfection effects on the buckling characteristics, few experimental data exist on axially compressed cylinders by spin-casting [10] and electro-deposition [11] techniques, therefore, it was of particular interest to evaluate these effects on the shear buckling characteristics.

In this paper, cylinders with vertical wrinkles made by press-forming were tested under static and pseudo-dynamic transverse loads. From the numerical analysis, it was clarified that a minimum buckling load was observed for a cylinder with wrinkles of the same circumferential wavelength as the buckling mode. From these test results and numerical analysis, the effect of initial shape imperfections on the shear buckling characteristics in static and dynamic buckling were clarified and the correction factor of the buckling load with imperfection was proposed.

TEST MODELS AND TEST SETUPS

Test models, having radius (R) of 250mm and 500mm, (R/t) of 125–200 and (L/R) of 1.0 as shown in Figure 1 and Table 1, were fabricated by rolling (nearly perfect) and by press-forming (imperfect cylinders) of austenitic stainless steel plates (SUS304) with a welded seam along the longitudinal direction. The first twelve models in Table 1 were for static tests and the last two models were for pseudo-dynamic tests, wherein the response displacement for a seismic ground motion was computer-controlled using the measured restoring force and solving the equation of motion numerically.

TABLE 1
Test model geometries and test conditions

Model No.	R mm	t mm	L/R	H/R	Wave No.	W im mm
R167-W0	250	1.47	1.0	1.2	—	0.62
R167-W12-T	250	1.47	1.0	1.2	12	1.32
R167-W12-2T	250	1.47	1.0	1.2	12	2.96
R167-W12-4T	250	1.50	1.0	1.5	12	5.72
R167-W6-2T	250	1.47	1.0	1.2	6	2.52
H200-W0	250	1.20	1.0	1.5	—	0.53
H200-W12-2T	250	1.20	1.0	1.5	12	2.32
H167-W0	250	1.51	1.0	1.5	—	0.47
H167-W12-2T	250	1.50	1.0	1.5	12	2.86
H167-W12-4T	250	1.50	1.0	1.5	12	5.84
H125-W0	250	1.94	1.0	1.5	—	0.49
H125-W10-2T	250	1.93	1.0	1.5	10	4.00
PH200-W0	500	2.44	1.0	1.5	—	0.51
PH200-W12-2T	500	2.45	1.0	1.5	12	4.82

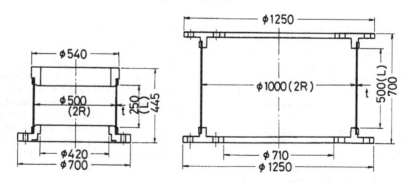

Figure 1. Configuration of test models

In these model numbers, test conditions (R:room temperature static, H:500° C static and PH: 500° C pseudo-dynamic) were indicated by the first symbol and the following numbers were R/t, imperfection modes (W0: nearly perfect and W6, W10, W12: wave numbers of 6, 10, 12) and imperfection peak-to-peak amplitudes normalized by its wall thickness. Shape imperfections tested here were wrinkles with 10 (R/t=125) or 12 (R/t=167 and 200) circumferential wave numbers, which were equal to the dominant wave numbers in the shear buckling mode, respectively. These imperfections were expressed in equation (1), where Wim is a peak-to-peak imperfection amplitude.

$$w = Wim/2 \cdot \cos(n\,\theta)\,\sin(\pi x/L) \qquad n: \text{circumferential wave number} \qquad (1)$$

The buckling test using a cylinder with 6 circumferential wave numbers was also made to evaluate the effect of wavelength on the buckling. Typical measured shape imperfections along the axial position are shown in Figure 2, where the shape imperfections were well controlled using templates with the desired wavelength and amplitude. The mechanical property tests using small test pieces cut out from imperfect cylinders showed that the yield stress and stress-strain relation were not changed compared with those for nearly perfect one.

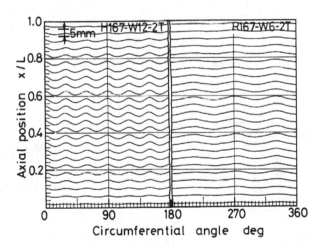

Figure 2. Distribution of measured initial imperfections

A seismic response wave, evaluated at a reactor support level of a half embedded reactor building, was used for pseudo-dynamic tests with the maximum acceleration input of 2495gal as shown in Figure 3. To clarify the difference of the post buckling characteristics for cylinders with and without imperfection, the maximum response displacements were expected to attain nearly two times of the buckling displacement under this acceleration input. The natural frequency of test models, which were nearly 1/16 scale models in dimension, was four times of a pool-type FBR main vessel (6.5 Hz), therefore, the time scale of the input ground motion was reduced by 1/4.

As shown in Figure 4, static test models were fixed horizontally and transverse shearing loads were applied by a screw jack in vertical direction. Pseudo-dynamic test models, on the contrary, were supported vertically and step by step transverse displacements (4000 steps) at the top of a cylinder were computer-controlled in the extended time duration using a hydraulic actuator.

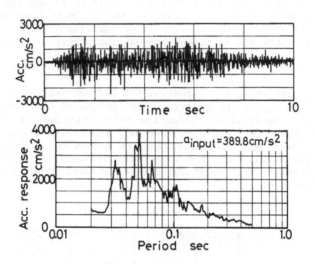

Figure 3. Input seismic motion

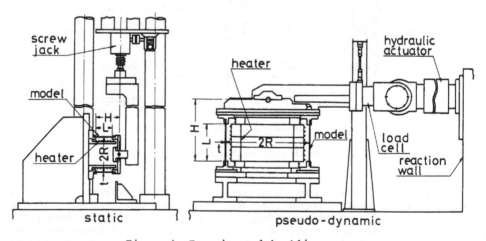

Figure 4. Experimental buckling setups

TEST RESULTS

Static Tests

Comparisons of typical load-displacement curves in static tests for nearly perfect and imperfect cylinders are illustrated in Figure 5 and buckling loads were summarized in Table 2. Buckling load reductions were 9% (Wim=T, RT), 20% (Wim=2T,RT) and 15%-18% (Wim=2T,HT), respectively, for imperfect cylinders with nearly the same wavelength as the buckling mode. The load-displacement curves in Figure 5 showed that the load carrying capacity after buckling for the imperfect cylinder decreased gently as expected by the numerical analysis. It was also clarified that the shear buckling was quite stable and less imperfection sensitive compared with axial one. In these tests, dominant buckling wave numbers were 10-12 (R/t=200), 9-11 (R/t=167) and 8-10 (R/t=125).

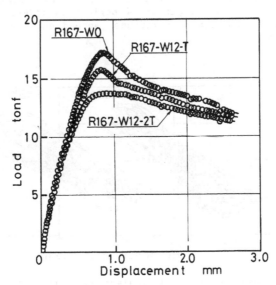

Figure 5. Comparison of load-displacement curves in static test

TABLE 2
Summary of experimental buckling loads

Model No.	σy kgf/mm²	Qexp tonf	δcr mm	Qp tonf	Qexp/Qp
R167-W0	30.5	17.3	0.84	14.9	1.16
R167-W12-T	30.5	15.8	0.83	14.9	1.06
R167-W12-2T	30.5	13.8	0.90	14.9	0.93
R167-W12-4T	30.6	12.7	1.06	15.4	0.83
R167-W6-2T	30.5	16.1	1.01	14.9	1.08
H200-W0	17.2	8.20	0.57	7.30	1.12
H200-W12-2T	17.2	6.95	0.54	7.30	0.95
H167-W0	17.0	11.5	0.83	10.0	1.15
H167-W12-2T	17.0	9.49	0.73	9.92	0.96
H167-W12-4T	17.0	8.42	0.84	9.92	0.85
H125-W0	16.4	14.5	0.94	13.3	1.09
H125-W10-2T	16.4	12.3	0.85	13.2	0.93
PH200-W0	17.2	31.6	2.30	29.9	1.06
PH200-W12-2T	17.2	27.0	2.03	30.1	0.90

Pseudo-Dynamic Buckling Tests

Cyclic load-displacement curves for nearly perfect and imperfect (PH200-W0 PH200-W12-2T) cylinders are shown in Figure 6. Buckling loads and displacements were 31.6ton, 2.30mm (PH200-W0) and 27.0ton, 2.03mm (PH200-W12-2T), respectively, and nearly four times of those in static tests (H200-W0, H200-W12-2T) as expected by the similarity rule. In these tests, maximum response displacements for the same acceleration input, which were 3.91mm (PH200-W0) and 4.52mm (PH200-W12-2T), were nearly equal each other and almost two times as large as the buckling displacements. The buckling load reduction of 15% (Wim=2T) in pseudo-dynamic tests was identical with those in static tests, and load-displacement skeleton curves under seismic loads were also similar to static test results. The total amounts of absorbed energy for nearly perfect and imperfect cylinders, calculated with experimental load displacement relations, agreed quite well regardless of the imperfection.

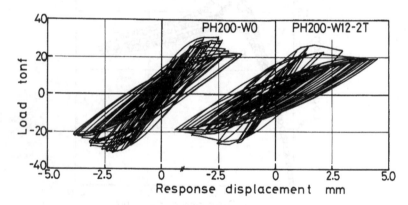

Figure 6. Comparison of load-displacement curves in dynamic test

DISCUSSIONS

Buckling Loads Estimations

Table 2 and Figure 7 show the comparison of experimental buckling loads Qexp in static and pseudo-dynamic tests with the evaluated load Qp given in equations (2) and (3), which were given in the interim buckling design guide [1][2].

$$Q_{el} = (\pi Rt) \ C_t \ Et/R \quad C_t = 0.8 \cdot 4.82(1+0.0239(L/\sqrt{Rt})^3)^{0.5}/(L/\sqrt{Rt})^2 \quad (2)$$

$$\frac{1}{Q_p{}^2} = \frac{1}{Q_{el}{}^2} + \frac{1}{Q_y{}^2} \quad \text{where} \quad Q_y = \pi Rt \ \sigma_y/\sqrt{3} \tag{3}$$

These formulae were based on the theoretical torsional buckling load with the capacity reduction factor of 0.8 and the quadratic interaction relation with the yield stress. As shown in Figure 7, it was clarified that some margin was recognized in Qp for nearly perfect cylinders and even for the imperfect cylinder whose amplitude was within a wall thickness. This margin was due to the capacity reduction factor mentioned above and conservative estimation of elastic shear buckling strength by torsional buckling equation. It was also clarified that the buckling load reduction was influenced significantly by the circumferential wavelength as shown in the test results of R167-W12-2T (Qexp/Qp=0.93) and R167-W6-2T (Qexp/Qp=1.08).

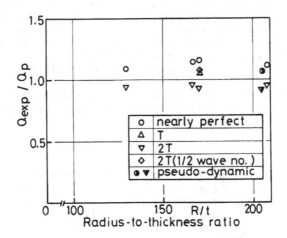

Figure 7. Comparison of buckling loads with estimated loads

Numerical Analysis

Numerical calculations using the doubly-curved thin shell elements in MARC (static tests) and nonlinear single-degree-of-freedom equation of motion (pseudo-dynamic tests) were carried out. In the static analysis, the numbers of element in circumferential and axial directions in 180° model were 38 and 9. The stress-strain curves were modeled by piecewise-linear relations and measured imperfections were expressed by the Fourier series expansion with the first 15 terms. From the numerical analysis, ratios of buckling load with initial imperfection to nearly perfect one are shown in Figure 8 as well as test results. It was clear that calculated buckling load reduction agreed well with test results and simulations were useful to evaluate the effect of imperfections on the buckling characteristics.

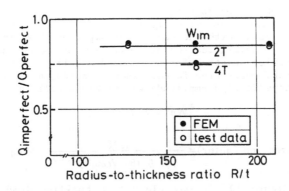

Figure 8. Comparison of buckling loads by FEM with test results

In the dynamic analysis, the equation of motion was numerically integrated using skeleton curves obtained by tests and empirical hysteresis rule. Figure 9 shows the relation between calculated maximum response displacements and the maximum acceleration inputs as well as test results. Calculated response displacements were 4.22mm (PH200-WO) and 4.27mm (PH200-W12-2T), these values agreed well with test results described before, and therefore, this nonlinear simulation method was verified. It was clarified that maximum response displacements were lower than linear response because of the energy absorption caused by plastic deformations before buckling and no immediate collapse occurred after buckling. This effect of response reduction associated with non-linearity factor (the ratio of an actual buckling displacement to a virtual linear buckling displacement) was incorporated in the interim buckling design guide [1][5].

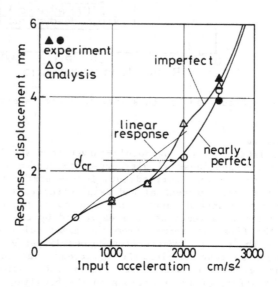

Figure 9. Comparison of response displacements with test results

Buckling Load Correction Factor

We proposed the following buckling load correction factor in the interim buckling design guide with the equivalent amplitude Weq, which was defined by the sum of the multiplication of An and Bn, where An and Bn were coefficients of Fourier spectrum for normalized buckling mode and measured imperfections. This correction factor can be considered as unity for the cylinder whose imperfection amplitude was within a wall thickness.

$$Weq = \Sigma \ An \cdot Bn \tag{4}$$

$$(Correction\ Factor) = \frac{1.18}{1 + 0.18\ (Weq/t)^{0.65}} \quad Weq/t \geqq 1.0 \tag{5}$$

As shown in Figure 10, it was clear that buckling load reduction with shape imperfections, obtained in this paper, were well explained using the above parameter Weq.

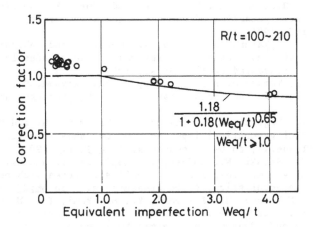

Figure 10. Correction factor of buckling load

CONCLUSIONS

Static and pseudo-dynamic buckling tests and numerical calculations were carried out to clarify the effect of geometrical imperfection on the shear buckling characteristics and the following conclusions were obtained.

1. The basic buckling characteristics, such as buckling loads and load-displacement relations, were found to be identical in the static and dynamic seismic loads.
2. In these tests, it was clarified that the shear buckling was less imperfection sensitive compared with axial one and quite stable even in the post buckling region.
3. Some margin of buckling load was recognized in the proposed equation for a nearly perfect cylinder and even for imperfect cylinders whose imperfection amplitude was within a wall thickness.
4. Buckling load reduction by FEM analysis and test results agreed quite well and calculated response displacements using the skeleton curves and empirical hysteresis rule agreed well with dynamic test results.
5. Maximum response displacements were lower than linear calculations because of the energy absorption caused by plastic deformation before buckling, and no immediate collapse occurred after buckling.
6. The buckling load was influenced by the imperfection amplitude and wavelength, therefore, we proposed here the buckling load correction factor using the equivalent imperfection amplitude.

ACKNOWLEDGEMENT

Authors gratefully acknowledge the helpful discussion of Prof. H. Akiyama and Prof. H. Ohtsubo (University of Tokyo) throughout the course of this work.

REFERENCES

1. Akiyama, H., Ohtsubo, H., Nakamura, H., Matsuura, S., Hagiwara, Y., Yuhara, T., Hirayama, H., Kokubo, K. and Ooka, Y., Outline of seismic buckling design guideline of FBR - A tentative draft, to be presented at 11th SMiRT, 1991.
2. Matsuura, S., Nakamura, H., Ogiso, S., Ooka, Y. and Akiyama, H., Buckling strength evaluation of FBR main vessels under lateral seismic loads, to be presented at 11th SMiRT, 1991.
3. Hagiwara, Y., Yamamoto, K., Nakagawa, M. and Akiyama, H., Seismic margin evaluation of FBR main vessels , to be presented at 11th SMiRT, 1991.
4. Murakami, T., Yoguchi, H., Hirayama, H., Nakamura, H. and Matsuura, S., The effect of geometrical imperfection on shear buckling strength of cylindrical shells, to be presented at 11th SMiRT, 1991.
5. Kawamoto, Y., Sasaki, N., Kodama, T., Hagiwara, Y. and Akiyama, H., The reduction of seismic response of shell structures with non-linear deformation characteristics, to be presented at 11th SMiRT, 1991.
6. Kokubo, K., Nakagawa, M., Kawamoto, Y., Murakami, T., Matsuura, S. and Hagiwara, Y., Corroboration of dynamic characteristics of FBR main vessels by pseudo-dynamic and dynamic buckling experiments, to be presented at 11th SMiRT, 1991.
7. Ogiso, S., Suzuki, M., Ooka, Y. and Matsuura, S., The state of the inelastic shear buckling analysis , and its application to low cycle fatigue estimation , to be presented at 11th SMiRT, 1991.
8. Galletly, G.D. and Blachut, J., Plastic Buckling of short vertical cylindrical shells subjected to horizontal edge shear loads , J. of Pressure Vessel Technology, 1985, Vol.107, pp.101-106.
9. Dostal, M., Austin, N., Combescure, A., Peano, A. and Angeloni, P., Shear buckling of cylindrical vessels benchmark exercise, Trans. of the 9th SMiRT, 1987, Vol.E, pp.199-208.
10. Tennyson, R.C. and Muggeridge, D.B., Buckling of axisymmetric imperfect circular cylindrical shells under axial compression, AIAA Journal, 1969 Vol.7, pp.2127-2131.
11. Wackel, N., Jullien, J.F. and Ledermann, P., Experimental studies on the instability of cylindrical shells with initial geometric imperfections, ASME PVP-Vol.89, 1984, pp.33-42.

Buckling of shells with local reinforcements

By Sigge Eggwertz, Bloms Ingeniörsbyrå, Sundbyberg, Sweden
Lars Å Samuelson, The Swedish Plant Inspectorate, Stockholm, Sweden

Abstract

The load carrying capacity of shells subjected to buckling is often considerably reduced due to initial imperfections and other disturbances. So far, current design codes address only the influence of initial imperfections and for instance local stress concentrations are left to the designer to consider. Short stiffeners attached to a thinwalled shell absorb some of the load carried by the shell and may give rise to severe stress concentrations. The present paper discusses the design of shells with such disturbances and a simple method of analysis is given. The results are checked through experiments and finite element calculations.

Introduction

The stability behavior of shells with a nearly uniform stiffness distribution and subjected to distributed loads is fairly well understood. Design recommendations are available for several cases in various design codes. Local disturbances, such as braces for support of external equipment, may lower the carrying capacity considerably. So far, very few investigations have been published on this subject which is of a fairly high practical importance. Local buckling in silo structures were noted by Bornscheuer, Ref [1]. A typical buckling pattern caused by a stiff door frame is shown in Fig. 1.

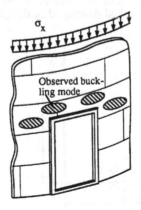

Figure 1. Local buckling pattern observed in a silo structure with a rigid door frame

A simplified approach to the buckling problem of shells with local stiffening was presented in Ref [2]. First the stress concentration in the shell caused by a stringer or partial ring was calculated by use of a simple method. This stress was then used to estimate the carrying capacity of the shell using the design rules for buckling under uniform loading or stress distributions.The method is similar to that proposed for design with respect to local loads according to Ref [3]. See Fig. 2. Since this method is based on the assumption that the maximum local membrane stress is to be used for the evaluation ot the buckling limit, it may be anticipated that the result is conservative.

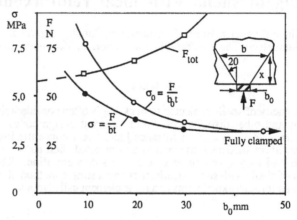

Fig. 2. Buckling loads and stresses of a circular cylindrical shell subjected to
local support loads, Ref.[3]

The validity of the simplified design method was checked by testing on small scale plastic cylinders. The cylinder buckling load was first determined without stiffeners to find the applicable reduction factor. Stringers with increasing stiffness and length were then attached to the shell and the buckling test was repeated. This was assumed to enable determination of the stiffening level where the disturbance starts to predominate over the initial imperfections, see Fig. 3. Since the wall thickness of the shell is only of the order of 0.2 mm it was not possible to measure the initial imperfections or the stresses in the shell. The stresses are, therefore, estimated by use of the simple method developed in Ref.[2] and by use of the finite element method (FEM).

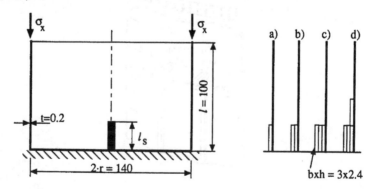

Figure 3. Stiffener arrangement for testing of the influence on the buckling load.

Basis for stability analysis

In shell stability analysis the membrans stress, caused by external pressure distributions, is used as design parameter. This may be expressed in the following manner:

$$\sigma_x \gamma_f \leq \frac{\sigma_u}{\gamma_m \gamma_n} \qquad (1)$$

So far, code design procedures only cover uniform, or nearly uniform stress distributions. If a stress concentration exists, local buckling may result, which may or may not lead to global collapse. It may be assumed, from the design point of view, that the local membrane stress should not exceed the carrying capacity of the shell as given for a uniformly distributed membrane stress field. The design criterion thus takes the form:

$$(\sigma_x + \Delta\sigma_x)\gamma_f \leq \frac{\sigma_u}{\gamma_m \gamma_n} \qquad (2)$$

where σ_u is the carrying capacity of the shell. Test results given in Ref.[3] indicate that the accuracy of Eq. (2) is reasonable if the stress field is dominated by the local stress $\Delta\sigma_x$. Application of the design value for a uniform stress distribution yielded a slightly conservative result. The smaller the area of the stress concentration the more conservative the result.

Estimation of the local membrane stress

A simple formula for the stress concentration $\Delta\sigma_x$ at the tip of a local stiffener of area A_s and length l_s was given in Ref.[2] for a cylindrical shell under axial compression. The formula was based on the assumption that a uniform stress distribution in the shell wall is disturbed by the stiffener and load is transferred to the stiffener through shear, see Fig. 4. It is rewritten in a slightly different form below:

$$\Delta\sigma_x = 2\sigma_x \left(1 - \frac{1}{1 + (A_s/l_s t)(E/G)} \right) \qquad \text{If } l_s < b_0 = 2\sqrt{rt}, \text{ set } l_s = b_0 \qquad (3)$$

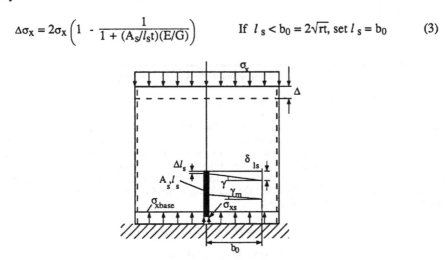

Figure 4. Assumptions regarding the stress redistribution in the vicinity of a local stiffener.

Since this is the stress level adjacent to the end of the stiffener, it is too severe to use this value in the design analysis. According to Ref.[3], the stress level at a distance $x_0 \approx 2\sqrt{rt}$ from the tip of the stiffener should be more relevant for design purposes. Application of the rules provided in Ref.[2] gives the following formula for the design stress:

$$\sigma_{eff} = \sigma_x + \Delta\sigma_x \frac{l_s - x_0}{l_s} \qquad (4)$$

Experimental investigation

A series of buckling tests was performed by use of two thinwalled plastic cylinders under axial compression, see Fig 3. The nominal geometries are identical for the cylinders but specimen No 2 had a significantly higher buckling load. This is probably because the initial deflections were smaller for specimen No 2.

The buckling load of the unstiffened cylinder was first determined through testing in a standard testing machine. Repeated loading showed that the buckling load was independent of previous loading history. A short stiffener, $l_s = l/4$ was then attached to the shell according to Fig. 3 and the buckling test was repeated. Subsequently the stiffener area was increased stepwise to a value $A_s = 12b_0t$ and the buckling load was determined for each configuration. The testing was repeated for different stiffener lengths $ls/l = 1/4$, 1/2 and 3/4. Some of the results are shown in Fig.5.

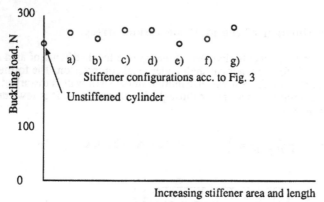

Figur 5. Test results test specimen No 1.

Although some of the buckling loads of test specimen No 1 with stiffening were sligthly below the value observed for the unstiffened shell, the test series seems to indicate that the load carrying capacity increases when local stiffeners are introduced. Test specimen No 2 gave a very similar result. Here only two stiffener configurations were tested both of length $ls/l = 1/2$. The second stiffener was made of aluminum with an in-plane stiffnes of approximately 6 times that of the plastic stringer. In both cases the buckling load of the cylinder remained practically the same as that of the unstiffened shell.

Numerical analyses

Linear stress distribution

A finite element model of the test cylinder, Fig. 6, was analyzed using $l_s/4$ and $l_s/2$ stiffener lengths respectively. The model, as shown in the figure, was subjected to a uniform axial displacement Δ_x, corresponding to an average axial stress σ_x =1.14 MPa. The membrane stresses in the shell are plotted in Fig. 7. The maximum stress caused by the short stiffener is 1.45 MPa indicating a stress concentration factor of 1.3. The corresponding figures for the longer stringer are 1.9 MPa and a stress concentration factor of 1.7.

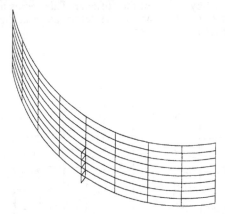

Figure 6. Finite element model of cylinder with local stiffeners.

Comparison with the results of Eq.(3) shows that the simplified formula yields a conservative estimate of the stress concentration factor and would thus be safe to use for design purposes.

Nonlinear collaps

Several parameters should be of interest in the analysis of the present problem. First of all the stability limit of a perfect shell with a local stiffener should be calculated in order to quantify a relevant knockdown factor. A second task would involve a parametric study of the interaction of the local stress concentration with the initial deflections of the shell. In the present investigation only the first task has been approached. Parametric studies of the interaction problem have been carried out by Teng and Rotter, Ref.[4] for a similar case. The results indicate that if the stress concentration coincides with an initial lateral deflection, a severe reduction in the load carrying capacity may result.

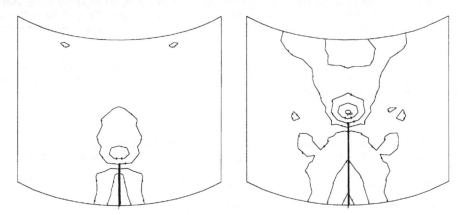

Figure 7. Stress concentrations in the two cylinders given in Fig. 6.

Non-linear collapse analyses were carried out for the two FE-models of Fig. 6. The collapse loads for the two cylinders were estimated at 450 and 560 N respectively which should be compared with the classical buckling load of the perfect cylinder of 720 N. The buckling pattern appeared to initiate at the boundary opposite to the stiffener and may originate from the nonuniform stress distribution caused by the tip of the stiffener. This seems to be in agreement with the observations during the tests where no visible deformation at the stiffener tip could be detected before the buckling load was reached.

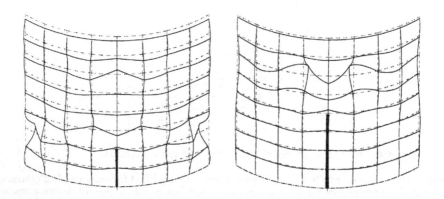

Figure 8. Typical results from nonlinear analyses of a cylinder with local stiffeners.

Suggested design method

The tests, Fig. 6, clearly show that adding local longitudinal stiffeners to the shell did not reduce the carrying capacity with respect to buckling. On the contrary, the local stiffeners seemed to increase the capacity even though the improvement is marginal. The reason for this behavior is that the shells may have had fairly large initial imperfections. The stress concentration caused by the stiffener did probably not coincide with any severe initial defect. In the second series the shell quality was higher but still no reducing effect of the stiffener on the capacity of the shell could be noted. These observations have led to the conclusion that the effect of the stiffener may be considered separate from that of initial imperfections in much the same way as for holes and cut-outs, see Refs.[5] and [6].

Assume that the shell is perfect and that a local stiffener causes a stress concentration according to Eq.(3). It is then reasonable to beleive that the buckling load of the cylinder is governed by the local stress at the tip of the stiffener:

$$\sigma_x + \Delta\sigma_x \le \sigma_{cr}$$

This implies that the function $\sigma_x /(\sigma_x + \Delta\sigma_x)$ may be regarded as a reduction factor for local stresses. Denoting $\sigma_x /(\sigma_x + \Delta\sigma_x) = \alpha_{ls}$, the design criterion would take on the following form:

$$\sigma_{u1} = 0.75 \, \alpha_{ls}\sigma_{cr} \qquad \text{Provided } \alpha_{ls}\sigma_{cr} \le f_y /2 \qquad (5)$$

$$\sigma_{u2} = 0.75 \, \alpha\sigma_{cr} \qquad \text{If } \alpha\sigma_{cr} \le f_y /2 \qquad (6)$$

Buckling in the plastic region may be handled by the interaction formula given in the ECCS Recommendations, Ref.[7],

$$\sigma_u = f_y[1 - 0.4123(\frac{f_y}{\alpha\sigma_{cr}})^{0.6}] \qquad (7)$$

The minimum value of σ_{u1} and σ_{u2} should be used in the design analysis. The procedure is shown in Fig. 9, where the results from the tests and analyses are included.

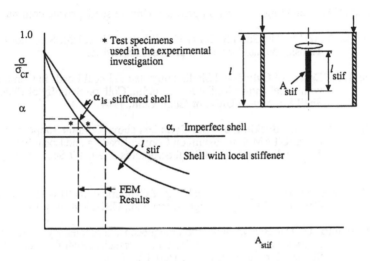

Figure 9. Suggested design procedure for shells with local stress concentrations.

Discussion

The carrying capacity of certain types of thin shells is extremely dependent on local disturbances such as initial imperfections, local forces (in-plane or lateral), support conditions, cutouts etc. Introduction of short stiffeners cause stress concentrations which may lead to premature buckling. The present study indicates that certain parameters valid for the design of cylindrical shells with local stiffening may be identified.

* Adding the stress concentration caused by the stiffener to the mean membrane stress and using the maximum value in the design condition, Eq.(2), clearly underestimates the carrying capacity of the shell with respect to buckling.

* It may be assumed that the local stress concentration provided by the stiffener reduces the load carrying capacity of the shell independently of the initial imperfections. Taking the reduction to be proportional to $\sigma_x/(\sigma_x + \Delta\sigma_x) = \alpha_{ls}$, the buckling loads of the two models analyzed above would be estimated with a reasonable accuracy.

* The stress concentration predicted by Eq.(3) is clearly higher for short stiffeners. This is in contrast to the FEM results where a longer stiffener yielded higher local stresses as should be expected. The collapse analyses, however, indicate that the shorter stiffener caused a lower critical load and use of the stress concentration factor from Eq.(3) may be reasonable.

Acknowledgements

The authors wish to express their gratitude to Miss Katarina Schyberg, The Royal Institute of Technology, Department of Materials Technology, who carried out the buckling tests and to Mr Carl von Feilitzen, The Swedish Plant Inspectorate for carrying out the numerical analyses.

References

[1] Bornscheuer, F.W., Buckling observed in practice, Unpublished, private communication.

[2] Samuelson, L. Å., (Ed.): The Shell Stability Handbook (in Swedish), Stockholm 1990. Translation into English in progress for Elsevier.

[3] Samuelson, L.Å.: Design of Cylindrical Shells Subjected to Local Loads in Combination with Axial and/or Radial Pressure. Int. Coll. on Stability of Plate and Shell Structures, Univ. of Gent, Gent, 1987.

[4] Teng, J. G., Rotter, J. M.: Buckling of Thin Cylinders above Column Support Terminations, IUTAM Symposium on Contact Loading and Local Effects in Thin-walled Plated and Shell Structures, Prag, 4-7 Sept. 1990, pp 589-585.

[5] Miller, C.D.: Experimental Study of the Buckling of Cylindrical Shells with Reinforced Openings. Chicago Bridge & Iron Co., CBI-5388, July 1982, 13 p

[6] Knödel,P., Schultz, U., Das Beulverhalten von biegebeanschpruchten Zylinderschalen mit grossen Mantelöffnungen. Berichte Versuchsanstalt für Stahl, Holtz und Steine, Karlsruhe, 4. Folge, Heft 12, 1985.

[7] ECCS: European Recommendations for Steel Construction. Buckling of Shells, ECCS Report No. 56, Brussels 1988.

ON THE PERFORMANCE OF DIFFERENT ELASTO-PLASTIC MATERIAL MODELS APPLIED TO CYCLIC SHEAR-BUCKLING

P. HORST[*], H. KOSSIRA[+] and G. ARNST[+]

[*] Deutsche Airbus GmbH, Hamburg, Fed. Rep. Germany

[+] Institut für Flugzeugbau und Leichtbau,

Technische Universität Braunschweig, Fed. Rep. Germany

ABSTRACT

The subject of this paper is the behaviour of cyclically shear-loaded panels, including buckling, plasticity and damage. The main objective of this paper is firstly to show how different the performance of material-laws may be, if cyclical buckling is existent and secondly how the resulting stresses and strains will influence the location of the initial crack and the main direction of crack-growth. Five different material-models have been tested; these are: a four-surface Mróz-model, a modified two-surface-model of the Dafalias-Popov-type, two models using a Phillips-Weng-type translational-law and a Tseng-Lee-model. Tests have been conducted with a specially designed test set-up.

INTRODUCTION

Thin-walled shear-panels are commonly used in lightweight structures. Also in cases of moderate plastic deformations, they exhibit a good-natured postcritical load-deflection behaviour. In emergency cases, if a multiple stress path is existing, this behaviour can be exploited, as one gets weight savings without a critical loss of stiffness. This kind of structural element has been examined by the authors for the last years. In this paper a comparison of five elasto-plastic constitutive relations is accomplished in order to extract essential characteristics of the material behaviour. All discussed experiments and calculations concern panels fabricated from the aluminium alloy 2024 T3 (resp. T351). Calculations have been conducted on the IBM 3090-600V computer of the Technische Universität Braunschweig.

MATERIAL MODELS

The material models used in this paper are based on the classical rate-independent plasticity theory, i.e. the normality rule and an additive decomposition of the strain tensor is employed. The application of models which include kinematic hardening is necessary, since the load direction changes and the pretreatment of the sheet metal causes a texture (L- and LT-direction). This texture should be taken into account by using an appropriate material model [1, 2].

There are a lot of different possibilities to define kinematic hardening material laws, as e.g. CHABOCHE shows in [3]. In case of the present study two- or multi-surface models have been used, since they represent a large group of different models, which may be compared to each other. The following five models have been used :

1. a multi-surface model like the one MRÓZ proposed in [4]

2. a two-surface model of the DAFALIAS-POPOV-type [5]

3. a two-surface model using a PHILLIPS-WENG-translational-rule [6]

4. a model of the preceding type, but with an additional deformation of the yield-surface, where the deformation of the yield-surface has been taken from [7]

5. a two-surface model of the TSENG-LEE-type [8]

According to the limited space, please look for further details on the test set-up and on the Mroz-model in [1] and on the Dafalias-Popov model in [2]. Details of model No. 3 and 4 are given in [9].

There are three essential distinctive criterions of the material models :

- the translational rule for kinematic hardening

- the way the elasto-plastic tangent-modulus is derived

- quantity and shape of the yield-surfaces.

As an universal translational rule of kinematic hardening has not yet been found, three different models have been used: first, the *Mróz*-rule which is employed in the models 1 and 2; second, the *Phillips-Weng*-rule which is employed in 3 and 4 and last but not least the *Tseng-Lee*-rule which is more or less a combination of the two preceding rules. Within the models 2, 3 and 4 the development of the elasto-plastic tangent-modulus is coupled with the kinematic rule through the direction of a certain distance measure. In model 5 exists a similar relation for the tangent-modulus, but the direction of the above mentioned distance measure is not connected with the applied kinematic rule. No direct connection is given in model 1. In general the translation direction of the yield-surface therefore has got an overwhelming influence on the stiffness-behaviour of the material. This statement comes true especially if nonproportional stress- or strain-pathes are concerned.

The shape of the yield surface has been chosen as follows :

All models base on the v.Mises type. The Mróz model has been modified according to Hill's anisotropic version, in order to include the effect of prestraining on the elastic limit in shear direction. In this case subsequent yield surfaces must all have the same anisotropic coefficients.

Model No.4 employs a model of REES which uses not only the second invariant of the stress-deviator, but also the third. The anisotropic coefficients are derived from plastic strains. The computational effort in this case is relatively high. Yield surfaces of 2024 T351 measured by ELLYIN ET AL. [10] can be simulated very accurate by this method, as shown in [9].

Parameters of the different material models have been determined from monotonic and cyclic stress-strain-curves of the material in an un-prestrained state and from four different stress-strain-curves in the prestrained state (L- and LT-direction in tension and compression). Additionally the yield surfaces from [10] have been used in order to obtain anisotropic parameters of model No. 4 and bialxial tests - accomplished at the Institut für Werkstoffkunde of the Technische Universität Braunschweig - have been made with given nonproportional stress-pathes in order to study the properties of 2024 T3 alloy.

Details about the mathematical description of the considered models are given in [1,2,9]. All of them are derived for plain stress, since a layered approach has been employed for the idealization of the elasto-plastic constitutive relation within the applied finite element code.

TEST SET-UP AND PARAMETERS OF SPECIMEN

The test set-up is shown in Fig. 1. The load - up to 250 kN - is applied by a hydraulic cylinder and is transformed into shear in the test panel. This set-up corresponds to a *picture-frame* test, but here edges are pin-jointed with the pins located exactly in the corners of the square plate. The test-panel is not penetrated as the pin-joints are parted in the middle. The frame is very stiff, in order to minimize deformations, thus gaining nearly pure shear and clamped boundary conditions which are of the same high quality in experiment and calculation.

The investigations were made with a panel-size of 500 x 500 mm². The thickness varies from 1.0 to 3.0 mm. This range is not only realistic in aerospace structures, it also excludes the case of real *plastic buckling*, where plastic deformations occur prior to buckling. It also circumvents the problem of a high imperfection sensitivity [1]. The shape of the entire test specimen is shown in [1]. Different parameters have been measured in order to characterize the buckling behaviour and the state of the elasto-plastic deformations. These are: the load P, the angle of shear γ, the out-of-plane deflection u_3 in one or more lines and the strain at different points of the panel.

RESULTS CONCERNING THE FIRST CYCLES

Due to the available space only three points shall be considered something more in detail: the behaviour of the panel during load-reversal, the strain-pathes and plastically deformed regions and the comparison of the investigated material models.

At first, some general aspects of the buckling behaviour have to be mentionened. An undamaged, perfect square-plate will buckle at a certain critical load. Under real circumstances this load is influenced by imperfections of different kind (e.g. geometrical, clamping etc.). As long as the aspect ratio of the panels does not exceed the given limits, the influence of these imperfections will decrease very soon and vanishes completely when the first plastic deformation takes place, as pointed out above. A nearly symmetrical main buckle appears within the diagonal of tension of the panel. More smaller buckles develop at higher load levels, parallel to the first buckle. There are two main regions of high plastic deformations, one on the concave side of the buckles and one at the edges, where the main buckle is constrained by the clamping. The remaining deformations influence the further buckling behaviour in a specific way.

Load-reversal
The range of load-reversal represents the most critical point of the behaviour of the cyclically shear-loaded panel, since remaining deformations from prior plastic loading act like geometrical imperfections each time when zero load is passed. These imperfections have a lasting influence on the further buckling behaviour. This influence of the load-induced imperfections is unpredictable due to small disturbances in the clamping area. Therefore it is not possible to predict the sign of the central deflection u_3^c after passing zero load. This is illustrated in Fig. 2, where the first load-cycle of two times four different test results are shown. All tests were accomplished with panels of a thickness of 1.4 mm. Each sub-figure includes two test results, one with and one without a change of the sign of u_3^c. The maximum load increases from Fig. 2a to 2d, but all other parameters are the same. These results support the statement of the last paragraph. Additionally to the test results, results from the finite element calculation, employing model No. 1 are given.

Two major statements may be derived from this investigations (Fig. 2):

o Since the remaining plastic deformations have to be assumed as nearly equal for the same maximum load, only disturbances from the edges can cause the 'random-like' behaviour after passing zero load.

o The amplitude of the remaining imperfections and therefore the disturbance of the developing buckle will grow with increasing maximum load.
As discussed in [1], intermediate states in the range of total unloading may be non-symmetrical and can lead to very rigorous snap-through effects.

Snap-through effects will occur in cases in which a real limit point of the equilibrum path has been reached by the system and no further increase of the load is possible without leaving this special path. The subsequent equilibrum state is reached by the snap-through. In the theoretical model the appropriate way to reach the next path will often not be a common quasi-Newton algorithm, since within the unstable part of the path plastic deformations may occur, which would not occur in the real snap-through case. This happens very likely if higher maximum loads are applied.

Although the snap-through behaviour is not perfectly reproduced in the theoretical model, as no time-dependent effects are included, it can be shown that the method of putting minimal disturbances on the system in order to change the equilibrum path acts very well. The comparison of the calculated value of the u_3^c-offset (deformation of the center of the panel in ϑ^3-direction) with that measured after a snap-through is an important clue in order to qualify the used material model, since u_3^c represents a measure of the total plastic deformation existing in the panel.

From all tests, having been conducted until now, it becomes obvious that the buckling behaviour of the second cycle is equal to the next cycles, as long as cracked parts of the panel are neglectable small. This fact is very usefull in case of constant amplitude tests.

Strain-Pathes and Plastically Deformed Regions

From theoretical computations - and this is valid regardless of the special material model that has been used - first plastic deformations will occur at about $\gamma = 0.2°$, as long as the panel dimensions are in the range mentioned above. The first plastic deformation takes place at a point exactly in front of both corners in the diagonal of tension. This spot is located at the concave side of the main buckle. At the opposite side of the plate plastic deformations occur just at the edge in the vicinity of the point where the plasticity starts. With increasing load this plastically deformed regions grow not only in the plain of the panel but also in thickness-direction.

It is an essential fact that in case of a cyclically shear-buckled panel wide parts of the panel will pass through a complete change of the stress- and strain-direction in the vicinity of total unloading. The description of the material properties in those cases of rapid changes in the direction of the stress or strain path is a critical point in the performance of the different models, which considerably influences the numerical solution of the buckling behaviour of the panels. From biaxial tests and their simulation with different material models it became obvious that none of the models is able to reproduce the stress-strain-relation of non-proportionally loaded test-specimens exactly; but there are severe differences which will be shown in the next chapter.

The regions of the panels where such a rapid change of the stress- and strain-direction takes place depend on the question whether a change of the sign of the central deflection takes place or not. These regions are given in Fig. 3 as a generic sketch. It gets apparent that the relevant regions of the panel are quite extensive, so the effects are not neglectible if the maximal loads are high enough.

Comparison of Different Material Models

For comparison of the five different models quoted above, eight tests shall be used. The results of these tests are shown in Fig.2. They allow a systematic examination of the influence of the maximal load and of the different kinds of load reversal behaviours. These tests are representative for some further tests with other panel thicknesses. In all cases the $\gamma - u_3^c$ - curves of the first and second load cycle have been considered in order to compare the buckling behaviours.

With respect to the qualification of the performance of the models the following examinations treat only the central deflection of the panel, which is more an integral measure of the buckling behaviour than the strains at special spots of the panel. But strain measurements widely support the following statements which are derived from Fig. 4 and 5.

Fig. 4 shows a $\gamma - u_3^c$ - curve in case of $\gamma_{max}/\gamma_{yield} = 2.0$, i.e. the angle of shear at maximal load exceeds the angle of shear at onset of yielding by a factor of two. For comparison the results from a test and of the considered five material models are plotted. Within the first monotonic loading the results of all models agree very well with test results. (A slight improvement of the calculated values can be obtained by using a mesh of more elements, but this would increase the computational effort to some extent.) During the first loading, stresses and strains show a mostly radial path from the origin of the stress- respectively strain-space. So the effects quoted above are not valid yet and the loading of the material almost correseponds to the tests that have been accomplished in order to determine the parameters of the models

At the negative maximum load two observations can be made; and these effects will become more extreme in cases of higher maximum loads:

○ The compliance of the material simulated with model 3 and 4 - which both employ the *Phillips-Weng* translation rule - is much higher than the compliance which is computed using the other models and even higher than that observed in the test. A higher compliance respectively a smaller hardening-modulus will result in a smaller central deflection, as the material is not forced to such an extent to get out of the mid-plane of the panel.

○ The second observation which can be made is that the local strain- and stress-pathes show rapid changes of their directions: After a high plastic deformation in one strain direction and an appropriate kinematic hardening of the material, a change of the external loading direction - even at the origin of the stress space - leads to a highly non-proportional loading of the material with respect to the origin of the translated yield surface.

As mentioned above, the translational rule and the applied evolution-rule of the the elasto-plastic tangent modulus are closely connected in the models 2, 3 and 4, which are obviously not able to reproduce the characteristics of the material as good as the models 1 and 5. The models 3 and 4 show a rather decreasing stiffness if non-proportional loading occurs, as the distance of the yield and the bounding surface is reduced very fast. In case of model 2, another effect, which is called *overshooting* in literature, occurs if high maximal loads are applied. This effect results in an instability of the calculation. Due to these facts, these models had to be abolished.

Fig. 5 shows the very good performance of the models 1 and 5 even if higher loads are applied. It cannot be decided yet which of them should be prefered, as both indicate results of similar quality for the global buckling behaviour. Model 1 is more or less insensible to non-proportional loading. Model 5 is especially designed for this case, but it got apparent from biaxial tests that even this model includes some deficiencies, which may explain the small differences between numerical and test results in Fig. 5.

CRACKINITIATION AND CRACKGROWTH

During the last years different tests have been conducted in order to evaluate the damage behaviour of panels at high loading cycles. The main objectives of these tests have been to determine the number of cycles until cracks have been initiated and the further crackgrowth behaviour.

The first crackinitiation occurs in a very small region; it is situated at the edge of the panel about 50 mm apart from one of the corners. There is no specific corner which is always the site of the crackinitiation, but it seems to be very likely that the crack will occur at an edge which is parallel to the L-direction of the pretreatment. The region of the crackinitiation is located where the main buckle interacts with the clamped boundary and therefore leads to high elasto-plastic and additionally non-proportional strains. All specimens were tested until now with $R = P_{min}/P_{max} = -1$.

Due to the fact that crackinitiation and crackgrowth occur in a region which is fully plastically deformed, it is not possible to apply ordinary linear elastic fracture mechanics and even the finite element mesh used in this study is not fine enough to cope with problems of crack analysis. But from finite element analysis not only the site of the crackinitiation but also the direction of the the crackgrowth may be interpreted. There are two main directions of crackgrowth: first, a crack beginning at the edge leading into the direction perpendicular to the diagonal of tension and second, along the main buckle. The first crack will not always show the greatest growth rate, but as within this region very high tension occurs, this crack will grow fast in the final stage and will result in the collapse of the panel. A typical crack path is shown in Fig.6. The initial crack occured in this specimen after cycling about 3800 times at a load amplitude of 80 kN. The crack propagation started slowly without significant changes in the load carrying capacity until at load cycle 4200 the accelerated crack propagation started and caused the total failure after some more cycles. Some typical results of these tests are shown in [1].

ACKNOWLEDGEMENT

The authors gratefully acknowledge the financial support of parts of their investigations by the Deutsche Forschungsgemeinschaft (Sonderforschungsbereich 319 regarding elasto-plasticity) and the Ministerium für Wissenschaft und Kunst des Landes Niedersachsen (regarding durability tests).

REFERENCES

1. Horst, P. and Kossira, H., Cyclic Buckling of Aluminium Panels, In Proc. of the 17th ICAS Congress, Stockholm, 1990

2. Kossira, H. and Horst, P., Cyclic Shear Loading of Aluminium Panels with Regard to Buckling and Plasticity, Thin-Walled Structures,1991, 11, 65-84

416

3. Chaboche, J.L., Time-independent constitutive theories for cyclic plasticity, Int.J.Plasticity, 1986, **2**, 149-188

4. Mróz, Z., On the description of anisotropic workhardening, J.Mech.Phys.Solids, 1967, **15**, 163-175

5. Dafalias, Y.F. and Popov, E.P., Plastic internal variables formalism of cyclic plasticity, J.Appl.Mech., 1976, **43**, 645-651

6. Phillips,A. and Weng, G.J., An analytical study of an experimentally verified hardening law, J.Appl.Mech., 1975, **42**, 375-378

7. Rees, D.W.A., The theory of scalar plastic deformation functions, ZAMM, 1983, **63**, 217-228

8. Tseng, N.T. and Lee, G.C., Simple plasticity model of the two-surface type, J.Engg.Mech., 1983, **109**, 795-810

9. Horst, P., Zum Beulverhalten dünner bis in den plastischen Bereich zyklisch durch Schub belasteter Aluminiumplatten, ZLR-Forschungsbericht 91-01, Braunschweig, 1991

10. Ellyin, F. and Neale, K.W., Effect of cyclic loading on the yield surface, J.Press.Vess.Tech., 1979, **101**, 59-63

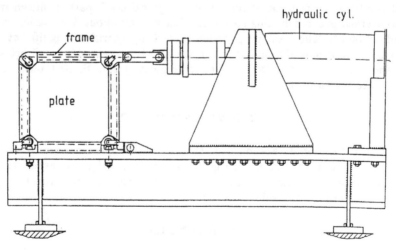

Fig. 1 Test set-up

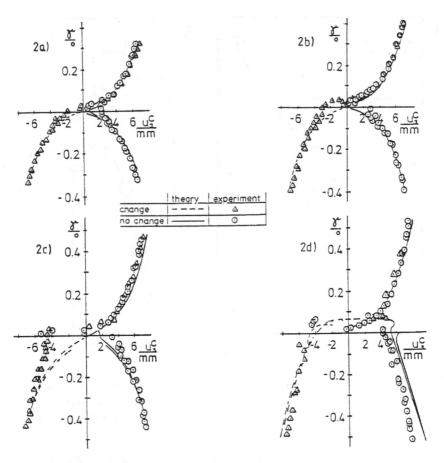

Fig. 2 Buckling behaviour during the first cycle with different maximum loads and different load pathes after unloading

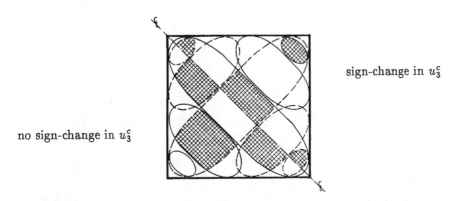

sign-change in u_3^c

no sign-change in u_3^c

Fig. 3 Regions with rapid change of strain directions depending on the load-deflection path

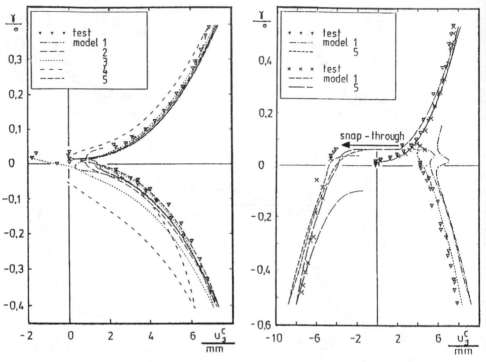

Fig. 4 Comparison of the different models with test results: angle of shear amplitude: $\hat{\gamma} = 0.42°$

Fig. 5 Comparison of the different models with test results: angle of shear amplitude: $\hat{\gamma} = 0.54°$

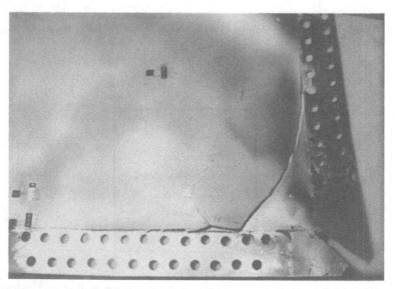

Fig. 6 Typical crack-pattern after total faliure

A SIMPLE GENERALIZATION FOR TIMOSHENKO'S MODEL PROBLEM OF THE STABILITY OF INFINITE BAR IN ELASTIC MEDIA AND SOME OF ITS APPLICATIONS

SERGEI GRISHIN
The Institute for Problems in Mechanics
of the U.S.S.R. Academy of Sciences,
prospect Vernadskogo 101, Moscow 117526, U.S.S.R.

ABSTRACT

The Timoshenko's model problem about the stability of an infinite bar under a longitudinal force in the media with Winkler's behaviour has been solved explicitly. Its analogue for a finite number of bars interacting trough a similar media is examined in this paper. In a particular case the formula generalizing the Timoshenko's one is obtained. In general case it is necessary to solve numerically a standard and well investigated eigenvalue problem for a family of Jacobian matrices. The studied problem may be interesting in composite structure design.

Timoshenko's scheme

In 1907, in connection with the problem of termostability of long rails S.P.Timoshenko formulated and solved the problem of the buckling of an infinite bar under a longitudinal force in the media which behaves as Winkler's basis. From mathematical point of view this problem is very simple. The equation for deflection υ is (in exposition we follow [1]):

$$EI\upsilon'''' + P\upsilon'' + W\upsilon = 0 \qquad (1)$$

where E is Young's modulus of bar material, I is a moment of inertia of bar cross-section, P is compressive longitudinal force, W is Winkler's coefficient. All these constants are positive. The prime denotes a derivative with respect to the longitudinal coordinate x. The characteristic equation

$$EI\lambda^4 + P\lambda^2 + W = 0 \qquad (2)$$

may be considered as biquadratic equation for λ, depending on the real parameter P which grows from zero. It is easy to see that the solutions υ bounded when x tends to infinity exist iff λ is pure imaginary. Denote $\mu = -i\lambda$, put this substitution into (2) and express P as function on real variable $\mu^2 \in (0, +\infty)$:

$$P = EI\mu^2 + W\mu^{-2} \qquad (3)$$

When $\mu^2 = \mu_x^2$ this function has an unique minimum P_x:

$$P_x = 2(WEI)^{1/2} ; \qquad \mu_x^2 = W^{1/2}(EI)^{-1/2} \qquad (4)$$

The last formulas are the immediate consequences from the condition of equality of roots for equation (2). Note that in practical engineering (4) is often used for estimation of critical values of forces buckling a finite bar in forms with many waves. Let us mention also that unlike our exposition, Timoshenko himself used the Fourier's series technique to a bit more general problem [2], but it does not matter for the following.

Generalization

The simplest natural generalization of Timoshenko's scheme is a buckling problem for a pack of bars connected by Winkler's media (Figure 1).

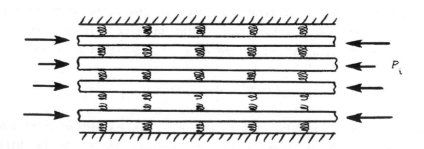

Figure 1.

The bars are enumerated by subscript $i=1,\ldots,N$. The subscript values $i=0$ and $i=N+1$ are associated with upper and lower boards respectively. We take the equation for deflection v_i of each bar in previous form:

$$E_i I_i v_i'''' + P_i v_i'' = q_i \qquad (i=1,\ldots,N) \qquad (5)$$

We calculate the surface loads according to Winkler's hypothesis:

$$q_i = w_{i-1,i}(v_{i-1} - v_i) + w_{i,i+1}(v_{i+1} - v_i) \qquad (i=1,\ldots,N) \qquad (6)$$

The substitution of (6) into (5) give a homogeneous system of ordinary differential equations with constant coefficients. It is convenient to write it in vector form:

$$\mathbf{A}v'''' + \mathbf{P}v'' + \mathbf{W}v = 0 \qquad (7)$$

Here as before the prime denotes a derivative with respect to longitudinal coordinate x; $\mathbf{v}$ is a vector of dimension N with v_i as components; $\mathbf{A}$ and $\mathbf{P}$ are diagonal matrices with the positive values $\alpha_i = E_i I_i$ and P_i respectively as entries; $\mathbf{W}$ represents the three-diagonal symmetrical matrix of following form:

$$\mathbf{W} = \begin{matrix} w_{0,1}+w_{1,2} & -w_{1,2} & & & \\ -w_{1,2} & w_{1,2}+w_{2,3} & -w_{2,3} & & \\ & -w_{2,3} & w_{2,3}+w_{3,4} & -w_{3,4} & \\ & & \cdots\cdots\cdots\cdots & & \\ & & & -w_{N-1,N} & w_{N-1,N}+w_{N,N+1} \end{matrix} \qquad (8)$$

We have the possibility to insert an homogeneous (even non-linear) compression of all pack from the side of rigid boards by good choice of Winkler's coefficients $w_{i,i+1}$. The functions $v_i(x)$ must be regarded as perturbations of principal homogeneous solution in this case. We may also eliminate the boards by putting $w_{0,1}$ and $w_{N,N+1}$ zero.

The characteristic equation for (7) must be written in the form:

$$\det(\mathbf{A}\lambda^4 + \mathbf{P}\lambda^2 + \mathbf{W}) = 0 \qquad (9)$$

The roots λ with non-zero real parts correspond to the unbounded solutions when x tends to infinity, therefore only pure imaginary λ are interesting. Putting $\mu=-i\lambda$, we obtain the real equation

$$\det(A\mu^4 - P\mu^2 + W) = 0 \qquad (10)$$

When all P_i are functions of common parameter ρ, growing from zero, the equation (10) can be solved step by step on ρ. For each step it is necessary to look for a real root μ^2 of a polynomial of degree $2N$ in a bounded interval. This interval may be taken for example as $(0; \max_{i=1\ldots N} P_i/a_i)$ because of the result of direct calculation: $\det W > 0$ and the fact that the positive values added to the leading diagonal of W make this determinant greater.

This is a general case, but the most nice solution can be obtained in the case of "positive proportionality":

$$P_i = \rho B_i \quad ; \; B_i > 0 \qquad (i=1,\ldots,N) \qquad (11)$$

Denote by B^{-1} the inverse matrix to B. It is also diagonal matrix with $1/B_i > 0$ as entries. After left- and right-multiplications of (7) (with the substitution (11)) by $\mu^{-1}B^{-1/2}$ and simple speculation we obtain the following equation - analogue to (3):

$$\det(AB^{-1}\mu^2 + B^{-1/2}WB^{-1/2}\mu^{-2} - \rho E) = 0 \qquad (12)$$

Here E represents the unit matrix. The last equation demonstrates that the buckling stress parameter ρ must be an eigenvalue of a single-parametric family of matrices

$$AB^{-1}\mu^2 + B^{-1/2}WB^{-1/2}\mu^{-2} \qquad (13)$$

with remarkable structure. Such matrices are called [3,4,5] Jacobian matrices. All their eigenvalues are real and different. When $\mu^2 \to +\infty$ the diagonal member in (13) becomes principal, all eigenvalues grow proportionally to μ^2. When $\mu^2 \to +0$ the Jacobian member in (13) becomes principal, the moduli of eigenvalues grow as μ^{-2}. The eigenvalues are continuous functions of μ^2 for any μ^2, therefore if some eigenvalues tend to $-\infty$ when $\mu^2 \to +0$, it have to be some μ^2 which gives $\rho = 0$. This is in conflict with reality because of (10) and (11). So in a quarter of plane of positive parameters μ^2, ρ, the equation (12) defines exactly N uncrossing curves $\rho_i(\mu^2)$, which are absolutely analogous to curve (3). The solution of the problem exists and is given by the following formula:

$$\rho_* = \min_{\mu \in (0, \infty)} \min_{i=1 \ldots N} \rho_i(\mu) \tag{14}$$

Plates

In Cartesian coordinates for a pack of plates under x-axis compression from the equation of the buckled plate [1] we obtain (7) where $\alpha_i = D_i$ ($D_i = \frac{1}{12} E_i h_i^3 / (1 - \nu_i^2)$ is flexural rigidity of plate with thickness h_i in material with moduli E_i, ν_i). The Winkler's coefficients $w_{i,i+1}$ for a soft linear elastic (E, ν) layer of thickness H can be calculated with the help of the formula: $w = H^{-1} E / (1 - \nu^2)$. The parameters P_i (or B_i) are completely defined by particular loading conditions.

The case for a pack of narrow annular plates under hydrostatic pressure (Figure 2, boards are not shown) is more interesting.

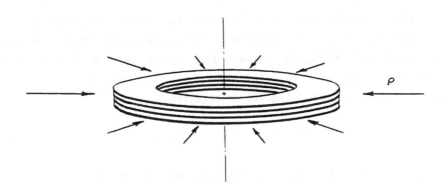

Figure 2.

We suppose the essential state of each plate to be planar. When inner and outer pressures give compressive integral hoop force and when the edges of plates are free or almost free, the rings can buckle out of their plane like bars. We suppose a priori the inner pressure to be equal to zero and we shall find the "bar-like" forms of buckling $v_i(\varphi)$. We take into account the pre-critical hoop compression and Winkler's normal reaction only by integrals over the widths of plates. After these assumptions the main equation of stability for each plate [6] becomes the following ordinary differential equation:

$$d_i D_i R_i^{-4}(v_i'''' + 4v_i'') + P_i R_i^{-2} v_i'' = q_i(\varphi) \qquad (i=1,\dots N) \qquad (15)$$

Here N is a number of plates, R_i are average radii, d_i are plate widths, the prime denotes a derivative with respect to circular coordinate φ. Unlike (5) the relationship (15) contains a supplementary term with v_i'', which appears due to bilaplacian.

The hoop forces P_i can be calculated as a result of a following static investigation . The soft rings, described by Winkler's model, transfer the lateral thrust to the stiffening ones. So P_i must be compensated by the loads ρ_i' which appear as a result of pressure ρ acting on the outer edges of stiffening and neighbouring soft half-thickness rings. Therefore we have

$$P_i = \rho_i'(R_i + {}^{1}\!/_{2} d_i); \quad \rho_i' = \rho(h_i + {}^{1}\!/_{2}(H_{i-1,i} + H_{i,i+1})) \qquad (i=1,\dots,N) \qquad (16)$$

Winkler's coefficients can be obtained for example by integration of basic coefficient $H^{-1}E(1-\nu^2)$ over the width of "interaction domain" between the neighbouring stiffening rings. This width may be defined only approximately, that is especially correct for the rings of different geometry. Finally we obtain the vector equation

$$\mathbf{A}v'''' + (\mathbf{B}\rho + \mathbf{C})v'' + \mathbf{W}v = 0 \qquad (17)$$

which differs from (7),(11) by the supplementary diagonal term $\mathbf{C} = \{4d_i R_i^{-4} D_i > 0\}$ which appears because of coordinate curvilinearity. The family of matrices analogous to (13) is

$$\mathbf{A}\mathbf{B}^{-1}\mu^2 - \mathbf{C}\mathbf{B}^{-1} + \mathbf{B}^{-1/2}\mathbf{W}\mathbf{B}^{-1/2}\mu^{-2} \qquad (18)$$

The spectral asymptotics at infinity ($\mu^2 \to \infty$) does not depend on the supplementary term, as opposed to the case $\mu < 2$. In this case the diagonal term in (18) decreases the leading diagonal terms in Jacobian part of (18). Therefore the eigenvalue $\rho = 0$ can be obtained for some μ by a special choice of matrix elements. This imperfection of the model may be explained by hypotheses, accepted above the equation (15). In great number of practically important cases, when these hypotheses are concluded verisimilitude, the behaviour of eigenvalues of the families (18) and (13) is analogous. It is new in comparison with (13) and very important that the number 2π should be one of the periods of the solution $\mathbf{v}$. It yields immediately that μ must be a natural number only: $\mu = 1, 2, \ldots$. Therefore (14) gives the lower estimate for an actual critical pressure

$$\rho_* = \min_{\mu=1,2,\ldots} \ldots \min_{i=1\ldots N} \rho_i(\mu) \tag{19}$$

Using (19) one has to compute only a finite number of eigenvalue problems because of the character of spectral asymptotics as $\mu^2 \to +\infty$.

Formula

In particular case when a Jacobian matrix is also a Toeplitzian one (or quasi-Toeplitzian) the roots of its determinant can be find explicitly [5]. We consider a symmetrical three-diagonal matrix **M** with some number $2Z$ in the leading diagonal and -1 in adjacent parallel diagonals. We solve the equation det **M**=0. Let D_n be the determinant of $n \times n$ matrix of the form **M**. Write the recursive relationship

$$D_n = 2ZD_{n-1} - D_{n-2}; \qquad D_1 = 2Z; \qquad D_2 = 4Z^2 - 1 \tag{20}$$

According to [4,7] we consider the quadratic equation

$$x^2 - 2Zx + 1 = 0 \tag{21}$$

If its roots

$$x_{(1,2)} = Z \pm (Z^2 - 1)^{1/2} \tag{22}$$

are different (hence $Z \neq \pm 1$), [4,7] gives the following formula:

$$D_n = C_1 x_1^n + C_2 x_2^n \tag{23}$$
$$C_1 = (D_2 - x_2 D_1)/(x_1(x_1 - x_2)); \quad C_2 = -(D_2 - x_1 D_1)/(x_2(x_1 - x_2))$$

If $Z = \pm 1$, $D_n \neq 0$ $\forall n$. We will look for the real roots of equation $D_n = 0$ such that $|Z| < 1$. Set $x = x_1$, $\bar{x} = x_2$. Noting that $|x| = 1$ we can write $x = e^{i\varphi}$, $Z = \cos \varphi$. Calculating the powers of x and substituting them into (23) we obtain:

$$D_n = \sin n\varphi \cot an \, \varphi + \cos n\varphi \tag{24}$$

Then it follows immediately:

$$\varphi = k\pi/(n+1) \quad (k = 0, \pm 1, \pm 2, \pm 3, \ldots) \tag{25}$$

When $k = m(n+1)$ $(m = 0, \pm 1, \pm 2, \ldots)$ $Z = \mp 1$ and the formula (23) fails. Due to the equality $Z = \cos \varphi$ for any other k from the sequence (25) Z takes only n different values

$$Z_k = \cos(k\pi/(n+1)) \quad (k = 1, \ldots, n) \tag{26}$$

Being a polynomial in Z, D_n does not have other roots.

The matrices (13),(18) become Toeplitzian in a particular case when their elements are independent of the number of the plate or of the bar. This case arises for example when the pack of Figure 2 consists of plates of only of two kinds: all the rigid ones are identical and so are all the soft ones. Dividing of each row of (18) by $wb^{-1}\mu^{-2}$ we obtain (for the eigenvalue problem) the matrix similar to $\mathbf{M}$ with

$$2Z = aw^{-1}\mu^4 - cw^{-1} + 2 - pw^{-1}b\mu^2 \tag{27}$$

Here the small letters a, w, c, b denote the entries of matrices $\mathbf{A}, \mathbf{W}, \mathbf{C}, \mathbf{B}$ (13),(18), respectively. For (13) $c = 0$ and for (18) $c = 4a$. Substituting Z_k from (26) in place of Z in (27) and solving (27) with respect to p we will obtain the function $p(\mu^2, Z_k)$. Its minimum p_* is attained for $Z_* = Z_1$ (26) and $\mu = \mu_*$:

$$p_* = b^{-1}(-c + (8aw(1-Z_*))^{1/2}); \quad \mu_* = (2a^{-1}w(1-Z_*))^{1/4} \tag{28}$$

It is easy to see that for a single bar (n=1, c=0) formulas (28) take the form (4). For a pack of rings ρ_* may be negative like the general case. The note preceding (19) for a rings remains valid.

Shells

We assume that the equations of the "technical theory" in the "mixed form" are acceptable to describing of behaviour of sandwich stiffening elements [8]:

$$D\nabla^4 \upsilon - \nabla_k^2 F - q - T_1\chi_1 - T_2\chi_2 - 2S\tau = 0$$
$$\nabla^4 F + Eh\,(k_1\chi_2 + k_2\chi_1 + \chi_1\chi_2 - \tau^2) = 0 \qquad (29)$$

Notations are standard: υ = deflexion; F = stresses potential; $T_1,\ T_2,\ S$ = stresses; $k_1, k_2, \chi_1, \chi_2, \tau$ = curvatures, their increments and torsion; q = surface load; ∇^4 = bilaplacian; ∇_k^2 = Vlasov's operator; D = flexural rigidity; h = thickness; E = Young's modulus. We shall describe in a few words a canonical cases of cylindrical, conical and spherical shells.

For a pack of cylindrical shells after linearization of equations barlikely perturbed near an homogeneous solution the following analogue of (15) can be obtained:

$$D_i R_i^{-4} \upsilon_i'''' + R_i^{-2} P_i \upsilon_i'' = q_i(\varphi) \qquad (30)$$

Here R_i = radii of cylinders; $\upsilon_i(\varphi), q_i(\varphi)$ = perturbations of displacements and surface loads; derivatives are taken with respect to angular coordinate φ. Expressing q_i in terms of υ_i by Winkler's formula (6) after usual speculation we obtain a matrices of type (13). For a closed cylinder μ can be natural, for a unclosed one - great.

For a pack of conical shells our scheme can be used only in the case of very narrow truncated cones - conical belts - with free edges. The equations (29) cannot be separated, therefore after all, for perturbations we take $\nabla_k^2 F=0$. We have the following analogue of (15):

$$d_i D_i R_i^{-4}(\upsilon_i'''' + 4\sin^2\gamma\ \upsilon_i'') + P_i R_i^{-2}\upsilon_i'' = q_i(\varphi) \qquad (31)$$

Here d_i = widths of belts along the generatrixes; γ = angle between the axis of cone and its generatrix; R_i = distances from the centers of belt sections to the axis of cone; another notations are usual.

For spherical belts the situation is still more complex. Unlike the cones the straight generatrixes, weak with respect to flexure, are absent in this case. The Vlasov operator remains a partial differential operator and the latter assumption, which give us the ordinary equation, is more strong.

For any type of shells the cases which lead to formulas (28) demand too specific organization of pack.

ACKNOWLEDGEMENTS

The author is grateful to the organizers of Colloquium for financial support, to professor V.M.Aleksandrov for useful discussions of the problem and to all the colleges who helped him in preparation of this paper.

REFERENCES

1. Rabotnov, Yu.N., Mechanics of deformable solids, Nauka, Moscow, 1988, pp. 132-4, 415-7 (in Russian).
2. Timoshenko, S.P., Stability of elastic systems, Gostekhizdat, Moscow, 1955, pp. 121-5 (in Russian).
3. Gantmakher, F.R., Krein, M.G., Oscillatory matrices and small vibrations of mechanical systems, Gostekhizdat, Moscow-Leningrad, 1941, pp. 74-82 (in Russian).
4. Mishina, F.P., Proskurjakov, I.V., High algebra, Fizmatgiz, Moscow, 1962, pp. 116, 25 (in Russian).
5. Il'in, V.A., Kuznetsov, Yu.I., Three-diagonal matrices and their applications, Nauka, Moscow, 1985, pp. 85-90 (in Russian).
6. Volmir, A.S., Stability of deformable systems, Nauka, Moscow, 1967, pp. 445-50 (in Russian).
7. Proskurjakov, I.V. Collection of problems in linear algebra, Nauka, Moscow, 1970, pp. 33-4 (in Russian).
8. Grigoljuk, E.I., Kabanov, V.V., Stability of shells, Nauka, Moscow, 1978, p. 47 (in Russian).

Finite-Elements for Finite-Rotation and Stability Analysis of Composite Shell Structures

Yavuz Başar, Yunhe Ding and Reinhild Schultz
Institut für Statik und Dynamik
Ruhr-Universität Bochum, FRG

Introduction

Composite laminates consist of an arbitrary number of layers made of the same or different orthotropic material. The material axes of each layer can be oriented differently with respect to the laminate coordinates in order to result in desirable structural properties such as large strength-to-weight ratio and desired directional strength. This beneficial aspect has led to a rapidly increasing use of composite laminates in a variety of structures and, accordingly, to intensive research activities which cover both theoretical and numerical modelling of composite laminates. A detailed survey about the advances in plate/shell theories and finite element models used in the analysis of composite laminates and literature available have been given e. g. in [1,2].

Most of the available finite element models are, however, restricted to linear analysis or deal with only plate structures or uses orthogonal Cartesian coordinates for the formulation. Finite element models based on consistent tensorial shell theories with curvilinear coordinates and applicable to strongly nonlinear composite shell problems with finite displacements and rotations are scarcely to find in literature. Thus, the development of reliable and efficient finite element models able to consider the above cited aspects presents still a challenging task for the scientists and engineers.

The objective of this paper is to report some of our research work in the finite-rotation and stability analysis of arbitrary composite shell structures by means of layered finite-element models with which orthotropic material properties varying in the thickness direction can be considered in a general manner. The investigations start from finite-rotation shell theories of Kirchhoff-Love [3] and Mindlin-Reissner types [4] which are, using an index notation, transformed into displacement-based [5] and mixed finite elements [6], respectively. In both kinds of element families the constraints existing for the so-called difference vector which describes the rotational movement of the normal vector are considered at the element level numerically. This procedure contributes significantly to the accuracy of the numerical models developed.

The constitutive relations are formulated in a general form for a shell structure with an arbitrary number of reinforced layers where the fiber direction may vary pointwise with respect to laminate coordinates. A computer-oriented procedure is presented permitting the transformation of the given physical orthotropic material constants into tensorial ones associated with curvilinear laminate coordinates.

The efficiency of the numerical models developed to deal with finite-rotation phenomena of composite shells is, finally, demonstrated by two strongly nonlinear examples.

Theoretical fundamentals

The development of mixed finite elements is based on a five-parametric finite-rotation shell theory [4] where the deformation state of the shell continuum is described by the middle-surface displacements v_i ($i = 1,2,3$) and the so-called difference vector w_i. The variable w_i is associated with the rotational movement of the unit normal vector $\overset{\circ}{a}_3$ into its deformed position a_3. By virtue of the zero strain condition $\gamma_{33} = 0$ throughout the thickness the difference vector w_i is subjected to a constraint [7] so that the tangential components w_α remain together with v_i as independent variables. The satisfaction of the constraint in question which is nonlinear in the unknown quantity w_3 causes in the numerical implementation difficulties. To avoid its use and to permit a clear determination of the deformed normal vector a_3 in every nonlinear range the variables w_α are replaced by new rotational degrees of freedom ψ_α ($\alpha = 1,2$) which fix the position of a_3 with respect of a global reference frame [8]. The definition of the variables ψ_α and the transformations relating them to w_i can be found in [6]. We note that these relations are satisfied at the element level as subsidiary conditions.

The internal work of the finite-rotation theory is given by the expression

$$\delta^* A^i = -\iint_F (\tilde{N}^{(\alpha\beta)} \, \delta\alpha_{(\alpha\beta)} + M^{(\alpha\beta)} \, \delta\beta_{(\alpha\beta)} + \tilde{Q}^\alpha \, \delta\gamma_\alpha) \, dF$$

relating the 2-D strain variables $\alpha_{(\alpha\beta)}$, $\beta_{(\alpha\beta)}$ and γ_α to their dual force variables: the pseudo-stress resultant tensor $\tilde{N}^{(\alpha\beta)}$, the moment tensor $M^{(\alpha\beta)}$ and the pseudo-shear stress tensor $\tilde{Q}^\alpha$. The variable γ_α describes a constant shear deformation throughout the thickness while $\alpha_{(\alpha\beta)}$ and $\beta_{(\alpha\beta)}$ are called the first and second middle surface strains. The kinematic relations to be satisfied by the strain variables are highly nonlinear ones

$$\alpha_{(\alpha\beta)} = \alpha_{(\alpha\beta)}(v_i), \quad \beta_{(\alpha\beta)} = \beta_{(\alpha\beta)}(v_i,w_\alpha), \quad \gamma_\alpha = \gamma_\alpha(v_i,w_\alpha),$$

and presented explicitly, i.e. in [4].

As the constitutive relations are given in earlier papers for isotropic linear elastic shell structures we now deal with their extension to arbitrary composite laminates which are supposed to consist of an arbitrary number n of unidirectionally reinforced layers where fiber direction may vary pointwise with respect to the laminate coordinates. It is assumed that these layers are perfectly bonded together, preventing any relative motion on the interfaces of the individual layers. Each of the layers possesses three perpendicular planes of material symmetry and therefore exhibits orthotropic material properties. The material behavior itself is assumed to be linear elastic. The middle surface of the structure selected as reference surface is described by curvilinear Gaussian coordinates Θ^α ($\alpha = 1,2$). The ply angle α is defined by the fiber direction and the coordinate line Θ^1.

Let $\bar{E}^{klmn}$ be the physical components of the orthotropic elasticity tensor defined in a Cartesian reference frame which coincides with the principal material axes. For orthotropic material $\bar{E}^{klmn}$ possesses nine independent coefficients [9]. The tensorial coefficients E^{ijrs} to be used in connection with the curvilinear coordinates are then given by:

$$E^{ijrs} = \frac{\partial\Theta^i}{\partial y^k}\frac{\partial\Theta^j}{\partial y^l}\frac{\partial\Theta^r}{\partial y^m}\frac{\partial\Theta^s}{\partial y^n}\bar{E}^{klmn},$$

where the transformation coefficients $\frac{\partial\Theta^i}{\partial y^k}$ ($i,j,...,n = 1,2,3$) can be calculated in any point of the shell continuum in terms of the fiber direction α and the angle φ enclosed by the coordinate lines Θ^α ($\alpha = 1,2$).

Using the zero stress condition $s^{33} = 0$ throughout the thickness coordinat Θ^3 the elasticity tensor E^{ijrs} can be transformed into the reduced one:

$$C^{\alpha\beta\rho\lambda} = E^{\alpha\beta\rho\lambda} - E^{\alpha\beta33}E^{33\rho\lambda}/E^{3333}, \quad C^{\alpha3\beta3} = E^{\alpha3\beta3}.$$

The constitutive relations needed for the development of layered finite elements can be derived according to the standard procedure [8]. The corresponding results are for the constant shear deformation theory of the form

$$\tilde{N}^{(\alpha\beta)} = \overset{0}{C}{}^{\alpha\beta\rho\lambda}\alpha_{(\rho\lambda)} + \overset{1}{C}{}^{\alpha\beta\rho\lambda}\beta_{(\rho\lambda)},$$

$$M^{(\alpha\beta)} = \overset{1}{C}{}^{\alpha\beta\rho\lambda}\alpha_{(\rho\lambda)} + \overset{2}{C}{}^{\alpha\beta\rho\lambda}\beta_{(\rho\lambda)},$$

$$\tilde{Q}^\alpha = \overset{0}{C}{}^{\alpha3\beta3}\gamma_\beta,$$

with the material tensors of the reference surface defined as:

$$\overset{k}{C}{}^{\alpha\beta\rho\lambda} = \sum_{i=1}^{n}\int_{Z_{ui}}^{Z_{oi}}C_i^{\alpha\beta\rho\lambda}(\Theta^3)^k\,d\Theta^3 \quad (k = 0,1,2)$$

$$\overset{0}{C}{}^{\alpha3\beta3} = \overset{n}{\underset{i=1}{\Sigma}} \int_{zu_i}^{zo_i} C_i{}^{\alpha3\beta3} \, d\Theta^3 \ .$$

The elasticity coefficients $C_i{}^{\alpha\beta\rho\lambda}$ and $C_i{}^{\alpha3\beta3}$ are supposed to be constant in each layer the top and bottom faces of which are determined by the values zo_i and zu_i of the coordinate Θ^3.

The development of displacement-based finite elements is achieved by means of a Kirchhoff-Love type shell theory [3] which has been obtained from the above formulation by the well-known orthogonality condition $\gamma_\alpha = 0$. Accordingly, the difference vector w_i becomes a purely dependent variable to satisfy, in addition to constraint discussed above, two further conditions [5]. Also, these constraints will be considered at the element level numerically. A detailed discussion of the procedure to be used for this purpose is given in [3].

Finite-Rotation Elements

The basic concepts used in the finite element procedure can be summarized as follows: Nonlinear shell equations are treated by an incremental-iterative solution strategy. The quantities of the fundamental state needed for the calculation of the tangential stiffness matrix and unbalanced forces are calculated according to exact nonlinear equations.

Table 1   Displacement-based finite-rotation shell elements

	Displacement-based Finite-Rotation Elements	
	KLFR 54	KLFR 48
Theoretical fundamentals	Kirchhoff-Love Theory: v_i, w_i with 3 subsidary conditions for w_i	
Degrees of freedom	v_i, $v_{i,\alpha}$, $v_{i,\alpha\beta}$ 3 x 18	v_i, $v_{i,\alpha}$, $v_{i,12}$ 4 x 12
Interpolation for v_i	quintic polynomials by Cowper	bicubic Hermite-polynomials
Integration points	21	16

Table 1 and 2 give a short description of the displacement-based and mixed finite-rotation shell elements developed. Details about these elements have been presented in [3,4] and will not be repeated here.

Table 2   Mixed finite-rotation shell elements

Mixed Finite-Rotation Elements				
SDFR 20		SDFR 45		
A	B	A	B	C
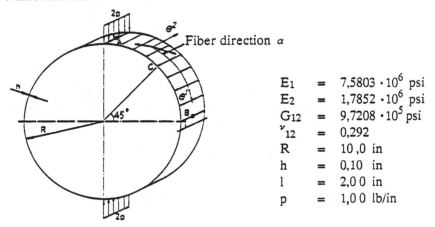				
Theoretical fundamentals	shear deformation shell theory: v_i, ψ_α			
Degrees of freedom	v_i, ψ_α 4 x 5		v_i, ψ_α 9 x 5		
Interpolations for v_i, ψ_α	bilinear polynomials		biquadratic polynomials		
Number of independent force parameters	8	14	32	36	38
Integration points	1 x 1	2 x 2	2 x 2	3 x 3	3 x 3

Numerical result

E_1 = $7{,}5803 \cdot 10^6$ psi
E_2 = $1{,}7852 \cdot 10^6$ psi
G_{12} = $9{,}7208 \cdot 10^5$ psi
ν_{12} = 0,292
R = 10,0 in
h = 0,10 in
l = 2,00 in
p = 1,00 lb/in

Fig. 1   Orthotropic circular ring under line loads

To demonstrate the efficiency of the finite elements developed for the analysis of composite shell structures undergoing finite displacements and, especially, finite rotations, two strongly nonlinear examples are given below.

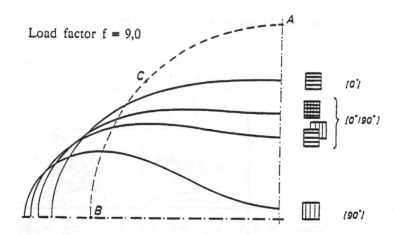

Fig. 2   Orthotropic circular ring: deformed configurations for f = 9,0

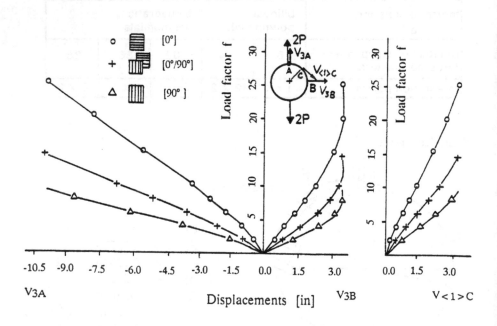

Fig. 3   Orthotropic circular ring: load-displacement diagrams

The first example is the geometrically nonlinear analysis of a circular ring. Geometry, finite element mesh and material properties are shown in Fig. 1. The structure is discretized by 32 quadrilateral displacement-based elements with 48 degreess of freedom. Three different cases of fiber directions have been considered. The load-displacement diagrams given in Fig. 3 show clearly that the structure possesses the largest stiffness when all the fiber directions concide with the principal load-carrying direction. Any deviation of the fiber directions from that reduces the stiffness of the system. When the critical load has been achieved for the circular ring reinforced in Θ^2 direction, the behavior of the structure reinforced in θ^1 direction remains still linear. The deformed configurations of the structure are shown in Fig.2 for the load factor $f=9{,}0$. By comparing the two variants for the 0°/90° lamination, the reduction of stiffness due to the coupling effect of stretching and bending in laminated composite structures can be observed obviously.

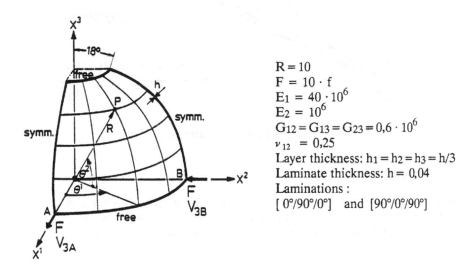

$$R = 10$$
$$F = 10 \cdot f$$
$$E_1 = 40 \cdot 10^6$$
$$E_2 = 10^6$$
$$G_{12} = G_{13} = G_{23} = 0{,}6 \cdot 10^6$$
$$\nu_{12} = 0{,}25$$

Layer thickness: $h_1 = h_2 = h_3 = h/3$
Laminate thickness: $h = 0{,}04$
Laminations:
[0°/90°/0°] and [90°/0°/90°]

Fig. 4   Hemispherical shell under point loads

As the second example, the hemispherical shell under two pairs of point loads shown in Fig.4 has been analysed by 16x16 quadrilateral mixed finite-rotation elements SDFR20-B. Two variants of layer fiber directions are considered. The load-displacement curves are plotted in Fig. 5. Compared with the displacements associated with the lamination [0°/90°/0°], the displacement at A in the case of lamination [90°/0°/90°] has been increased while the displacement at B has been reduced. The influence of fiber directions on the nonlinear behavior of the strucure is not very obvious in this case.

436

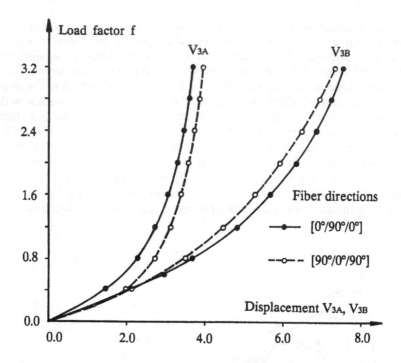

Fig. 5   Hemispherical shell under point loads

Acknowledgements

This work has been carried out within a research project "Nonlinear Analysis of Laminated Shell Structures" at the Ruhr-University. The financial support of the DFG- German Research Council is greatly acknowledged.

References

[1]   Reddy, J.N.: On Refined Computational Models of Composite Laminates. Int. J. Num. Meth. Engng. Vol. 27, 361-382, 1989.

[2]   Pandya, B.N. , Kant, T.: Higher-Order Shear Deformable Theories for Flexure of Sandwich Plates - Finite Element Evaluations. Int. J. Solids & Structures Vol. 24, No. 12, 1267-1286, 1988.

[3]   Başar, Y., Ding, Y.: Finite-Rotation Elements for the Nonlinear Analysis of Thin Shell Structures. Int. J. Solids & Structures, Vol. 26, No. 1, 83-97, 1990.

[4]   Başar, Y.: A Consistent Theory of Geometrically Nonlinear Shells with an Independent Rotation Vector. Int. J. Solids & Structures, Vol. 23, No. 10, 1405- 1415, 1987

[5] Ding, Y.: Finite-Rotations-Elemente zur geometrisch nichtlinearen Analyse allgemeiner Flächentragwerke. Dissertation am Institut für Statik und Dynamik, Ruhr-Universität Bochum, 1989.

[6] Başar, Y. and Ding, Y.: Theory and Finite-Element Formulation for Shell Structures undergoing Finite Rotations. Advances in the Theory of Plates and Shells , edited by G. Z. Voyiadjis and D. Karamanlidis, Elsevier Science

[7] Ramm, E.: Geometrisch Nichtlineare Elastostatik und Finite Elemente, Habilitationsschrift Bericht Nr. 76-2, Institut für Baustatik der Universität Stuttgart, 1976.

[8] Başar, Y. and Krätzig, W.B.: Mechanik der Flächentragwerke. Friedr. Vieweg, Braunschweig, 1985.

[9] Jones, E.M.: Mechanics of Composite Materials, McGraw-Hill Book Company, New York , 1975.

MIXED FORMULATION FOR SHELLS
AND MESH REFINEMENT

Iñigo A. Arregui[1,2] - Philippe Destuynder[1] - Michel Salaün[1]

[1] Institut Aérotechnique (I.A.T.) - 15 rue Marat - 78210 Saint Cyr l'Ecole - FRANCE

[2] Depto. de Matematica Aplicada - E.T.S.I. Minas - Rios Rosas, 21 - 28003 Madrid - SPAIN

ABSTRACT

A mixed formulation for a Koiter thin shell model is presented here. As many authors did, the rotation rate of the normal appears as a new unknown, but the Kirchhoff-Love condition is taken into account by using duality techniques : the Lagrange multiplier associated to this relation is the transverse shear stress. Moreover, in order to study the non linear phenomena (crash or stamping, e.g.), we have introduced in our formulation a geometrical approximation of the middle surface of the shell. Then, we are allowed to make correction of the geometry at each step of such non-linear processes. Finally, automatic refinement of the mesh has been introduced in our code. An application of our approach is given at the end of this paper : it is the classical case of the pinched cylinder.

INTRODUCTION

The mechanical model we consider is one of the thin shell models developed by W. T. Koiter [11], in which Kirchhoff-Love assumptions for displacements are needed.

The main drawbacks of direct formulations is that a conformal finite element approximation needs C^1 elements, which are expensive and not always very efficient for singular or non-linear problems. So many authors have introduced the rotation rate of the unit normal as an independent unknown of the problem, the link between this quantity and the displacements (membrane and deflection) being obtained by various techniques : penalty methods (QUAD 4 of McNeal [12] e.g.), discrete methods (DKT, DKQ, cf. Batoz-Dhatt [2] e.g.), for example.

Moreover, those formulations present bending locking phenomenon [3], which appears particularly when the geometry of the shell middle surface is approximated by flat elements.

First of all, we recall some basic aspects concerning shell theory. In a second step, the mixed formulation and the geometrical approximation are presented. In a third part, we explain the technique used for mesh refinement. Finally, an example is given.

I. THE KOITER MODEL FOR SHELLS

I.1 Notations

We consider a thin elastic shell, the medium surface of which is denoted by ω. This surface is described by a mapping, say ϕ, which transforms a plane open set $\widehat{\omega}$ onto ω; this mapping ϕ is supposed to be at least $C^3(\overline{\widehat{\omega}})$. It is also necessary to assume that the boundary γ of ω corresponds to the one of $\widehat{\omega}$, denoted by $\widehat{\gamma}$.

Hence a point m on ω is defined by $m = \phi(\xi^1, \xi^2)$, where (ξ^1, ξ^2) are the coordinates in the plane containing $\widehat{\omega}$. Then, the *tangent plane* to ω at each point m is spanned by the two vectors $\underline{a}_\alpha = \phi_{,\alpha}$, supposed to be linearly independent.

Remark 1

We note $._{,\alpha}$ for $\dfrac{\partial \cdot}{\partial \xi^\alpha}$. Furthermore, greek indices are supposed to belong to the set $\{ 1 , 2 \}$. Finally, implicit summation over repeated indices is assumed from 1 to 2. ∎

Thus the *unit normal* to ω is defined by : $N = \dfrac{\underline{a}_1 \wedge \underline{a}_2}{\|\underline{a}_1 \wedge \underline{a}_2\|}$.

The *curvature* of ω is characterized by the tensor : $b_{\alpha\beta} = (N, \underline{a}_{\alpha,\beta})$, where $(.,.)$ is the scalar product in $\mathbb{R}^3$ and $\underline{a}_{\alpha,\beta}$ is the partial derivative of $\underline{a}_\alpha$ with respect to ξ^β.

The *metric tensor* is : $g_{\alpha\beta} = (\underline{a}_\alpha, \underline{a}_\beta)$, and its inverse is : $g^{\alpha\beta} = (\underline{a}^\alpha, \underline{a}^\beta)$, where $\{\underline{a}^\alpha\}$ is the dual basis of $\{\underline{a}_\alpha\}$ defined by : $\underline{a}^\alpha \cdot \underline{a}_\beta = \delta^\alpha_\beta$ (Kronecker symbol).

The *Christoffel symbols*, used in covariant derivatives, are : $\Gamma^\lambda_{\alpha\beta} = g^{\lambda\mu}(\underline{a}_\mu , \underline{a}_{\alpha,\beta})$. Let us notice that : $b_{\alpha\beta} = b_{\beta\alpha}$, $\Gamma^\lambda_{\alpha\beta} = \Gamma^\lambda_{\beta\alpha}$, as a consequence of : $\underline{a}_{\alpha,\beta} = \phi_{,\alpha\beta} = \underline{a}_{\beta,\alpha}$.

Then, the definition of the integral of a function f on ω is :

$$\int_\omega f = \int_{\widehat{\omega}} f \circ \phi(\xi^1, \xi^2) \sqrt{|g|}$$

where : $|g| = g_{11} g_{22} - g_{12}^2$, is the *determinant of the metric tensor*.

I.2. The shell model

We consider a fully consistent shell model, i.e. satisfying the rigid body motion invariance of the internal energy (cf. [11]). Such models are described by two tensors. The first one represents the *change of metric on the medium surface* due to the deformation : it is denoted by $\gamma_{\alpha\beta}$. The second one is the *change of curvature of the medium surface* that we name $\rho_{\alpha\beta}$. If v is a displacement vector field on the surface $\omega : v = v_\alpha \, \underline{a}^\alpha + v_3 \, N$, then $\gamma_{\alpha\beta}$ and $\rho_{\alpha\beta}$ are respectively given by :

$$\begin{cases} \gamma_{\alpha\beta} = \frac{1}{2} \left(v_{\alpha|\beta} + v_{\beta|\alpha} \right) - b_{\alpha\beta} \, v_3 \, . \\ \rho_{\alpha\beta} = \frac{1}{2} \left(\theta_{\alpha|\beta} + \theta_{\beta|\alpha} \right) + \frac{1}{2} \left(b_\alpha^\lambda \, v_{\beta|\lambda} + b_\beta^\lambda \, v_{\alpha|\lambda} \right) - b_\alpha^\lambda \, b_{\lambda\beta} \, v_3. \end{cases} \tag{1}$$

where (Kirchhoff-Love relation) :

$$\theta_\alpha = - b_\alpha^\lambda \, v_\lambda - v_{3,\alpha} \, , \; \text{with} : \; b_\alpha^\lambda = g^{\lambda\beta} \, b_{\beta\alpha} \, . \tag{2}$$

and the covariant derivation being given by :

$$v_{\alpha|\beta} = v_{\alpha,\beta} - \Gamma_{\alpha\beta}^\lambda \, v_\lambda \, . \tag{3}$$

Let us just recall for sake of clarity that we set : $v_t = v^\alpha \, \underline{a}_\alpha = v_\alpha \, \underline{a}^\alpha$ with $v_\alpha = g_{\alpha\beta} \, v^\beta$. Then we can formulate Koiter shell model for linear elasticity as :

$$\begin{cases} \text{Find a displacement field } u \text{ lying in a functional space } V \text{ such that :} \\ u = u_t + u_3 \, N, \text{ with } u_t = u^\alpha \, \underline{a}_\alpha \, , \text{ and :} \\ \forall \, v \in V \, , \, a_0 \, (u,v) + \varepsilon^2 \, a_2 \, (u,v) = l \, (v) \; ; \end{cases} \tag{4}$$

where the bilinear forms a_0 and a_2 are respectively defined, for u and v in V , by :

$$\begin{cases} a_0(u,v) = \displaystyle\int_\omega R^{M\alpha\beta\mu\lambda} \, \gamma_{\alpha\beta} \, (u) \, \gamma_{\mu\lambda} \, (v) \\ a_2(u,v) = \dfrac{1}{3} \displaystyle\int_\omega R^{F\alpha\beta\mu\lambda} \, \rho_{\alpha\beta} \, (u) \, \rho_{\mu\lambda} \, (v) \end{cases} \tag{5}$$

where $R^{M\alpha\beta\mu\lambda}$ and $R^{F\alpha\beta\mu\lambda}$ are respectively the membrane and the bending stiffness tensors. The linear form $l\,(.)$ represents the mechanical loading. For sake of brevity, we set :

$$l(v) = \int_\omega f_3 \, v_3 + \int_\omega g^{\alpha\lambda} \, f_\lambda \, v_\alpha \, . \tag{6}$$

But one could include in the model more general loading with, for instance, boundary terms.

Let us now define the space V . Splitting between tangential and normal components, we set : $V = V_t \times V_3$. For example, we take concerning boundary conditions :

$$\begin{cases} V_t = \left\{ v_t = v^\alpha \, \underline{a}_\alpha \, , \, v^\alpha \in H^1 \, (\omega) \, ; \, v^\alpha = 0 \; \text{on} \; \gamma_0 \cup \gamma_1 \right\}, \\ V_3 = \left\{ v \in H^2 \, (\omega) \, , \, v = 0 \; \text{on} \; \gamma_0 \cup \gamma_1 \, ; \, v_{,\alpha} = 0 \; \text{on} \; \gamma_0 \right\}, \end{cases} \tag{7}$$

where γ_0 and γ_1 are two portions of the boundary γ of ω : the shell is assumed to be clamped on γ_0 and simply fixed on γ_1 . Let us recall finally that the existence and uniqueness of a solution to the Koiter model is classical [4] .

Remark 2

In the expression of the change of curvature $\rho_{\alpha\beta}$, the term : $\left(b_\alpha^\lambda\, v_\lambda\right)_{|\beta} = b_{\alpha|\beta}^\lambda\, v_\lambda + b_\alpha^\lambda\, v_{\lambda|\beta}$, is involved (in the derivatives of θ precisely) . It contains first order derivatives of the curvature tensor b_α^λ ; hence third order derivatives of the mapping ϕ are required. ■

II. A MIXED FORMULATION FOR THIN ELASTIC SHELLS

For sake of simplicity, we will assume in the sequel that both the whole edge and the free edge of the shell are connected. But, as it is proved in [9] or [15], it is not a major restriction.

II.1. The continuous formulation

Let us recall the definition of the bending moments $m^{\alpha\beta}$ and the resultant transverse shear stress, say $Q = Q^\alpha\, \underline{a}_\alpha$:

$$m^{\alpha\beta} = R^{F\alpha\beta\mu\lambda}\, \rho_{\mu\lambda}(u)\ ,\quad Q^\alpha = \frac{1}{\sqrt{|g|}}\left(m^{\alpha\beta}\,\sqrt{|g|}\right)_{,\beta} + \Gamma_{\xi\lambda}^{\alpha}\, m^{\xi\lambda}\ . \tag{8}$$

One can prove (cf. [9] or [15]) that there exist two potential functions ϕ and ψ such that :

$$Q = \operatorname{grad}\phi + \operatorname{rot}\psi\ , \tag{9}$$

with the notations :

$$\operatorname{grad}\phi = \frac{\overline{\partial\phi}}{\partial m} = g^{\alpha\lambda}\,\phi_{,\lambda}\,\underline{a}_\alpha\ ;\ \operatorname{rot}\psi = \frac{1}{\sqrt{|g|}}\left(-\psi_{,2}\,\underline{a}_1 + \psi_{,1}\,\underline{a}_2\right). \tag{10}$$

Furthermore, we need to have ϕ constant on the part $\gamma_0 \cup \gamma_1$ of the boundary. As ϕ is defined up to an additive constant, we choose for instance $\phi = 0$ on $\gamma_0 \cup \gamma_1$.
Let us now introduce the following functional spaces :

$$\left|\ \begin{aligned} &L_0^2(\omega) = \left\{\psi \in L^2(\omega) / \int_\omega \psi = 0\right\},\\[4pt] &V_3 = \left\{v_3 \in H^1(\omega) / v_3 = 0 \ \text{ on } \gamma_0 \cup \gamma_1\right\},\\[4pt] &W_t = \left\{\mu = \mu_\alpha\,\underline{a}^\alpha / \mu_\alpha \in H^1(\omega),\ \mu_\alpha = 0 \ \text{ on } \gamma_0 ,\ \mu_s = \mu_\alpha\,\tau^\alpha = 0 \ \text{ on } \hat{\gamma}_1\right\} \end{aligned}\right. \tag{11}$$

(τ^α being the components of the unit tangent vector to $\hat{\gamma}$) .

The functional spaces used for the mixed formulation are then :

$$\mathcal{V} = V_t \times V_3 \times W_t\ ,\qquad \mathcal{M} = V_3 \times L_0^2(\omega) \tag{12}$$

and it is worth noticing that the potential ϕ and the displacement v_3 lie in the same space V_3 .

Finally, the two bilinear forms used in the mixed formulation are :

$$
\begin{cases}
A(X,Y) = \displaystyle\int_\omega R^{M\alpha\beta\mu\lambda}\,\gamma_{\alpha\beta}(X)\,\gamma_{\mu\lambda}(Y) + \frac{\varepsilon^2}{3}\int_\omega R^{F\alpha\beta\mu\lambda}\,\rho_{\alpha\beta}(X)\,\rho_{\mu\lambda}(Y) \\[2mm]
B(\Xi,Y) = \dfrac{\varepsilon^2}{3}\left\{ \displaystyle\int_\omega g^{\alpha\lambda}\,\varphi_\lambda\!\left(\mu_\alpha + b^\beta_\alpha v_\beta + v_{3,\alpha}\right) \right. \\[3mm]
\qquad\qquad \left. - \displaystyle\int_\omega \frac{\psi}{\sqrt{|g|}}\left(\mu_{2,1} + \left(b^\beta_2 v_\beta\right)_{,1} - \mu_{1,2} - \left(b^\beta_1 v_\beta\right)_{,2}\right)\right\}
\end{cases}
\tag{13}
$$

with the notations : $Y = \left(\underline{v}_t = v^\alpha \underline{a}_\alpha\,,\, v_3\,,\, \mu = \mu^\alpha \underline{a}_\alpha\right)$, $\Xi = (\varphi,\psi)$.

The mixed formulation (equivalent to the Koiter model we used) consists in finding an element (X,Λ) in the space $\mathcal{V}\times\mathcal{M}$ such that :

$$
\begin{cases}
\forall\, Y \in \mathcal{V},\ A(X,Y) + B(\Lambda,Y) = F(Y), \\
\forall\, \Xi \in \mathcal{M},\ B(\Xi,X) = 0.
\end{cases}
\tag{14}
$$

where $F(.)$ is the extension to $\mathcal{V}$ of the linear form $1(.)$ defined in (6).
The existence and uniqueness of a solution to this problem is based on F. Brezzi Theorem [5] ; it has been proved in [9] or [15].

Remark 3
The main advantages of this formulation are the following:
a) A conformal approximation needs only C^0 elements.
b) It avoids transverse shear locking effects.
c) Though we assume the mapping discribing the middle surface to be C^3, we explicitely use only the two first derivatives of ϕ. As a matter of fact, the third derivatives of ϕ would appear in the bilinear form B (see (13)) but a simple computation leads to :

$$
\left(b^\beta_2 v_\beta\right)_{,1} - \left(b^\beta_1 v_\beta\right)_{,2} = b^\beta_2 v_{\beta,1} - b^\beta_1 v_{\beta,2} + \left(\Gamma^\beta_{\mu 2}\, b^\mu_1 - \Gamma^\beta_{\mu 1}\, b^\mu_2\right) v_\beta
\tag{15}
$$

and the third order derivatives disappear ! This is very convenient if one has to approximate the geometry of the middle surface. ∎

II.2.The discrete formulation

We introduce two finite dimensional subspaces of $\mathcal{V}$ and $\mathcal{M}$, say $\mathcal{V}^h$ and $\mathcal{M}^h$, such that :

$$
\mathcal{V}^h \to \mathcal{V} \qquad \mathcal{M}^h \to \mathcal{M} \qquad \text{when } h \to 0.
$$

(h being a small parameter linked to the mesh size).
In a practical way, we use first degree displacements fields, but we must add a bubble function for the rotation rate, in order to satisfy Brezzi condition for the discrete model. (see Figure 1).

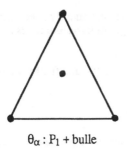 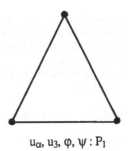

$\theta_\alpha : P_1 + \text{bulle}$ $u_\alpha, u_3, \varphi, \psi : P_1$

Figure 1

Then we prove that there exists $\left(X^h , \Lambda^h \right)$ uniquely defined in $\mathcal{V}^h \times \mathcal{M}^h$ such that :

$$\begin{cases} \forall\, Y \in \mathcal{V}^h , \quad A\left(X^h , Y\right) + B\left(\Lambda^h , Y\right) = F\left(Y\right) \\ \forall\, \Xi \in \mathcal{M}^h , \quad B\left(\Xi , X^h\right) = 0 \end{cases} \tag{16}$$

As the finite elements suggested above, in the construction of $\mathcal{V}^h$ and $\mathcal{M}^h$, use linear approximation of the fields components, we obtain (cf. [15]) the following error estimate between (X, Λ) and (X^h, Λ^h) :

$$\left\| X - X^h \right\|_{\mathcal{V}} + \left\| \Lambda - \Lambda^h \right\|_{\mathcal{M}} \leq c\, h \tag{17}$$

where c is a constant dependent on the solution (X, Λ) and on the mapping ϕ but not on h (as soon as h is small enough). The use of second degree elements would permit to obtain $O(h^2)$ as it has already been shown for plates [6,7] .

II.3. The geometrical approximation

Now, we assume that the middle surface ω of the shell is approximated by flat elements, at the vertices of which we know the exact normal vector (the vertices are on ω !).

Then we can build a local (i.e. depending on the element) basis of the tangent plane at each vertice of the mesh. By interpolation of these "nodal" bases, we obtain an approximation of the different geometrical tensors (cf. I.1.) and we prove that the additional error, due to this approximation, is consistent with the error due to the finite elements interpolation (cf. [15]) .

III. MESH REFINEMENT

Self-adaptive mesh refinement techniques always present three different aspects that can be treated independently :
a) The determination of the elements that will be subdivided.
b) The refinement strategy.
c) The resolution of the system of equations arising from the discrete formulation associated with the new mesh.

In this paragraph, we will deal with the two first aspects.

The usual way to determine the elements that will be refined is to compute an error estimate on each element of the current grid (Babuska and Rheinboldt [1], Kelly et al [10]).

However, we will not do any error estimate : we base our adaptive technique for the refinement of elements on the greatest stresses and/or changes of curvature fields associated to:
a) the displacement fields of the shell.
b) the buckling eigenmodes.

First of all, we will do an approximation of buckling modes, and finally we will explain the refinement strategy.

III.1. A buckling criterion

Let us consider the linear problem $A u = f$. This problem is equivalent to search a set of eigenvalues λ_i and eigenvectors w_i such that :

$$u = \sum_{i \geq 1} \alpha_i w_i \text{ , where : } \alpha_i = \frac{(f, w_i)}{\lambda_i} \tag{18}$$

If the first eigenvalue is close to zero (say λ_1), we will have :

$$\alpha_1 \gg \alpha_j \text{ , } \forall j \neq 1 \tag{19}$$

and the solution will be colinear with the corresponding eigenvector, i.e. :

$$u \approx \frac{(f, w_1)}{\lambda_1} w_1 = \alpha_1 w_1 \text{ (if } (f, w_1) \neq 0) \tag{20}$$

When $A u = 0$, buckling phenomenon appears. So, in order to approach the problem, let us consider the matrix A , which is factorized in Crout form : $A = L . D . L^t$ (L is a lower triangular matrix, and D is diagonal) .

Let us notice by d_1 the smallest value on the diagonal matrix D . If d_1 is close to 0, we introduce the vector Z which is such that : $Z_i = \delta_{i1}$ (Kronecker symbol) . It is clear that :

$$D . Z \approx 0 . \tag{21}$$

Then, if we set :

$$L^t . w_1 = Z , \tag{22}$$

we obtain :

$$A . w_1 = L . D . L^t . w_1 = L . D . Z \approx 0 \tag{23}$$

The vector w_1 , solution of the triangular system (22) , is then a "buckling field". It is a displacement field and we can associate to it the corresponding stresses and change of curvature tensor. We can treat these fields in the same way that the original fields and use them as indicators for refinement.

III.2. Refining strategy

Let us denote by "c" the value of a field that we have computed on each element of the grid (stress, curvature tensor, e.g.). In order to refine elements with the greatest values of "c", we suggest two different criteria :

1) We compute the maximum value "c_{max}" in the mesh and refine elements verifying :

$$c \geq \gamma \cdot c_{max} \qquad (24)$$

where $0 < \gamma < 1$. In order to accelerate the refinement process, a small value of γ can be used ($\gamma \approx 0.25$ e.g.). An illustration of this technique is given on Figure 2.

2) We assume the distribution of the field "c" on the shell has a Gauss normal distribution. We denote by $\bar{c}$ its mean value and σ its mean-square value. Then, elements such that :

$$c \geq \bar{c} + \gamma \cdot \sigma \qquad (25)$$

are selected to be refined.

The second strategy is a little bit more expensive, but new grids are more regular than those obtained by the first criterion (see Figure 3). If we want to enlarge progressively the refined area, we can repeat the second process in a local way, i. e. on non selected elements which are in contact with selected elements (see Figure 4). Examples of meshes obtained by these different methods are shown on Figures 2 , 3 and 4 (for the example of the pinched cylinder).

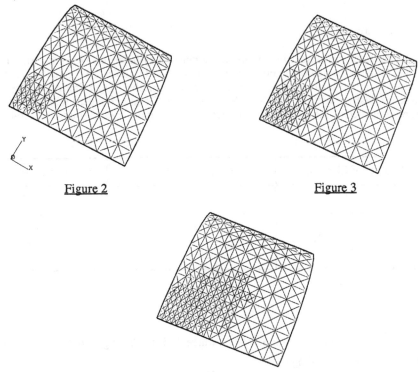

Figure 2   Figure 3

Figure 4

To refine the triangles, we use here an algorithm proposed by Rivara [14]. Each triangle is divided in four sub-triangles in the following way : we connect the middle point of the longest side with its opposite vertex and with the middle points of the other two sides. Once a first refinement is achieved, a second one is needed to assure the conformity of the new mesh : if the non-conforming nodes are lying on the longest side of the non-conforming triangle, they are connected with their opposite vertex. However, if they are not lying on the longest side, they are connected with the middle point of this last one, as in the following figures (cf. Figure 5). This process is repeated until total conformity is obtained. Let us remark that the final grid has angles bounded away from zero (cf. [14]).

Figure 5

III.3. An example

We consider the classical example of the pinched cylinder (cf. [13] e.g.) for which an analytical solution (using double Fourier series) is known. In Table 1, we give the value of the deflection w at the loading point for different rafined meshes. The criterion we used is the first one (cf. last paragraph) with Von Mises stresses. Good agreement with the analytical solution is obtained as the mesh is refined (let us recall that the "exact" solution is 164.24)

Number of Elements	Number of Vertices	Deflection
1250	676	132.90
1846	984	146.68
3885	2020	159.48

Table 1

Remark 4

As a matter of fact, we give the ratio $\dfrac{E\,t\,w}{P}$, where E is the Young modulus, t the thickness of the shell and P the loading. ∎

CONCLUSIONS

This paper presents a first approach in mesh refinement for shell structures. Many points have to be solved and/or improved, for example :
a) Introduction of hierarchical finite elements, in order to use initial mesh as a preconditioner of the refined meshes.
b) Study of good criteria for refinement.
c) Geometrical approximation, the aim being to add vertices which are really (i.e. as close as possible) on the medium surface of the shell when the mapping is not exactly known.

REFERENCES

[1] BABUSKA I., RHEINBOLDT, W.C. [1978], *A posteriori error estimates for the finite element method,* Int. J. Num. Meth. Eng., 12, p. 1597-1615.

[2] BATOZ J.L., BEN THAR M., DHATT G.S. [1982], *Les éléments DKT et DKQ et l'analyse des plaques et coques minces,* Conference "Tendances Actuelles en Calcul des Structures", Sophia-Antipolis.

[3] BELYTSCHKO T., STOLARSKI M. [1983], *Shear and membrane locking in curved* C^0 *elements,* Comput. Methods in Appl. Mech. and Eng., 41, p. 279-297.

[4] BERNADOU M., CIARLET P.G. [1976], *Sur l'ellipticité du modèle linéaire de coques de W.T. Koiter,* Lecture notes in Applied Sciences and Engineering, vol. 34, Springer-Verlag, Berlin, p. 89-136.

[5] BREZZI F. [1974], *On the existence, uniqueness and approximation of saddle-point problems arising from Lagrange multipliers,* RAIRO, série rouge, 8, p. 129-151.

[6] DESTUYNDER Ph., NEVERS Th. [1988], *A new finite element scheme for bending plates,* Comp. Meth. Appl. Sci. Eng., 68, p. 127-139.

[7] DESTUYNDER Ph., NEVERS Th. [1989], *Some numerical aspects of mixed finite elements for bending plates,* Comp. Meth. Appl. Sci. Eng., 78, p. 73-87.

[8] DESTUYNDER Ph., NEVERS Th., SALAÜN M. [1989], *Développement d'un élément fini de plaque mince en flexion,* La Recherche Aérospatiale, 4, p. 65-80.

[9] DESTUYNDER Ph., SALAÜN M. [1990], *Une formulation variationelle mixte pour les problèmes généraux de coques minces,* C. R. Acad. Sci. Paris, t. 310, Série I, p 215-220.

[10] KELLY D.W., GAGO J.P., ZIENKIEWICZ O.C., BABUSKA I. [1983], *A posteriori error analysis and adaptive processes in the finite element method,* Int. J. Num. Meth. Eng. 19, p. 1593-1656.

[11] KOITER W.T. [1970], *On the foundations of the linear theory of thin elastic shells,* Proc. Kon. Ned. Akad. Wetensch, B73, p. 169-195.

[12] MAC NEAL R.H., [1978], *A simple quadrilateral shell element,* Comp. Struc., 8, p.175-183.

[13] MEEK J.L., TAN H.S. [1986], *A faceted shell element with loof nodes,* Int. J. Num. Meth. Eng. 23, p. 49-67.

[14] RIVARA M.C. [1984], *Algorithms for refining triangular grids suitable for adaptive and multigrid techniques,* Int. J. Num. Meth. Eng. 20, p. 745-756.

[15] SALAUN M. [1991], *Formulation mixte théorique et numérique d'un modèle de coque mince de W.T. Koiter,* Thesis.

UPON THE DIFFERENT THEORIES OF PLASTIC BUCKLING
ELEMENTS FOR A CHOICE

A COMBESCURE CEA DMT Saclay (FRANCE)

ABSTRACT

This paper gives the different theories that can be used to predict plastic buckling of metallic structures and shows by the application on two typical examples the interest and the limitations of each of the approach.

1 THREE DIFFERENT THEORIES OF PLASTIC BUCKLING

We suppose in this part that a fully incremental non linear analysis including plasticity and large displacement has been performed and that we want now to check the stability of the equilibrium states obtained by the incremental analysis just performed.We have then to evaluate the tangeant stress strain matrix so that the tangeant stiffness can be evaluated on the actual configuration.

Let us define the following notations:

E is the Young's modulus

E_t is the tangeant modulus

E_s is the secant modulus

σ_y is the conventionnal yield stress

ν is the Poisson's ratio

h is the plastic tangeant modulus h $= E \times E_t / (E-E_t)$

D   is the elasticity matrix

D_t is the tangeant stress strain matrix.

To construct the tangeant stress strain matrix, D_t, we can use different theories:

The first and most simple way is to use Ungesser's approach which defines the D_t matrix by the following formula:

(1) $$D_t = \frac{E_t}{E} \times D$$

In this simple equation we simply replace the Young's modulus by the tangeant modulus when the point is plastic.

The second is the Von Karman theory which consists to replace E_t by E_r in equation (1). E_r is the reduced Von Karman modulus given by the following equation:

(2) $$E_r = 4 E \times E_t / \left(\sqrt{E} + \sqrt{E_t} \right)^2$$

The third is the most common and consists of using the classical flow theory to construct the tangeant D_t matrix. In this case the matrix is given by the following formula:

(3) $D_t = (I - C) \times D$

(4) $C = \dfrac{\left(\dfrac{\partial F}{\partial \sigma}\right) \times \left(\dfrac{\partial F}{\partial \sigma}\right)^t \times D}{\left(\dfrac{\partial F}{\partial \sigma}\right)^t \times D \times \left(\dfrac{\partial F}{\partial \sigma}\right) + h}$

In this equation F is the Von Mises equivalent stress given by equation (5):

(5) $F = \sqrt{\sigma_1^2 + \sigma_2^2 - \sigma_1 \times \sigma_2 + 3 \times \sigma_{12}^2}$

In this equation σ_i is the component of the stress tensor. In the case of buckling of axisymmetric structures loaded axisymmetrically the prebuckling stresses are axisymmetric so that the term (3,3) of the tangeant matrix D_t is certainly the elastic shear modulus G.

The fourth expression is the evaluation of the tangeant matrix D_t by the deformation theory of plasticity. The inverse of the D_t matrix is evaluated by the following formula for axisymmetric prebuckling:

(6) $(D_t)^{-1} = \begin{pmatrix} A^{-1} & 0 \\ & 0 \\ 0 \ 0 & 1/G_s \end{pmatrix}$

G_s is the secant shear modulus given by the following equation:

$$(7) \qquad G_s = E_s / (2 \times (1.+ \nu))$$

A is a 2 × 2 matrix given by equation (8):

$$(8) \qquad A = \left(\frac{1.}{E_t} - \frac{1.}{E_s} \right) \times N + \frac{1.}{E} \times K + \left(\frac{1.}{E_s} - \frac{1.}{E} \right) \times H$$

In equation (8) N K and H are three 2 × 2 matrix given by equations (9) (10) and (11):

$$(9) \qquad N = \frac{1.}{\sigma^{*2}} \begin{pmatrix} s_1^2 & s_1 s_2 \\ s_1 s_2 & s_2^2 \end{pmatrix}$$

where s_1 is the deviatoric stress state, and σ^{*2} the equivalent Von mises stress.

$$(10) \qquad K = \begin{pmatrix} 1 & -\nu \\ -\nu & 1 \end{pmatrix}$$

$$(11) \qquad H = \begin{pmatrix} 1 & -.5 \\ -.5 & 1. \end{pmatrix}$$

We shall refer in the following the four approaches with the four following names respectively:
The tangeant modulus theory shall refer to Ungesser's model.
The Von Karman theory shall refer to the second approach.
The flow theory predictions shall refer to the third behavior.
The deformation theory shall refer to the last one.
All the computation in the following using these models have been done with the INCA code of the CASTEM system.

2 PLASTIC BUCKLING OF A CYLINDER UNDER COMBINED TRACTION AND EXTERNAL PRESSURE

A thin cylinder submitted to a tensile axial load and to a simultaneous external pressure is tested against buckling.The cylinder has a nominal radius of 75mm and a length of 150mm.The material is a nickel alloy with a Young's modulus of 160000.Mpa a Poisson's ratio of .3 a conventional yield stress σ_y of 150.Mpa and an elastic limit σ_1 of 60.Mpa.The cylinder is clamped at its two extremities.The elastic external pressure buckling with zero axial load is called the Yamaki value and is equal to .0356Mpa.We can observe on figure 1 that the predictions of plastic buckling start to differ drastically when plasticity takes place.Von Karman prediction are very unconservative when the plasticity is small which is obvious from the formula given by equation (2).When the plasticity is important it approaches the tangeant modulus theory.The tangeant modulus theory predicts a collapse under pressure which is smaller and smaller when the tension increases.This is not observed experimentally.It can also be well understood from the equation (1) which shows that the material remains isotropic when plasticity developps which is not the case with flow theory or deformation theory.For these two modelisations one can observe that flow and deformation theories give nearly the same result up to an axial stress of 1.5 times the proportionnal stress σ_e.The flow theory gives here the best prediction of the observed experimental buckling pressures.The deformation theory seems to bee a bit too conservative .A full description of the work can be seen in ref (1).

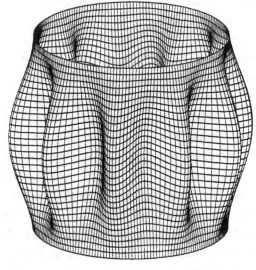

BUCKLING MODE OF THE CYLINDER

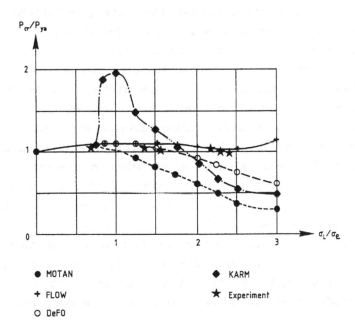

- MOTAN   ◆ KARM

+ FLOW   ★ Experiment

○ DeFO

Fig. 1 – Interaction diagram between tension σ_L/σ_e and external pressure

3 TORIPHERES UNDER EXTERNAL PRESSURE.

The torispherical head having the geometry defined on figure 2 has been studied.It is subjected to an uniform internal pressure.The material is steel having a young's modulus of 200000.Mpa a poisson's ratio of .28,it is perfectly plastic with a yield stress of 310.Mpa.The structure is clamped at its top.A detailed description of the experimental data can be found in ref (2) as well as prediction done with BOSOR 5 code.For dome 2 having a diameter to thickness ratio of 500,the predictions of plastic buckling are the following:with the tangeant modulus one obtains buckling for a pressure of .60Mpa;the buckling pressure is .84Mpa if one uses the deformation theory and no buckling is predicted if one uses the flow theory.The experiment shows a first threshold buckling for.67Mpa and a chart recorder pressure of .97Mpa.One can consider that the predictions are rather good with tangeant modulus for the threshold pressure and with deformation theory for the chart recorded pressure.The flow theory is not sure because in spite of its full mathematical sound basis it perdicts no buckling when one observes a buckling.One could suggest that the real structure has a small imperfection and that this is the reason of the experimental buckling.We have then used the imperfect shell element COMU with different amplitude of imperfections to simulate this effect.The imperfection are chosen parallel to the plastic buckling mode.When one plots the maximum load obtained on the imperfect torisphere one obtains the curve given on figure 4.From this curve one sees that an imperfection of only one thousandth of the thickness shall produce the buckling.From this study one concludes that the flow theory is not a sure tool to predict

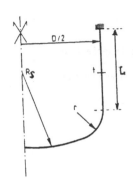

cylinder diameter D
cylinder high L   L/D = 0.5
torus radius r   r/D = 0.15
sphere radius R_s R_s/D = 1
thickness t   D/t = 500

FIGURE 2 Geometry of the
internally pressurised
torisphere

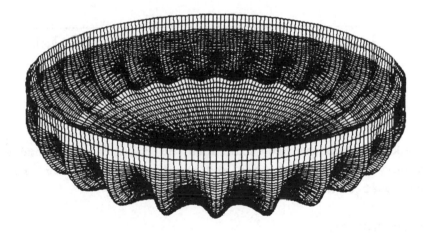

FIGURE 3 Plastic buckling
with deformation theory

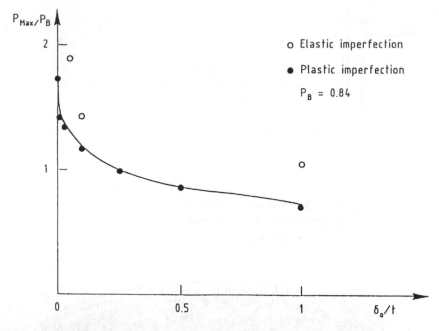

FIGURE 4 Influence of initial imperfection
on buckling

buckling because it tolerates no imperfection. Deformation
theory is more realistic and the tangeant modulus theory is
the most conservative approach.

4 CONCLUSIONS

This paper shows that flow theory should not be used to
predict plastic buckling if one wants to avoid in
certain cases the imperfection analysis. The tangeant
modulus leads to the most conservative predictions but, when
the stress state is bidirectionnal with a tensile and a
compressive component, it can be a bit too conservative. The
deformation theory leads to reasonnable results in most
cases but tolerates only imperfections which are not bigger
than one tenth of the thickness.

(1) CHRAIBI Said

Flambage de coques cylindriques sous combinaisons de chargements mecaniques et thermiques cycliques.

THESE INSA LYON obtained on june 22 1990

(2) G D GALLETLY J BLACHUT university of liverpool

Intrenally pressurised machine dome ends-a comparison of the plastic buckling predictions of the deformation and flow theories.

Journal of mech. eng. science vol 204 1990 p 169-186

Buckling and Postbuckling Behavior of Cracked Plates in Tension

by

E. RIKS [1], C.C. RANKIN [2]

and

F.A. BROGAN [2]

1)   Aero/Astronautics, Delft University, Delft, The Netherlands

2)   Lockheed Missiles & Space Company, Palo Alto, California.

Extended Abstract

1. Introduction

There are two categories of problems that concern through cracks in thin-walled structures which are known to be influenced by a nonlinear geometrical effect. The first category comprises through cracks in thin-walled pressurized vessels. Here, the nonlinearity presents a stiffening effect. It restrains the crack opening displacements and the intensity of the stress singularity at the crack front so that the latter is *lower* than the value that one obtains when the elasticity solutions are linearized. It is noted that a longitudinal through crack in a pressurized fuselage belongs to this category. This is an important practical problem in aircraft design which received much attention recently, e.g. (Refs. 1, 2 and 3, 4 and 5).

In this paper we focus on a related problem. A cracked sheet *in tension* buckles locally around the crack faces if the loading exceeds a certain critical value This a problem that belongs to the second category. Here the nonlinear effect is a softening effect. It tends to amplify the intensity of the stress concentration around the crack front so that it is *larger* than the value that one obtains when the elasticity solutions are linearized. The buckling phenomenon and its consequences in this case are well known among experimenters in the field of fatigue cracking. The problem has been considered experimentally, analytically and computationally at various instances in the past. We can mention for example (Refs.6 - 17).

When buckling takes place in the situation sketched above, the initially flat cracked plate specimen will develop out of the plane displacements in the vicinity of the crack. While this occurs, the strength of the crack front singularity will further increase, but this at a faster rate than one would have observed had the plate not buckled. Clearly, the principal issue is here to determine in what way the crack front singularity increases once buckling is taking place. Let it be emphasised that this is precisely the issue that we intend to consider in this presentation.

2. Computational Tools

To be able to solve the problem as sketched above, one is in need of:

(i) A finite element discretization procedure that adequately represents the nonlinearity of the cracked plate buckling problem, (Refs.1, 2, 3).

(ii) A bifurcation and continuation method that enables the analyst to compute the complete pre- and postbuckling states including the bucklingpoint, cq. load, (Refs. 18, 19, 20).

(iii) A method that is capable to compute the energy release rates of the cracks in the pre- and post buckled states, (Refs. 1,2,3 & 21, 22, 23).

With the recent additions described in in (Ref.3), the three basic requirements (i) to (iii) are now a regular part of the STAGS code (Ref.24). In the presentation, we will first give a review of the basic principles of these methods. Then we will proceed to present some new results for the centrally cracked plate that we obtained with the help of these methods. In this note, it suffices to summarize the most important part of the results of the calculations.

3. Summary Results

The plate dimensions that we considered are : Thickness = t = 1 mm; Width = 2b = 400 mm: Length = 2l = 800 mm; (central) Crack-length = 2a. The loaded edges are fully clamped, the unloaded edges are free. For the elasticity constants we took: Youngs modulus $E = 70000$ N/mm^2 , Poissons ratio $v = .3$. With these model definitions, the following data were obtained for the buckling loads which belong to the so-called symmetrical buckling mode, e.g. (Refs.12, 13, 17).

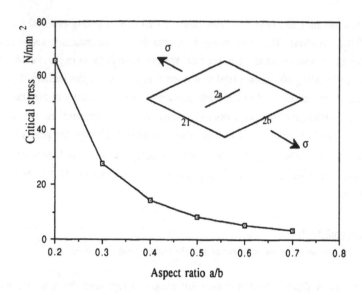

Figure 1 **Results Buckling Loads**

It is noted that we believe that our results are accurate. They compare well with findings reported in the literature for the larger aspect ratios a/b > .4.

Next, we computed the postbuckling behavior of the plate for three choices of aspect ratios a/b from the set given in Figure 1, a/b = 0.2, 0.4, 0.6. The most interesting case of this set is the third choice, the case determined by a/b = 0.6. We present first a summary of the postbuckling solutions obtained by STAGS (Fig.2) & (Fig.3):

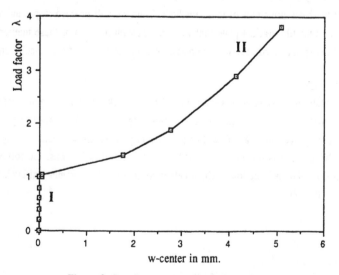

Figure 2 **Load vs. center displacement w_c**

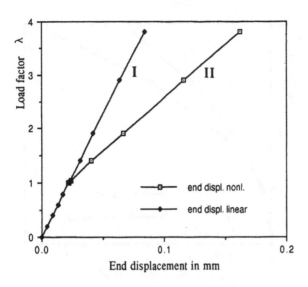

Figure 3 Load vs end displacement

The first figure (Fig.2) depicts the load vs. the displacement normal to the plate at the center of the crack. The second figure (Fig.3), gives the load vs. the end displacement Δu at the loaded edges. These relations show a stable buckling phenomenon as one would expect on the basis of physical reasoning. The load-displacement response of the cracked plate in tension is thus very much similar to that of a simply supported plate in compression.

To complete the analysis we also computed the energy release rates of the crack as a function of the load intensity in the pre- and postbuckling state. These results are presented in the last two figures (Fig.4 & Fig. 5).

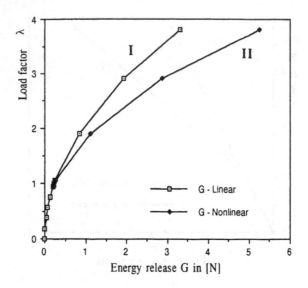

Figure 4 **Energy Release Rate vs. Load**

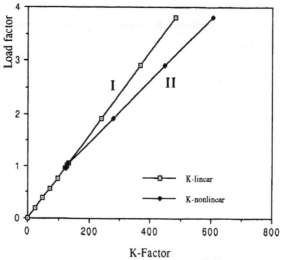

Figure 5 **K- Factor** (defined by $K = \sqrt{GE}$) vs. load

It is noted, that all our results for the other cases not shown here are qualitatively in agreement with the form of the solution given in these figures. The transformation of (Fig. 4) into (Fig. 5) is given here because the energy release rate G of the crack in the linear case is given by:

$$G = \frac{K^2}{E} \qquad \text{(Ref.25)} \qquad (1)$$

where K is the mode 1 stress intensity factor. Before buckling, the stress intensity factor in our case of the centrally cracked plate is accurately determined by the formula:

$$(K)_{_1} = \sigma_\infty \Theta(a,b) = \sigma_\infty \sqrt{\pi a} \sqrt{[\sec\{\frac{\pi a}{2b}\}} \qquad \text{(Ref.25)} \qquad (2)$$

where the subscript I outside the brackets denotes that this value corresponds to the pre-buckling state. The graph in (Fig.5) can be seen as an extrapolation of the relation (1) into the postbuckling range. The result of this extrapolation seems to suggest that the postbuckling solutions for K along the bifurcation branch II are still linearly dependent on the load factor $\lambda = \frac{\sigma_\infty}{\sigma_{\infty cr}}$ and this in turn suggests that the K-factor at buckling as given by (2), can be modified to:

$$(K)_{_{II}} = \sigma_\infty \kappa \Theta(a,b) = \sigma_\infty \kappa \sqrt{\pi a} \sqrt{[\sec\{\frac{\pi a}{2b}\}]} \qquad (3)$$

where κ is a positive constant, smaller than unity, which depends on the specific plate geometry, boundary and loading conditions. We note further that if the previous conjecture is correct this constant κ must be proportional to the drop in stiffness of the plate specimen at buckling.

4. Conclusion

Recent enhancements of the STAGSC1- code alleviate the analysis of cracks in thin-walled shell structures. Applied to the problem of a centrally cracked plate, we were able to obtain with this code the stress singularity characterization of the crack along the postbuckling states of the plate. We believe that these results are new in the sense that they have never been produced before.

REFERENCES

1. Riks E., den Reijer P.J., "A Finite Element Analysis of Cracks in a Thin Walled Cylinder under Internal Pressure", NLR TR 87021 U, National Aerospace Lab. NLR, the Netherlands, Jan. 1987.

2. Riks E., "Bulging Cracks in Pressurized Fuselages: A Numerical Study", NLR MP 87058 U, National Aerospace Lab. NLR, The Netherlands, Sept. 1987.

3. Riks E., Brogan F.A., & Rankin C.C., "Bulging Cracks in Pressurized Fuselages: A Procedure for Computation", December Meeting ASME, San Francisco, 1989. In: Analytical and Computational Models of Shells, Proceedings Winter Annual Meeting ASME (A.K. Noor, T. Belytschko, J.C. Simo eds.), C.E.D. Volume 3, The American Society of Mechanical Engineers, 1989.

4. Ansell, H., "Bulging of Cracked Pressurized Aircraft Structure," Ph.D. Thesis, Linkoping Institute of Technology, Sweden, April 1988.

5. Dong, Chen., " Bulging Fatigue Cracks in a Pressurized Aircraft Fuselage", PhD. Thesis, Delft University, Dept. of Aeronautics & Astronautics, Delft, The Netherlands, January , 1991.

6. Petyt M., "The Vibration Characteristics of a Tensioned Plate Containing a Crack", J. Sound Vib. 8, 377 (1968).

7. Dixon R. and Strannigan J. S.,. "Stress distribution and Buckling in Thin Sheets with Central Slits", Proc. 2nd. Int. Conf. on Fract., Brighton, 105 (1969).

8. Dyshel M. S., "Stability under Tension of Thin Plates with Cracks", Soviet Appl. Mech. 14, 1169 - 1173 (1978)

9. Dyshel M. S., "Stability of Thin Plates with Cracks under Bi-axial Tension", Soviet Appl. Mech. 18, 924 - 928 (1982)

10. Guz A. N., Kuliev G. G. and Tsurpal I.A.,"On Fracture of Brittle Materials from Loss of Stability near a Crack", Engng Fract. Mechanics, 10, 401 (1978).

11. Dal, Y. M., " Local Bending of a Stretched Rectangular Plate with a Crack", Soviet Appl. Mech. 17, 120- 125 (1978)

12. Markström K., Storåkers B.,"Buckling of Cracked Members under Tension", Int. J. Solids Structures, Vol. 16, pp 217 - 229, 1980.

13. Rossmanith H. P., Troger H., Tschegg E., " Beulen und Reissen von Gezogenen Dunnen Blechen mit Innenrissen", Z. Flugwiss. Weltraumforsch. 5, (1981), Heft 1.

14. Fujimoto T. and Sumi S., "Local Buckling of Thin Tensioned Plate Containing a Crack", The Memoirs of the faculty of Engineering, Kyushu University, Vol. 12, pp. 355-370 (1982)

15. Fujimoto T. and Sumi S., "Elastic buckling of Center Cracked Plates under Tension", J. J. S. M. E. 52, 1582-1586 (1985)

16. Sih G. C. and Lee Y. D., " Tensile and Compressive Buckling of Plates Weakened by Cracks", Theor. Appl. Fracture Mech. 6, 129-138 (1986)

17. Shaw D., Huang Y. H. "Buckling Behavior of a Central Cracked Thin Plate under Tension" Engrg. Fracture Mechanics, Vol. 35. no 6. pp 1019 - 1027, 1990.

18. Riks E., "Some Computational Aspects of The Stability Analysis of Nonlinear Structures", Comp. Meth. in Appl. Mech. and Engng. 47. 219-259, 1984.

465

19. Thurston G. A., Brogan F.A., Stehlin P., " Postbuckling Analysis Using a General Purpose Code", AIAA paper No. 85-079-CP. Presented at the AIAA/ASME/AHS 26th Structures, Structural Dynamics and Materials Conference, Orlando,Florida, April 15- 17, 1985.

20. Rankin C. C., Stehlin P. and Brogan F. A., "Enhancements to the STAGS Computer Code", NASA CR-4000, 1986.

21. Parks D.M., "A Stiffness Derivative Finite Element Technique for Determination of Elastic Crack Tip Stress Intensity Factors", Int. J. Fracture 10, No. 4, 487-502, December 1974.

22. Hellen T.K., "On the Method of Virtual Crack Extensions", Int. J. Num. Meth. Engng. 9, No.1, 187-207 (1975).

23. LeFort P., DeLorenzi H.G., Kumar V., German M.D., "Virtual Crack Extension Method for Energy Release Rate Calculations in Flawed Thin Shell Structures" , Journal of Pressure Vessel Technology, Vol. 109/101-107, February 1987.

24. Almroth B.O., Brogan F.A., Stanley G.M.,"Structural Analysis General Shells", Vol.2 Users Instructions for STAGSC-I, LMSC D633873.

25. Broek D., " Elementary Engineering Fracture Mechanics" Sijthof & Noordhoff, Alphen a/d Rijn - The Netherlands. (1987).

NUMERICAL CALCULATION OF LIMIT LOADS FOR SHELLS OF REVOLUTION WITH PARTICULAR REGARD TO THE APPLYING EQUIVALENT INITIAL IMPERFECTIONS

GERHARD SPEICHER and HELMUT SAAL
Institut für Stahlbau und Holzbau, Universität Stuttgart
Pfaffenwaldring 7, D-7000 Stuttgart 80

ABSTRACT

Two concepts for the numerical calculation of the limit capacities for shells of revolution are presented. In concept C, the plastic buckling load has to be determined for the perfect geometry and in conservative consideration of the non-elastic material behaviour. The real limit load is obtained by the application of a reduction factor, which includes all the effects of the geometrical and the structural imperfections to this buckling load. A complete numerical method is given with the concept D. Here, the limit loads are computed by the computer program, which has to be able to consider an imperfect geometry with the imperfection amplitude w_0 in the shape of the eigenform due to the lowest eigenvalue and the geometrical and the physical nonlinearity. The application of the proposed formulae is shown to 2 stiffened shells of revolution whose experimental limit loads are known and it demonstrates the conservativity of the procedures.

INTRODUCTION

Today, Finite-Element-programs are available to nearly all the consulting engineering offices. Regarding this fact in all fields of civil engineering a lot of efforts to find regulations for the use of computer programs for numerical computations can be noticed. This is also valid for the field of buckling of shells. Shell structures incline to buckling failure because of their very decided sensitivity to several parameters. Therefore it is very important to safeguard the quality of the numerical computations.

Already in 1976, a task group of the "European Convention For Constructional Steelwork ECCS" dealt with the inclusion of numerical methods to achieve the load carrying capacity of shell structures [1]. At present, a task group of the "Deutscher Ausschuß für Stahlbau" is developing new rules for buckling design of shell structures [3]. These rules should deal with the design procedures for shell structures not considered in the DIN 18800 part 4 [2] and should also regulate the use of numerical design methods for calculating the limit loads of shell structures.

The draft of "DASt-Richtlinie 017" [3] proposes several rules and design methods to be used in conjunction with computer programs to obtain limit loads. The aim of our investigation is first, to check these proposed methods by application to some shell structures, and second, to investigate, how to apply the equivalent initial imperfections to obtain the limit capacity of shell structures. The draft [3] offers several possibilities to calculate the limit loads by numerical investigations. Four different concepts already presented in [4], can be used alternatively to determine the limit loads.

Concept A :   Numerically calculated ideal buckling load of the perfect shell; total reduction factor for imperfections and non-elastic material behaviour

Concept B :   Numerically calculated ideal buckling load of the perfect shell; reduction factor according to the stress components for imperfections and non-elastic material behaviour; interaction criterion

Concept C :   Numerically calculated non-elastic (elastic-plastic) buckling load of the geometrically perfect shell; total reduction factor for imperfections

Concept D :   Numerically calculated limit loads of the imperfect shell

NUMERICAL INVESTIGATION

Concept A and B
The concepts A and B correspond to the design procedures in [2] with the only difference that the appropriate slenderness $\overline{\lambda}_s$ has to be determined with a computationed ideal buckling stress. Therefore both concepts are not considered here.

Concept C
In relation to A and B, the concept C is of essential importance for the computer based design. It is only allowed to apply it, if the working line used in the program corresponds to the real material behaviour or is a conservative approximation. This elastic-plastic, geometrically nonlinear computation gives a buckling load which must be still reduced with a factor regarding the practical imperfections. These required reduction factors are known only for very few cases, where a sufficient number of experimental results exists. For other shell buckling problems, [3] requires a conservative approximation. In the following the procedure will be described to get these required reduction factors for the design due to concept C.

A broad variation of the geometric dimensions showed that the plastic buckling loads calculated for axially loaded, perfect cylindrical shells only depend upon the R/T-ratio (R = radius, T = wall thickness). Figure 1 shows the non-elastic limit stress σ_{pl_calc} related to the yield strength f_y as a function of the appropriate slenderness $\overline{\lambda}_{sx}$ [2].

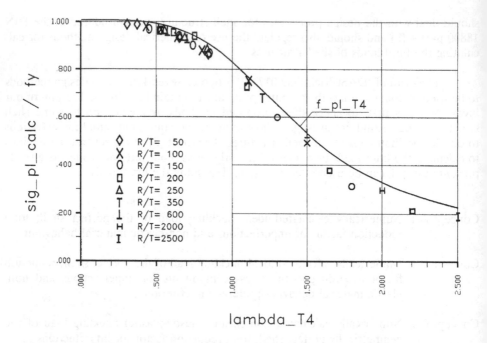

Figure 1. Distribution of the calculated $\sigma_{pl\ calc}/f_y$-values for cylinders with different R/T-ratios and presentation of the reduction curve f_{pl_T4}

In figure 1 all results fit one smooth curve. This curve representing the relation between the appropriate slenderness $\bar{\lambda}_{sx\_T4}$ and the plastic reduction f_{pl_T4} is given by the following formulae.

$$f_{pl\_T4} = 1.0 \qquad \text{if} \qquad \bar{\lambda}_{sx\_T4} \leq 0.25 \quad (2)$$

$$f_{pl\_T4} = -0.219 \cdot \bar{\lambda}^3_{sx\_T4} + 0.018 \cdot \bar{\lambda}^2_{sx\_T4} + 0.032 \cdot \bar{\lambda}_{sx\_T4} + 0.994 \qquad \text{if} \quad 0.25 < \bar{\lambda}_{sx\_T4} \leq 1.00 \quad (3)$$

$$f_{pl\_T4} = -0.59 \cdot \bar{\lambda}_{sx\_T4} + 1.415 \qquad \text{if} \quad 1.00 < \bar{\lambda}_{sx\_T4} \leq 1.50 \quad (4)$$

$$f_{pl\_T4} = 1.043 \cdot (\bar{\lambda}_{sx\_T4})^{-1.67} \qquad \text{if} \quad 1.50 \leq \bar{\lambda}_{sx\_T4} \quad (5)$$

These formulae allow the calculation of the "real" buckling stress $\sigma_{u,r}$ of an arbitrary axisymmetric shell structure in combination with $\sigma_{pl\ calc}$ given by a computer program which is able to consider the non-elastic material behaviour and the geometrical nonlinearity.

$$\sigma_{u,r\_T4} = f_{el\_T4} \cdot f_{pl\_T4} \cdot f_y = f_{el\_T4} \cdot \sigma_{pl\_calc} \qquad (6)$$

On the other hand according to [2] the "real" buckling stress σ_{u,r_T4} is given by the product

$$\sigma_{u,r\_T4} = \kappa_2 \cdot f_y \tag{7}$$

where κ_2 is a reduction factor which is determined from the lower bound curve to a huge number of experimental buckling loads of axially compressed circular cylindrical shells. The comparison of (6) and (7) gives the reduction factor

$$f_{el\_T4} = \kappa_2 / f_{pl\_T4} \tag{8}$$

accounting for imperfections only. The curves for κ_2, f_{el_T4} and f_{pl_T4} are shown in figure 2.

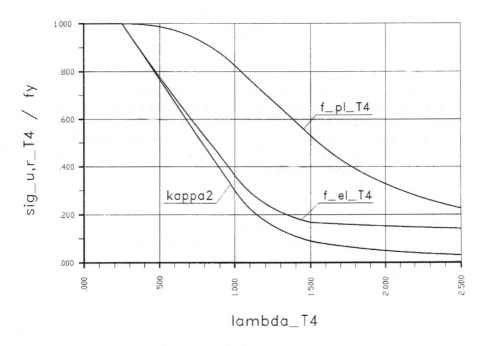

Figure 2. Splitting of the reduction factor κ_2 according to [2] in f_{pl_T4} including the plastic reduction and f_{el_T4} including the effects of the imperfections

The experimental verification of this reduction factor requires that yield strength and Young's modulus have been determined and documented for the test specimens. Only 11 test results of axially compressed circular cylindrical shells could be considered to satisfy this condition. In figure 3 their experimental buckling loads are plotted to check their agreement to the κ_2-curve. The test no.'s 1-3, 6 and 11 are very close to the design curve.

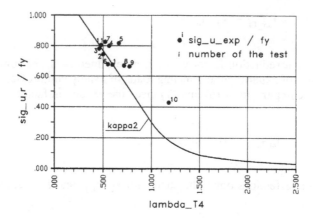

Figure 3. Distribution of the experimental limit loads in relation to the κ_2-curve according to [2]

In table 1, the ratios of the calculated buckling stresses $\sigma_{u,calc} = f_{el\ T4} \cdot \sigma_{pl\_calc}$ to the limit stresses $\sigma_{u,exp}$ from the tests are tabulated. For the tests no.'s 1-3, 6 and 11, which are very close to the κ_2-curve according to [2], this ratio is very close to one. For the other test specimens the deviation of the calculated buckling stress $\sigma_{u,calc}$ from the experimental values is larger but always to the safe side as expected. If the numerical values $\sigma_{u,calc}$ are related to the real limit stress $\sigma_{u,r}$ given by [2], all the deviations are less then 10% as also represented in table 1.

TABLE 1

R/T-ratio, Young's modulus, yield strength, numerical limit stress related to experimental and real limit stress and references

test no.	R/T	f_y (kN/cm^2)	E (kN/cm^2)	$\dfrac{\sigma_{u,calc}}{\sigma_{u,r\_T4}}$	$\dfrac{\sigma_{u,calc}}{\sigma_{u,exp}}$	ref.
1	206	20.7	19900	0.990	0.971	[5]
2	146	21.9	21200	0.968	1.016	[5]
3	105	26.0	21400	0.968	1.019	[5]
4	225	17.4	20600	0.980	0.874	[6]
5	158	34.3	20600	0.970	0.731	[6]
6	189	19.9	20600	0.986	1.048	[6]
7	159	21.2	20600	0.977	0.885	[6]
8	257	24.5	20400	0.954	0.808	[7]
9	256	28.6	20400	0.949	0.737	[7]
10	372	45.6	20150	0.905	0.386	[7]
11	133	20.7	20000	0.928	0.910	[8]

Concept D

This concept represents a complete numerical method. The limit loads are calculated by a computer program which includes the geometrical and the material nonlinearity as well as an imperfect geometry. To obtain reliable results it is necessary to consider all the deteriorating effects, which will be represented by the geometrical equivalent initial imperfections. If these equivalent initial imperfections are known it is possible to perform conservative buckling load analyses for all cases not regulated in [2].

In a first step, the limit loads have been numerically calculated for cylindrical shells subjected to axial compression (the same tests as for concept C, see table 1) and with equivalent initial imperfections in the shape of the eigenform due to the lowest eigenvalue given by the program. The value w_0 of the imperfection amplitudes was varied, until the obtained limit loads agreed with the reduced ideal buckling load due to curve κ_2 in [2]. The test no.'s 1-3, 6 and 11 were regarded particulary because of their very small deviation from the limits loads due to [2]. This procedure guarantees the same safety level as [2].

The values w_0 of the imperfection amplitudes were described according to [3] as multiples of the wall thickness T and of the measuring length L_m for the check of the initial imperfections given by [2]. The amplitudes w_0 were varied for the 11 test cylinders and 4 other cylinders with arbitrary chosen dimensions and parameters. Figure 4 shows the results of this variation, exemplary for test no. 1.

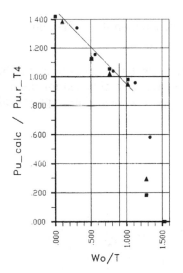 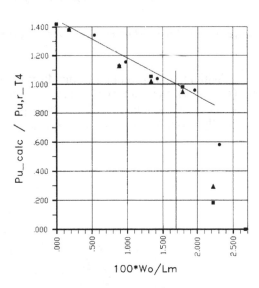

Figure 4. Distribution of the ratios of the calculated to the limit loads given by [2] for varied imperfection amplitudes w_0, which are related to the wall thickness T and the measuring length L_m according to [3], for test no. 1

In addition to the variation of w_0 the effects of the applied circumferential wave number m of the imperfections - the eigenvalues of the different eigenforms are situated very close together - and the load step width were also investigated. In figure 5 the load-axial-shortening-curves for test no. 3 are plotted as function of the w_0/T-ratio. The numerical and the experimental limit loads are given too.

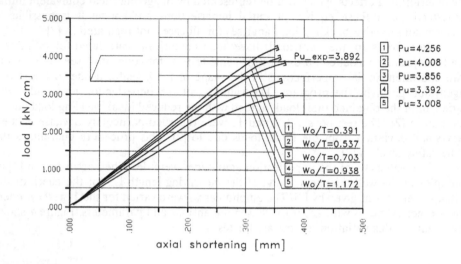

Figure 5. Load-axial-shortening-curves of test no.3 depending on the w_0/T-ratio

With an approximation to the distribution of the calculated values by a straight line (see fig. 4), the demanded values w_0/T and $100 \cdot w_0/L_m$ were obtained for the different ratios of the numerical limit loads p_{u_calc} to p_{u,r_T4}. The values are represented graphically in fig. 6 as a function of $\overline{\lambda}_{sx\_T4}$.

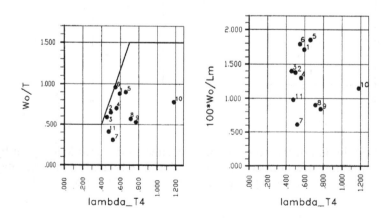

Figure 6. Plot of the values w_0 of the imperfection amplitudes related to T and L_m for $p_{u\_calc}/p_{u,r\_T4} = 1.0$ as function of the approriate slenderness $\overline{\lambda}_{sx\_T4}$

In contrast to the very large scatter of the values depending on L_m, the w_0/T-values increase nearly linear for $\bar{\lambda}_{sx}$-values between 0.4 and 0.7 to a maximum. On this basis we suggest as a conservative approximation the amplitudes w_0 for the equivalent initial imperfections to be taken into account in a buckling analysis of arbitrary shells of revolution according to equations (9) to (12).

$$\max w_0 \quad = 2.0 \cdot L_m/100 \tag{9}$$

$$
\begin{array}{llll}
\max w_0/T & = 0.5 & \text{if} & \bar{\lambda}_{sx\_T4} \le 0.4 & (10) \\
\max w_0/T & = 3.333 \cdot \bar{\lambda}_{sx\_T4} - 0.833 & \text{if} & 0.4 < \bar{\lambda}_{sx\_T4} \le 0.7 & (11) \\
\max w_0/T & = 1.5 & \text{if} & 0.7 < \bar{\lambda}_{sx\_T4} & (12)
\end{array}
$$

The governing value max w_0 is given by the maximum of both conditions.

APPLICATION OF THE CONCEPTS C AND D TO 2 EXAMPLES

Finally we will check the proposed reduction factor f_{el_T4} and imperfection amplitude w_0 for their conservativity by comparing the calculated to the experimental limit loads of 2 selected shell structures which are not covered by [2].

$\sigma u,exp = 10.27$ kN/cm^2 [8]

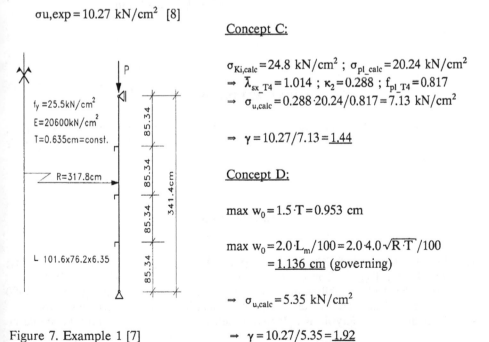

Concept C:

$\sigma_{Ki,calc} = 24.8$ kN/cm^2 ; $\sigma_{pl\_calc} = 20.24$ kN/cm^2
$\Rightarrow \bar{\lambda}_{sx\_T4} = 1.014$; $\kappa_2 = 0.288$; $f_{pl\_T4} = 0.817$
$\Rightarrow \sigma_{u,calc} = 0.288 \cdot 20.24/0.817 = 7.13$ kN/cm^2

$\Rightarrow \gamma = 10.27/7.13 = \underline{1.44}$

Concept D:

$\max w_0 = 1.5 \cdot T = 0.953$ cm

$\max w_0 = 2.0 \cdot L_m/100 = 2.0 \cdot 4.0 \sqrt{R \cdot T}/100$
$\qquad = \underline{1.136}$ cm (governing)

$\Rightarrow \sigma_{u,calc} = 5.35$ kN/cm^2

Figure 7. Example 1 [7]

$\Rightarrow \gamma = 10.27/5.35 = \underline{1.92}$

In figure 1:
$f_y = 25.5$kN/cm^2
$E = 20600$kN/cm^2
$T = 0.635$cm=const.
$R = 317.8$cm
$L\ 101.6 \times 76.2 \times 6.35$

85.34
85.34
85.34
85.34
85.34
341.4cm

P

$P_{u,exp}=320.0$ kN   [9]

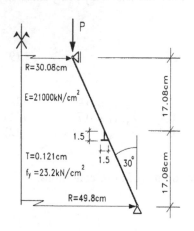

$R=30.08$cm

$E=21000$kN/cm^2

1.5

$T=0.121$cm

$f_y=23.2$kN/cm^2

$R=49.8$cm

$30°$

17.08cm

17.08cm

Figure 8. Example 2 [9]

Concept C:

$\sigma_{Ki,calc}=47.68$ kN/cm^2 ; $\sigma_{pl,calc}=22.77$ kN/cm^2

$\Rightarrow \bar{\lambda}_{sx\_T4}=0.698$; $\kappa_2=0.582$; $f_{pl\_T4}=0.951$

$\Rightarrow \sigma_{u,calc}=0.582 \cdot 22.77/0.951=13.93$ kN/cm^2

$\Rightarrow P_{u,calc}=13.93 \cdot 2\pi \cdot 30.08 \cdot 0.121 \cdot \cos 30°$

$=275.98$ kN

$\Rightarrow \gamma=320.0/275.98=\underline{1.16}$

Concept D:

max $w_0=1.5 \cdot T=\underline{0.1815 \text{ cm}}$ (governing)

max $w_0=2.0 \cdot L_m/100$

$=2.0 \cdot 4.0 \cdot \sqrt{R \cdot T/\cos 30°}/100=0.164$ cm

$\Rightarrow P_{u,calc}=321.3$ kN

$\Rightarrow \gamma=320.0/321.3=\underline{0.996}$

ACKNOWLEDGEMENT

The numerical analyses were performed with the computer program "HORUS" [10], which was placed at our disposal from Prof. Dr.-Ing. W. Wunderlich. This support is gratefully acknowledged.

REFERENCES

1. Task group 9 of the ECCS, Shells. In Second International Colloquium on Stability; Introduction Report, Tokyo 1976, Liège 1977, Washington 1977, pp. 275-297
2. DIN 18800 Teil 4, Stahlbauten - Stabilitätsfälle, Schalenbeulen, Ausgabe Nov. 1990
3. Deutscher Ausschuß für Stahlbau, DASt-Richtlinie 017, Entwurf Mai 1990
4. Schmidt, H. and Krysik, R., Beulsicherheitsnachweis für baupraktische stählerne

475

Rotationsschalen mit beliebiger Meridiangeometrie - mit oder ohne Versuche ? In Festschrift H. Duddeck, ed. J. Scheer, H. Ahrens, H.-J. Bargstädt, TU Braunschweig, 1988

5. Schulz, U., Zylinderschalen mit und ohne Innendruck im elastisch-plastischen Beulbereich. Der Stahlbau, 60 (1991), H.4, pp. 103-110
6. Lindenberger, H., Bericht über Druckversuche an Kreiszylindern. In Fortschritte im Stahlbau, Vortragstagung 07.02.1958, ed. Deutscher Stahlbau-Verband Köln, pp. 53-64
7. Miller, C.D., Buckling of axially compressed cylinders. Journal of the Structural Division, Vol.103, NO. ST3, march 1977, pp. 695-721
8. Bornscheuer, F.W., Flächige, gekrümmte Bauteile - Beulsicherheitsnachweise für isotrope Schalen. In Stahlbau-Handbuch 1, Stahlbau-Verlags-GmbH Köln, 1982, pp.552-566
9. Klöppel, K. and Motzel, E., Traglastversuche an stählernen, unversteiften und ringversteiften Kegelstumpfschalen. Der Stahlbau, 45 (1976), H.10, pp. 289-301
10. Wunderlich, W., HORUS - User's manual

THE PRECRITICAL DEFLECTIONS OF COOLING TOWERS ORIGIN AND INFLUENCE

By Dr W. AFLAK, Prof. J.F. JULLIEN
Concrete and Structures Laboratory INSA LYON (FRANCE)
and M. BOLVIN
EDF/SEPTEN LYON (FRANCE)

I - INTRODUCTION

Cooling towers are thin-walled shell structures susceptible to buckling when subjected to permanent or short term loading. This type of structure, like all Civil Engineering construction, is designed for a long life. It is possible that the deterioration of the structure with time can lead to failure and this then poses the problem of their obsolescence. This latter is perhaps not only linked with the deterioration of reinforced concrete as a material, but also with the possibility of buckling.

Because of the lack of knowledge of the mechanism which governs this aging process, very strict design rules have been applied in order to achieve adequate security. The new generation of cooling towers are hence designed with a greater thickness than the old ones and also have two layers of reinforcement in place of a single layer at the centre of the section. The new designs have been accompanied, for the last four years, by research coordinated by EDF (France). This is aimed at identifying the physical parameters of aging and of failure of cooling towers, particularly those towers that are more than 30 years old. It is intended to produce simple design rules once the principles which govern the aging process are established. It is in this context that the present study was undertaken. It is an attempt to analyse and interpret the observed in-situ deteriorations of several cooling towers, by means of numerical calculations, using the results obtained in laboratory on the buckling of circular cylinders.

II- SYNTHESIS OF OBSERVED IN-SITU RESULTS [1]

The analysis and comparison of the results of the survey (1) of deformation and cracking on several cooling towers of EDF (France) (2), led to the following conclusions :

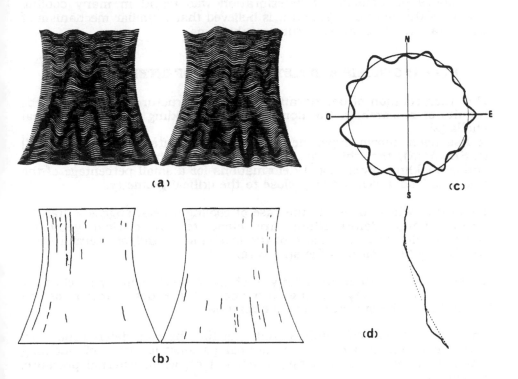

Fig.1 - *Pont sur Sambre 2 cooling tower*
  a/ actual deformed shape (exageration factor : x 18)
  b/ crack distribution.
  c/ Horizontal section
  d/ Vertical section

- There are radial imperfections of large amplitude in the shape of the shell (Fig.1- a).

- The geometry of these imperfections is modal and characterised by both meridional (Fig. 1 - d) and circumferential undulations (Fig. 1 - c).

- The circumferential wave number is more significant than the meridional one.

(1) *The geometric deformations were measured by photogrammetry and the cracks were identified by means of a video camera.*

(2) *The cooling towers studied were :*
Pont sur Sambre 2 and 3, Ansereuilles 3 and Bonchain 1.

- The preponderant local circumferential mode for each horizontal section varies with height.

- The cracks are exclusively meridional, i.e. there are no horizontal cracks (Fig. 1-b).

The above phenomena of deterioration was found in many cooling towers of the older design and it is believed that a similar mechanism of deterioration effects all structures of this type.

III - HYPOTHESIS AND METHODOLOGY OF INTERPRETATION

The interpretation of deteriorations of these structures is based on the results of research in the domain of the buckling of thin cylindrical shells [2].
It has been shown experimentally, and validated numerically [3-7] (INSA LYON), that this type of shell under loading which is likely to cause buckling, gives modal deformations for a small percentage of the buckling load which are very close to the critical geometry.

This result when applied to the case of cooling towers, suggests that the actual modal deformations are close to the critical geometry corresponding to one or more of the mechanical and/or thermal loads which may be applied to the structure.

These loads lead to an instability of shape if their intensity reaches the critical value, always taking into account the deterioration of the properties of the structure and the material.

The interpretation and understanding of the observed deterioration can thus be accomplished by a numerical parametric study of buckling under different types of loading : self weight, wind, external pressure, thermal gradient.

The comparison of the calculated and observed deformations support the hypothesis that the loads are the origin of the behaviour. In addition to the structural mechanism, there are the parameters relating to the behaviour of reinforced concrete. The modelling of the latter takes into account its heterogenity and anisotropy and also the different phenomena resulting from creep and shrinkage.
The work presented in this paper describes the qualitative interpretation of the aging phenomena and the validation of the results based on the assumption of buckling. We assume, for simplification, that reinforced concrete is a homogeneous, isotropic, perfectly elastic material with symmetric behaviour under tension and compression.
The calculations are carried out using a finite element program, INCA, developed by C.E.A. (France). This program permits a two dimensional modelling of shells, the application of a non axisymmetric loading, and allows for the introduction of geometrical imperfections by the use of a Fourier series.
The results given below are based on cooling tower n° 2 of the Pont sur Sambre Power Station (Fig. 2).

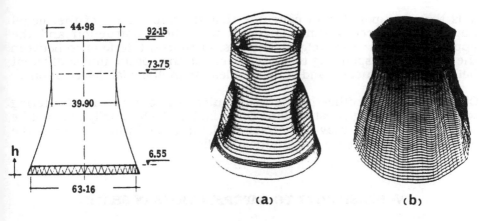

Fig. 2 - Overall dimensions of Pont sur Sambre 2 cooling tower (m).

Fig. 3 -   *a/ buckling mode under self weight*
          b/ actual deformed shape

IV- ORIGIN OF GEOMETRICAL IMPERFECTIONS

Amongst all the numerically obtained critical deformations under the elementary loads considered, that corresponding to self weight gives deformations which are similar to those observed (Fig.3).

This result emphasises the importance of the action of gravity and leads to the conclusion that the initiation of the geometrical imperfections are caused by this effect at the time of construction. This hypothesis is confirmed by the good agreement between the buckling modes calculated for different levels of construction (h) less than the total height (H), and the observed in-situ deformations at the same levels (table 1)

h/H	Measurments		INCA Calculation	
	Predominant mode	Next important mode	Critical mode	Next modes
0.20	9	10	6	9
0.23	10	9	6	7, 8
0.25	10	9	9	7, 8, 10, 11
0.29	9	10	8	7, 9, 10
0.32	9	10	7	6, 8
0.35	9	6	6	7, 8
0.45	6	4	6	5
0.50	6	4	5	6
0.75	6	5	5	
1.00	4	6	5	

- Table 1-

It is therefore possible to conclude that the appearance of the modal imperfections occurs simultaneously with the construction and the application of the self weight. There is, of course, interference between the modes corresponding to these effects as they occur under different heights of construction and thus they lead to multimodal deformations.

Clearly, there are other geometrical faults due to errors in shuttering which occur during construction. These faults which initially may not be modal in character, may be become modal in form during the life of the structure as in the case of cylindrical shells [3].

V- SENSITIVITY TO IMPERFECTIONS IN SHAPE

The term sensitivity here includes the effect of geometric imperfection on the static state and on the stability of the shell.

V-1 Choice of imperfection

The geometric imperfection chosen for study was determined by the following considerations :

- the imperfection must be representative of the real deformation but must also be capable of being described by a numerical procedure.

- the most critical imperfection vis-a-vis stability is that one which has the same shape as the buckling mode.

- the critical mode must correspond to the predominant loading, i.e. the self weight as given in the previous conclusions of the study.

- The imperfection must be multimodal, since the mode of the real defect varies with height.

For the cooling tower Pont sur Sambre 2, we have retained a geometric imperfection made up of the superposition of buckling modes 5,6 and 9 which were generated over the complete surface. These latter are the most frequent modes given by the elastic bifurcation calculation at the various height of the shell under self weight.
The choice is justified by the global similarity which is obtained between the numerical calculation, as it is described by these modes (Fig.4), and the real imperfection.

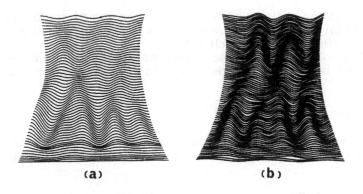

Fig. 4 - a/ Initial imperfection obtained by calculation
b/ actual deformed shape (modes 5,6,9 only)

This gives a method for the definition, in cooling towers in general, of the geometric imperfections which are created during construction and which cause deterioration.

V- 2 Static behaviour

The calculations show that the presence of a multimodal imperfection has two important consequences for the static behavior of cooling towers :

- a modification in the radial rigidity of the shell

This is raised or lowered according to the position of the defect and the nature of load (Fig.5).

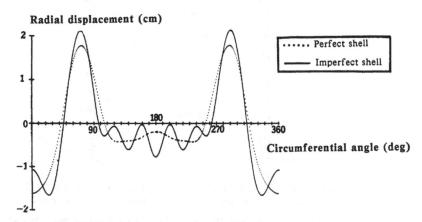

Fig. 5 - Radial displacement of perfect and imperfect shells under the combination of self weight, wind and external lateral pressure.

The decrease in the radial rigidity is able to accommodate the growth of the small initial imperfections by the effect of the creep of the concrete with time, particularly under self weight.
This result is collaborated by the observation that the amplitude of the modes on the bottom two thirds of the cooling towers are more than the average.

- a pertubation on the stress field

Under self weight, the meridional and circumferential membrane stresses are little effected by the existence of a multimodal defect (Fig.6).

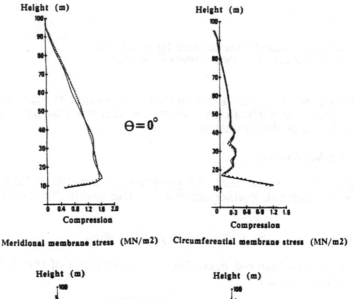

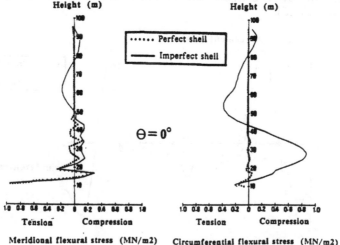

Fig. 6 - Stress distributions in the perfect and imperfect shell under self weight.

The bending stresses, unimportant in the perfect shell, become important in the imperfect shell. This applies particularly to the circumferential bending stresses (Fig.6).

This result follows the same relationship as the preponderence of the circumferential undulations over the meridional undulations.

The excess of the circumferential flexural stress in the presence of an imperfection gives an explanation of the origin of the meridional cracks observed.

V-3 Buckling behaviour

The critical load factor for buckling λ_{cr} is calculated for two combinations of load, namely :

I) λ (G+V+P)

II) λ (3G+V+P)

where   G : self weight
                         V: wind load (regulation NV 65)
                         P: external pressure.

In combination (I), it is assumed that the three loads contribute in the same way to instability. The same notion is retained in the second combination but the self weight load is multiplied by 3.

The application of this coefficient permits, in an approximate manner, the combination of permanent load (which acts with a reduced modulus of elasticity), with those loads of a different nature such as wind whose duration may only be of several hours (and which acts with an instaneous modulus of elasticity). The results are summarised in Table 2 and give rise to the following conclusions :

Shell Combination		Perfect shell		Imperfect shell	
		λ_{cr}	n	λ_{cr}	n
I	λ (G + V + P)	8.3	5	11.05	5
II	λ (3G + V + P)	3.3	7	4.4	5

-Table 2-

- The presence of the imperfection does not lead to any reduction of the critical load. The physical significance of this result is discussed later.

- The delayed effect (combination II) considerably reduces the critical load of cooling tower.

VI- LIMITATIONS AND TRENDS OF THE CALCULATIONS

Some doubts must remain about the physical significance of the increase in the critical load of the imperfect shell in comparison with the perfect one (Table 2).

In effect, the presence of geometric imperfections favours the deterioration of the material and hence the appearance of cracks.
The existence of the latter, which is not taken into account in the calculations, modify the behaviour of the structure. The cracks diminish the rigidity of the shell and, as a consequence, lower the critical load of the cooling tower as compared to the calculated load. This latter refers to an uncracked shell.
A more realistic estimation of the margin of security and its variation with time has to consider, in addition to the geometric imperfections, the instantaneous nonlinearties and the time - dependent effects of reinforced concrete.
The large flexural effects caused by the modal imperfections makes it necessary to consider the flexural behavior with its resulting cracks and nonlinearities as well as the membrane characteristics.

VII- CONCLUSION

The same deterioration observed in many cooling towers establishes the mechanism of aging in this type of structure.
The interpretation of these mechanisms on a basis of buckling allows the deformation of cooling towers to be considered in the same way as the observed precritical deformation of cylindrical shells.

The study emphasises the predominant role which is played by the self weight in the behaviour of this type of structure.

This loading is at the origin of the geometric imperfections observed, and the latter have their origin in the original stages of construction. The existence of these imperfections increase the local radial flexibility of the shell and causes the vertical cracking by the creation of circumferential bending under the action of different loads.

The results enable the formulation of design rules for the dimensioning of cooling towers which take into account their aging.

The results quantify the phenomena whereby the deformations are created and give a qualitative assessment of their amplitudes. They must be developed to take account of time.
This development requires a viscoplastic model of the material which allows for damage. Such a study is being undertaken at the present time.

REFERENCES

[1] AFLAK W., JULLIEN J.F.- "Analyses des paramètres réagissant les déformations des coques des aérorégrigérants atmosphériques", Rapports INSA, n° 1.714/001 - 013 (13 rapports).

[2] JULLIEN J.F., REYNOUARD J.M. - "Reflexion on the origine of deflections observed on the cooling towers of Pont sur Sambre and Ansereuilles", 3rd international symposium on natural draught cooling towers, Paris, April 1989, pp. 595-604.

[3] WAECKEL N. - "Stabilité des plaques et des coques. Influence des défauts géométriques initiaux", Thèse de Doctorat d'Etat, INSA DE LYON, MARS 1984, 348 p.

[4] AFLAK W. - "Flambage de coques cylindriques sous compression axiale. Influence des imperfections géométriques et des imperfections de conditions aux limites", Thèse de Doctorat, INSA de LYON, Juillet 1988, 256 p.

[5] DEBBANEH N. - "Flambage de coques de révolution à méridienne brisée, sous pression latérale externe. Influence des conditions aux limites", Thèse de Doctorat, INSA de LYON, Juillet 1988, 245 p-

[6] WAECKEL N., JULLIEN J.F.- "Experimental studies of the instability of cylindrical shells with initial geometric imperfections", Pressure vessels and piping conference and exhibit, San Antonio (Texas), June 1984, Preprints ASME, special publication PVP, vol 89, p. 34-48.

[7] AFLAK W., DEBBANEH N., WAECKEL N., JULLIEN J.F. -"Buckling of cylindrical shells", 9th- SMIRT, Lausanne, 1987, p. 157-165.

EXTERNALLY-PRESSURISED CFRP DOMES

by
J. Blachut, G.D. Galletly and F. Levy-Neto
Department of Mechanical Engineering
University of Liverpool

ABSTRACT

The results of external pressure tests on twenty-five 0.23 m dia. woven CFRP torispherical shells are given in the paper. Two values of D/t (diameter-to-thickness ratio) were used for the models (i.e. D/t ≈ 80 and 120) and the r/D - ratio was varied from 0.2 to 0.5 (r = knuckle, or toroidal, radius). The test results are compared with theoretical predictions of the failure pressures of the models. Both bifurcation buckling and material failure (FPF = first ply failure) were considered as possible failure modes and both stress and buckling analyses of the composite domes were carried out using the BOSOR 4 shell program. The FPF pressures were obtained from the stresses and various material failure criteria.

All the models failed in the material failure mode. The ratios of the experimental collapse pressures to the FPF predictions were in the range 0.85-1.30. As the models were made by the hand lay-up/vacuum bag technique, and as the experimental values for the compressive strength of the CFRP had a fair amount of scatter, this agreement can be considered satisfactory.

The test results also confirm a recent theoretical prediction that there is an optimum value of r/D insofar as dome strength is concerned. The optimum torisphere is both lighter and stronger than the comparable hemisphere. A reason for the occurrence of these optimum torispheres is given in the paper.

INTRODUCTION

The use of fibre-reinforced plastic (FRP) end closures in aerospace and marine structures [1] is attractive from the weight-saving point of view. However, it is only recently that test results on externally-pressurised GFRP and CFRP hemispherical and torispherical domes (see Fig. 1) have been reported in the literature [2,3,4,5]. For the thinner shells (D/t > 300, say), the failure mode was often bifurcation buckling or axisymmetric collapse (snap-through). With the thicker shells (D/t ≈ 100), the failure mode was usually first ply failure (FPF).

Another recent result of interest is the theoretical prediction in [6] that the collapse strengths of externally-pressurised composite torispheres having r/D - ratios of about 0.35 are considerably higher than the strengths of comparable hemispheres. These torispheres are also lighter than the hemispheres (ratios of ≈ 0.9:1.0).

In view of the few test results available on externally-pressurised woven CFRP domed ends, it was decided to carry out additional tests on them. The shapes chosen for investigation were torispherical and hemispherical shells having moderate values of the diameter-to-thickness ratio, i.e. D/t ≈ 80 and ≈ 120. The aims of these tests were:

(i)   to check the repeatability of the experimental failure pressures,

(ii) to determine how well theory and experiment agreed insofar as failure
      pressures, failure modes and failure locations were concerned, and

(iii) to verify that there were torispherical FRP domes which were stronger
      and lighter than their hemispherical counterparts.

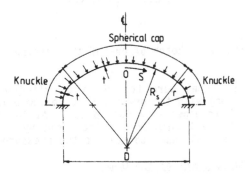

Fig. 1   Geometry of externally-pressurized torisphere.

EXPERIMENTAL DETAILS

a)   CFRP Model Domes

All the dome closures were $[0°/60°/-60°]_s$ 6-ply CFRP shells and had a
diameter of ≈ 0.225m.   One series of domes had a diameter-to-thickness
(D/t) ratio of ≈ 80 and the other series had D/t ≈ 120. Both torispherical
and hemispherical domes were tested. The R_s/D - ratio of the torispheres
was 0.55 and their r/D - ratios were in the range 0.2-0.5. The CFRP shells
were made from dry cloth which was then impregnated with resin.   Male
moulds were used in the manufacture of the domes and the final
consolidation and cure was carried out using the vacuum bag technique.   For
the D/t ≈ 80 shells, the carbon cloth was 0.56mm thick 8 end satin whereas,
for the D/t ≈ 120 shells, it was 0.30mm 5 end satin.

b)   Test Chambers

Two external pressure facilities were employed in the investigation.
The D/t ≈ 120 domes were tested in an open-ended pressure chamber (see Fig.
2) which was designed to receive the base ring at the open end. The chief
advantage of this arrangement is that the inside (concave) surface of the
test dome is easily visible. Elastic buckles may be seen and the dome is
accessible for deformation measurements.   Any onset of 'weeping' when
cracks appear in the matrix can be detected readily. The pressure chamber
was filled with oil and its pressure was increased by means of a hand pump.
A strain-gauged transducer and a Bruel & Kjaer strain indicator were
employed to measure the pressure.

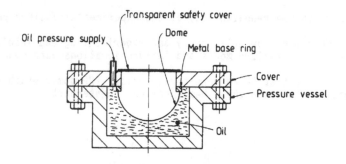

Fig. 2   Sketch of one of the testing rigs.

The $D/t \approx 80$ domes were tested in a cylindrical pressure chamber 0.30m dia. and 1m long with the pressurising medium being water. The domes were attached at their edges to a base plate. As for the $D/t \approx 120$ domes, the pressure in the test chamber was increased by a hand pump; a B & K strain indicator was again used to measure the pressure.

c)   Thickness Measurement of the Domes

The thicknesses of the models were measured along 8 or 12 equispaced meridians, at every 10mm meridionally from the base of the domes. The only exception to this procedure was in the knuckle region of the torispheres, where measurements were made at every 5mm. The thickness usually increased from the pole to the base of the dome and a typical variation of thickness in a dome is as shown in Table 1 of [5].

d)   Mechanical Properties of CFRP

The relevant elastic material properties of the quasi-isotropic composites used in the present paper are two orthogonal moduli of elasticity, E_1 and E_2, Poisson's ratios, ν_{12} and ν_{21}, and the in-plane shear modulus, G_{12}. The relevant strength properties are the ultimate tensile strengths X_t and Y_t, the ultimate in-plane shear strength, S, and the ultimate compressive strengths, X_c and Y_c.

The tensile and shear strengths were determined according to ASTM procedures D 3039-76 and D 3518-76 [7]. For the compressive strengths, the IITRI compression test [8] was used (IITRI = Illinois Institute of Technology and Research Institute) - see Fig. 3.

At least five tensile and five compressive test specimens were prepared, with the fibre directions at 0° and 90° to the applied load and another five specimens were prepared with the fibre directions at -45° and +45°. The test specimens were prepared using the hand lay-up technique; the consolidation of the layers, and the cycles of cure and post-cure, were identical to those used in the manufacture of the domes.

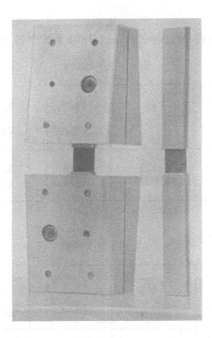

Fig. 3   CFRP flat specimen with mild steel tabs; also a specimen in the
jaws of the compressive strength rig.

The average results (and the standard deviations) obtained in the
mechanical property tests are summarised in Table 1.

THEORETICAL PREDICTIONS OF COLLAPSE PRESSURES

Two failure modes of the CFRP domes were considered in the
theoretical analysis, i.e. elastic bifurcation buckling and material
failure (FPF = first ply failure). The BOSOR 4 program [9] was used to
find the buckling pressures of the composite domes. The FRP is treated as
a layered orthotropic material and the orthotropic properties are given
relative to the meridional directions of the shell of revolution - see also
[2], [3] and [5].

The direct forces N and bending moments M at the middle-surface of
the FRP dome can be obtained using the stress analysis option in BOSOR 4
(INDIC = 0). The stresses in the various plies throughout the shell can
then be determined using classical laminate theory (see [2], [3], [10]).
In order to determine whether material failure is likely to occur, one
needs a failure criterion. Several of these have been suggested in the
literature and seven of them were utilised in this investigation. In this
short paper, only the results of the Tsai-Wu [11] and Owen [12] criteria
will be given. More information is available in [5].

For all the domes considered in this paper, the FPF pressures were
lower than the bifurcation buckling pressures. Hence, in the theoretical
results which follow, only the FPF pressures will be given. However,

TABLE 1
Material properties obtained for CFRP domes.

D/t ≈ 80 Domes

$E_1 = E_2$ (kN/mm²)	G_{12} (kN/mm²)	ν_{12}	$X_t = Y_t$ (N/mm²)	$X_c = Y_c$ (N/mm²)	S (N/mm²)
55.856 (0.636)*	3.438 (0.036)	0.125 (0.0014)	625.11 (70.2)	503.4 (58.0)	64.51 (0.72)

D/t ≈ 120 Domes

$E_1 = E_2$ (kN/mm²)	G_{12} (kN/mm²)	ν_{12}	$X_t = Y_t$ (N/mm²)	$X_c = Y_c$ (N/mm²)	S (N/mm²)
66.50 (2.4)*	5.0 (0.19)	0.048 (0.002)	618.0 (41.0)	465.0 (41.0)	78.0 (6.2)

* = standard deviations in brackets

interested readers can find the buckling pressures of the D/t ≈ 120 domes in Table 4 of [5].

As the thickness of the CFRP domes was variable, in both the meridional and the circumferential directions, two simplifications were used to expedite the analyses. These were:

(i)   method A: an overall average thickness of the dome was determined and this constant thickness was used in the calculations, and

(ii)   method B: average thicknesses for the spherical cap and the knuckle were found and used, as constant thicknesses, in the relevant shell segments.

With the thicker domes (D/t ≈ 80), an average thickness, which varied along the meridian, was also employed in the analyses. This is denoted by method C herein.

RESULTS

The experimental failure of the domes was sudden, brittle and accompanied by a loud bang. The collapse pressures of the D/t ≈ 120 domes are given in Table 2 and those of the D/t ≈ 80 domes in Table 3. Other features of the tests, such as the location along the meridian of the FPF, which layer cracked first, etc. may be found in [5].

The numerical results for the FPF pressures of the various domes are also listed in Tables 2 and 3. With the D/t ≈ 80 domes, the failure criterion used was that of Tsai-Wu in stress space. With the D/t ≈ 120 domes, several failure criteria were studied with method B; the minimum pressures were found with either the Owen or the Tsai-Wu (in strain space) criteria. With method A applied to the thinner domes, Tsai-Wu in stress space was employed.

DISCUSSION OF RESULTS

a)   D/t ≈ 120 Domes

The repeatability of the experimental pressures listed in Table 2 is quite good. For all models except dome 13 (a hemisphere), the pressures

TABLE 2
Experimental Collapse Pressures and Theoretical FPF Pressures for the D/t ≈ 120 Domes (R_s/D = 0.55 throughout).

Dome No.	r/D	D/t_{av}	P_{expt} (N/mm^2)	P_{FPF} (N/mm^2) Method A	P_{FPF} (N/mm^2) Method B	$\dfrac{P_{expt}}{P_{FPF}}$ Method A	$\dfrac{P_{expt}}{P_{FPF}}$ Method B
1	0.2	117	5.09	4.90	5.56	1.04	0.95
2		116	5.51	4.92	5.31	1.12	1.04
3		120	5.26	7.69	6.10	0.685	0.80
4	0.3	120	5.28	7.71	6.18	0.685	0.855
5		119	5.67	7.73	6.16	0.735	0.92
6		120	5.59	7.70	6.12	0.725	0.915
7	0.37	114	7.04	8.33	7.18	0.845	0.98
8		115	6.81	8.25	7.53	0.825	0.905
9	0.40	121	7.12	7.12	6.98	1.00	1.02
10		124	7.23	6.97	7.12	1.04	1.015
11	0.425	116	6.69	6.92	7.12	0.97	0.94
12		115	6.37	6.97	7.16	0.91	0.89
13		122	6.80	5.68	5.92	1.20	1.15
14	0.5	116	5.80	5.98	6.85	0.97	0.85
15	(Hemi-	119	6.11	5.84	6.70	1.05	0.91
16	sphere)	118	6.04	5.88	6.04	1.03	1.00

were repeatable to within 10%. With dome 13, the model started to leak at 5.72 N/m² - this pressure is close to the failure pressures of the other hemispheres.

As may be seen from Table 2, the agreement between the test collapse pressures and the theoretical FPF pressures is also fairly good. In general, the agreement is best with method B although, for the hemispheres, there is not much difference between methods A and B.

With regard to the variation of collapse strength with a dome's r/D - ratio, this is illustrated in Fig. 4. As may be seen, the average strength of the $^r/_D$ = 0.4 torispheres was 20% higher than that of the corresponding hemisphere. Also, there is reasonable agreement between test and theory and the variation of strength with r/D is broadly confirmed.

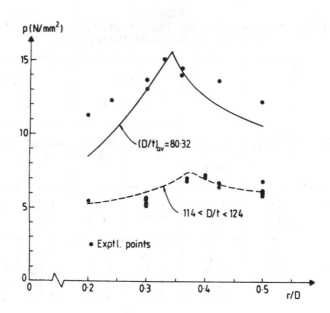

Fig. 4   Variation of the FPF pressure vs. the r/D parameter (R_s/D = 0.55 throughout).

b)   <u>D/t ≈ 80 Domes</u>

For this series of models, there were duplicate tests only for the domes with r/D = 0.30 and 0.37. As may be seen from Table 3, their collapse pressures were repeatable to within 7%.

The agreement between the experimental collapse pressures and the theoretical FPF pressures (based on the Tsai-Wu criterion in stress space) was best when method C was employed. In this method, the thickness varied along the meridian and the ratios of p_{expt}/p_{FPF} were in the range 0.89 to 1.28.

The variation of the collapse pressures versus the r/D - ratio is shown in Fig. 4. As may be seen, the maximum strength occurs for a

TABLE 3
Experimental collapse pressures and theoretical FPF pressures for the D/t $\approx$
80 domes ($R_s/D = 0.55$ throughout).

Dome No.	r/D	D/t_{av}	P_{expt} (N/mm^2)	P_{FPF} (N/mm^2)		P_{expt}/P_{FPF}
				Method A (average thickness)	Method C (variable thickness)	
1	0.20	84.19	11.31	8.09	8.87	1.40-1.28
2	0.24	79.41	12.28	10.54	10.89	1.17-1.13
3	0.30	79.04	13.57	13.47	13.75	1.01-0.99
4	0.30	84.74	13.04	12.34	13.41	1.06-0.97
5	0.33	79.72	15.04	15.11	15.63	1.00-0.96
6	0.37	75.70	14.00	15.61	15.81	0.90-0.89
7	0.37	79.58	14.48	14.75	15.70	0.98-0.92
8	0.425	79.58	13.59	12.18	13.33	1.12-1.02
9	hemisphere	80.94	12.17	10.22	11.47	1.19-1.06

torisphere having $r/D \approx 0.35$ and its strength is about 25% higher than that of the comparable hemisphere. In general, the test results confirm the variation in collapse strength which was predicted theoretically.

So far, only the influence of a single shape parameter, i.e. r/D, on the collapse strength of a dome has been examined herein. However, the R_s/D ratio can also influence the collapse strength - see Fig. 5. The effect of that, for $r/D = 0.2$ and 0.4, is shown in Table 4.

Table 4 also gives the weight W of the test domes, normalised with respect to the weight W_H of the reference hemisphere. The ratios of the volume V enclosed by a dome, divided by the volume enclosed by the hemisphere, V_H, are also tabulated.

It may be seen, from Table 4, that if the shape of the dome is not chosen with care, then one can obtain a shell which is only one-quarter the strength of another shell which has a comparable weight and volume, viz. domes 4 and 7.

The weight W, the volume V and the test pressure $\bar{p}_{expt}$ are frequently combined into the so-called performance ratio η (i.e. $\eta = \bar{p}_{expt}$ V/W). Table 4 contains the relevant values of η and they vary from 0.26 to 1.19 of the performance of the comparable hemisphere.

TABLE 4
Tabulation of shell weight W, volume V and performance ratio η. Quantities corresponding to the reference hemisphere denoted by W_H, V_H and η_H ($\eta = p$ V/W).

Dome No.	$\dfrac{r}{D}$	$\dfrac{R_s}{D}$	$\dfrac{D}{t}$	$\dfrac{W}{W_H}$	$\dfrac{V}{V_H}$	$\bar{p}_{expt}$ (N/mm^2)	$\dfrac{\eta}{\eta_H}$
1	0.20	0.55	116.5	0.807	0.691	5.30	0.759
2	0.30	0.55	119.8	0.839	0.777	5.45	0.844
3	0.37	0.55	114.5	0.924	0.845	6.93	1.060
4	0.40	0.55	122.5	0.884	0.878	7.18	1.192
5	0.425	0.55	115.5	0.956	0.906	6.53	1.034
6	0.20	1.00	120.0	0.711	0.547	2.03	0.261
7	0.40	1.00	120.0	0.894	0.863	1.72	0.278
8	0.20	0.55	84.18	0.768	0.691	11.31	0.836
9	0.24	0.55	79.41	0.836	0.724	12.28	0.874
10	0.30	0.55	81.89	0.845	0.777	13.31	1.006
11	0.33	0.55	79.72	0.863	0.805	15.04	1.153
12	0.37	0.55	77.64	0.937	0.845	14.24	1.055
13	0.425	0.55	79.58	0.954	0.906	13.59	1.061

AN EXPLANATION FOR THE OCCURRENCE OF OPTIMUM TORISPHERES

For both D/t $\approx$ 80 and D/t $\approx$ 120, the experimental and theoretical results show that some torispheres are stronger than the comparable hemispheres. This would be an unusual finding for metallic shells (see, for instance, [13]); however, the failure mode for these CFRP models is not buckling but first ply failure (i.e. a material failure mode). It is of interest to see if an explanation can be found for these results.

Considering the sixteen D/t $\approx$ 120 models, the average global thickness was t_{gl} = 1.867 mm (this is close to 225/120 = 1.875 mm). Using the above constant thickness for all the models, and the stacking sequence $[0/60^{\circ}/-60^{\circ}]_s$, stress calculations were carried out assuming that the models were subjected to the same external pressure. The value of the latter was taken as 6.15 N/mm^2, which was the average experimental failure pressure of the D/t $\approx$ 120 models.

The quantities determined in the above calculations were:

(i)   the maximum value of the failure index,
(ii)   the meridional location of the most critical section (i.e. s/s_{total}),
     and
(iii) the stress resultants (M_ϕ, M_θ, N_ϕ, N_θ) at this critical section.

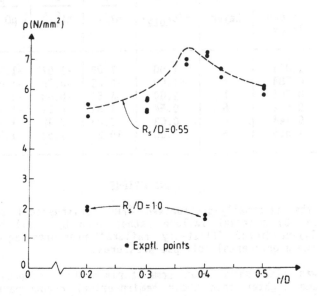

Fig. 5   Comparison of dome strength vs. the R_s/D ratio for torispheres
having the same diameter and $D/t \approx 120$.

The only quantity which varied in the calculations was the toroidal, or knuckle, radius of the models ($0.2 \leq r/D \leq 0.5$). The failure criterion used in (i) above was that of Owen.

A summary of the numerical results obtained is given in Table 5. From this Table it may be seen:

a)   For r/D = 0.50, 0.425 and 0.40, N_ϕ, N_θ and s/s_{total} do not vary with r. However, M_ϕ, M_θ and the failure index decrease when r/D changes from 0.50 to 0.40. For this reason, the r/D = 0.40 torisphere is the strongest shape of the three.

b)   For r/D = 0.37 and 0.30, the critical section moves away from the clamped edge to a region where N_θ is significantly higher than its value for 0.4 < r/D < 0.5. M_ϕ and M_θ also change sign and the most highly stressed layer shifts from layer 1 (exterior) to layer 6 (interior).

c)   For r/D = 0.20, M_ϕ and M_θ reach their highest absolute values for all the models.

d)   As far as the strength of these domes is concerned, M_ϕ seems to be the most important parameter.

TABLE 5

A Comparison of the Stress Resultants and Failure Index for $0.2 \leq r/D \leq 0.5$ ($p = 6.15$ N/mm^2).

| r/D | Most Critical Section | | | $\dfrac{M_\phi}{pt_{g1}^2}$ | $\dfrac{M_\theta}{pt_{g1}^2}$ | $\dfrac{4N_\phi}{pD}$ | $\dfrac{4N_\theta}{pD}$ |
	Failure Index	Layer	s/s_{total}				
0.50	1.084	1	1.00	-7.09	-1.07	-1.00	-0.31
0.425	0.788	1	1.00	-5.20	-0.78	-1.00	-0.31
0.40	0.701	1	1.00	-4.51	-0.68	-1.01	-0.31
0.37	0.642	6	0.54	1.31	0.31	-1.21	-1.11
0.30	0.744	6	0.63	2.08	0.42	-1.18	-1.17
0.20	1.566	6	1.00	10.2	1.53	-1.01	-0.31

CONCLUSIONS

All the externally-pressurised CFRP torispherical shells in this study failed by material failure rather than by shell buckling. The theoretically-predicted "first-ply failure" pressures agreed reasonably well with the experimental collapse pressures.

It was also shown that some torispherical composite shells are stronger and lighter than their hemispherical counterparts. A recent theoretical prediction concerning this was verified experimentally in the process.

REFERENCES

1. Smith, C.S., "Design of Marine Structures in Composite Materials", Elsevier Appl. Sci., London/NY, 1990.

2. Blachut, J., Galletly, G.D. and Gibson, A.G., "CFRP Domes Subjected to External Pressure", J. Marine Struct., vol. 3, 1990, 149-173.

3. Levy-Neto, F., Galletly, G.D. and Mistry, J., "Buckling of Composite Torispherical and Hemispherical Domes", in Proc. of 'Composite Materials Design and Analysis - CADCOMP 90, Brussels', (eds.) W.P. de Wilde and W.R. Blain, Springer-Verlag, 1990, pp. 375-393.

4. Galletly, G.D. and Blachut, J., "Collapse Strength of Composite Domes Under External Pressure". Proc. of the 2nd Inter. Conf. on 'Advances in Marine Structures', (eds.) C.S. Smith and R.S. Dow, Elsevier Appl. Sci., 1991, pp. 708-732.

5. Blachut, J., Galletly, G.D. and Levy-Neto, F., "Towards Optimum CFRP End Closures". Submitted for publication in Proc. I.Mech.E.

6. Blachut, J. and Galletly, G. D., "A Numerical Investigation of Buckling/Material Failure Modes in CFRP Dome Closures", in Proc. of 'Composite Materials Design and Analysis - CADCOMP 90, Brussels', (eds.) W.P. de Wilde and W.R. Blain, Springer-Verlag, 1990, pp. 395-411.

7. American Society for Testing Materials, Annual book of ASTM Standards, Part 36, Plastic Fibre Composites, 1977.

8. Hofer, K.E. and Rao, P.N., "A New Static Compression Fixture for Advanced Composite Materials", J. of Testing and Evaluation, 4, 5, 1977, 181, 278-283.

9. Bushnell D., "BOSOR 4 - program for stress, buckling and vibration of complex shells of revolution", in 'Structural mechanics software series', (eds.) N. Perrone and W. Pilkey, University Press of Virginia, 1977, pp. 11-143.

10. Jones, R.M., "Mechanics of Composite Materials", McGraw-Hill Book Company, New York, 1975; also Hemisphere Publishing Corporation, 1990.

11. Tsai, S.W., "A Survey of Macroscopic Failure Criteria for Composite Materials", J. Reinforced Plastics and Composites, 3, (1984), 40-62.

12. Owen, M.J., "Biaxial Failure of GRP - Mechanisms, Modes and Theories", in Composite Structures 2, (ed.) Marshall I.H., Elsevier Appl. Sci., 1983, pp. 21-39.

13. Blachut, J., Galletly, G.D. and Moreton, D.N., "Buckling of Near-Perfect Steel Torispherical and Hemispherical Shells Subjected to External Pressure", AIAA J., 11, 28, 1990, 1971-1975.

BUCKLING TESTS WITH AXIALLY COMPRESSED UNSTIFFENED CYLINDRICAL SHELLS MADE FROM CFRP

B. Geier, H. Klein, R. Zimmermann
Deutsche Forschungsanstalt für Luft- und Raumfahrt
Institut für Strukturmechanik
Flughafen, D-3300 Braunschweig, Germany

ABSTRACT

The results of buckling tests with axially compressed unstiffened cylindrical shells made from carbon fiber reinforced plastic are reported and evaluated. The set of specimens comprises ten shells, the laminate set-ups of which lead to maximum, minimum and intermediate buckling loads. Their design by mathematical optimization and their manufacture, as well as the test equipment and the test procedure are dealt with. A comparison of experimental and computed buckling loads is performed on the basis of knock-down factors. The results admit the conclusion that imperfection effects do not overwhelm the influence of the individual nominal design on the buckling load, and that the small strain, moderate rotation theory supplies a good foundation for the prediction of buckling loads of laminated composite cylindrical shells subjected to axial compression, even if edge effects are disregarded.

INTRODUCTION

Buckling tests are indispensible in the development of appropriate methods to predict the buckling behaviour of shells. The results of such tests with axially compressed unstiffened circular cylindrical shells made from carbon fibre reinforced plastic (CFRP) are reported and evaluated.

Laminated fibre reinforced shells offer an increased freedom to the designer compared to structures made of homogeneous isotropic material. For the same number of elementary plies he has the choice between different fibre orientations varying from ply to ply. By using mathematical programming methods the best and the worst of all possible configurations for a laminate with a given number of plies can be found. The result is startling: The buckling load of the best cylinder, the optimum, may be as high as nearly 2.8 times that of the worst one which we shall designate as 'pessimum'.

499

The test series was intended to answer the following questions:

- Is the computed large range between optimum and pessimum confirmed by experiments?
- Which simplifications in modelling the structure are admissible for the prediction of buckling loads with sufficient trustworthiness?

Ten circular cylindrical shells were tested in axial compression, among them the optima and pessima for three and five angle plies. In this paper the following items are addressed:

- Design of the test specimens,
- manufacturing, test device and test procedure,
- test results and comparison with computations.

TESTS

Design of the Test Shells

The number of plies and the associated orientations are different among the test shells; the radius and the length are fixed to 250 mm and 510 mm, respectively. The laminate of each shell consists of n_p units called angle ply, each of which is built by two plies oriented under $+ \Theta$ and $- \Theta$, respectively, with regard to the x-coordinate which runs in the direction of the cylinder axis. The plies are tapes, 0.125 mm in thickness, with fibres going in parallel. The nominal in-plane stiffness properties of the tapes are (in N/mm^2):

$$Q'_{11} = 124446 , \quad Q'_{12} = 2802 , \quad Q'_{22} = 8771 , \quad Q'_{33} = 5695 .$$

(Subscripts '1' and '2' indicate the fibre direction and the in-plane direction normal to it, respectively, and '33' means shear).

Each shell is designed by variation of the fibre orientations, mainly aiming at maximization or minimization of the buckling load for a given number of angle plies n_p (e.g. a fixed mass). This has been achieved by a systematic optimization procedure using a specific mathematical programming system (CADOP, cf. [1]). In such a procedure the buckling load has to be evaluated very often, and a quick method for its computation is a must. So the simplifications leading to classical buckling load have been presumed. Details are presented in the chapter on comparison of tests and analysis. Despite its shortcomings, the classical load is used here for the optimization runs because of its efficiency with respect to computer time. More information on the optimization and its results is given in ref. [2]. A list of the test designs is presented in Table I.

Manufacturing of the Test Shells

The fabrication of the test shells is performed by tape laying upon a heated core, the radius and the length of which are 250 mm and 550 mm, respectively. The tapes are cut according to the desired fibre orientations (Table 1); 90° orientations are made by winding. Having applied a single layer, deficiencies like unevennesses, gaps between fibres, and gas bubbles are manually removed by pressing with a rubber roll ('smoothing'). After the application of all the layers of the laminate, a tear-off-fabric is put on and overwound by a well defined amount of glass fibres with a definite fibre stress. Then the laminate hardens at 140°C for 12 hours. After cooling down, the tear-off-fabric and the glass windings, which have absorbed surplus resin, are removed, and the CFRP laminate can be pulled from the core. The ends are now edged such that they are plane and normal to the shell axis. The length will be 530 mm. The structure is then equipped with end plates to assure circular cross sections. The edges of the laminate are fitted into 10 mm deep grooves of the end plates, filled with a mixture of epoxy resin and quartz powder. By hardening of the resin the CFRP structure is bonded to the plates. The effective length of the shell now comes to 510 mm.

TABLE 1
Compilation of the Test Shells

Shell-No.	Laminate Built-Up, Orientations in °	Remarks
Z11	$\pm60/0_2/\pm68/\pm52/\pm37$	Complem. angles of the optimum for $n_p = 5$
Z12	$\pm51/\pm45/\pm37/\pm19/0_2$	Pessimum for $n_p = 5$
Z14	$\pm51/90_2/\pm40$	Optimum for $n_p = 3$
Z17	$\pm30/90_2/\pm22/\pm38/\pm53$	Optimum for $n_p = 5$
Z18	$\pm37/\pm52/\pm68/0_2/\pm60$	Reverse stacking to Z11
Z21	$\pm39/0_2/\pm50$	Complem. angles of the optimum for $n_p = 3$
Z22	$\pm49/\pm36/0_2$	Pessimum for $n_p = 3$
Z23	$\pm60/0_2/\pm68/\pm52/\pm37$	Repetition of Z11
Z24	$\pm51/\pm45/\pm37/\pm19/0_2$	Repetition of Z12
Z25	$\pm30/90_2/\pm22/\pm38/\pm53$	Repetition of Z17

Test Equipment and Test Procedure

The tests were performed in the buckling test device of the Institute for Structural Mechanics of DLR. Figure 1 illustrates details of the test installation.

The upper plate can be moved in vertical direction in order to adapt the test device to various lengths of test specimens. It is fixed during the tests and reacts the force that is applied to the movable lower drive plate by a servo-hydraulic cylinder. The drive plate acts against three load cells on which there is a stout cylindrical structure that is meant to distribute the three concentrated forces coming from the load cells, into a smooth force distribution at its upper surface. The test specimen is placed between that load distributor and the fixed upper plate. All parts of the test device are extremely stiff, the most flexible parts being the three load cells which are shortened by 0.15 mm at their nominal loads of 100 kN.

Although the test device and test specimens are manufactured with extreme care we do not expect that the fixed upper plate and the load distributor are perfectly plane and parallel to each other, nor do we expect the end plates of the test specimens to be perfectly plane and parallel. To make sure that the test specimens will be uniformly loaded we apply, between the end plates of the test specimens and the adjacent parts of the test device, thin layers of a kind of epoxy concrete, i.e. epoxy reinforced with a mixture of sand and quartz powder. This has the side effect of securing the test specimens against lateral displacement at the lower end. At the upper end the epoxy

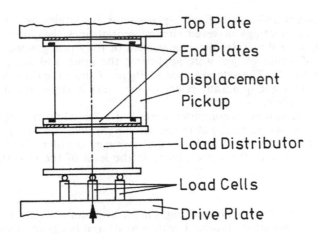

Figure 1. Test Equipment

layer is separated from the upper plate of the test device by a thin foil. The separation is required to achieve a well defined load-free state.

Three displacement transducers are used to measure the axial shortening of the cylinders during the tests. Their signals are recorded and, moreover, used for control purposes. Hence, the test device is displacement-controlled. According to the particular arrangement of the transducers the elastic deformation of the test device does not influence the control at quasi-static loading.

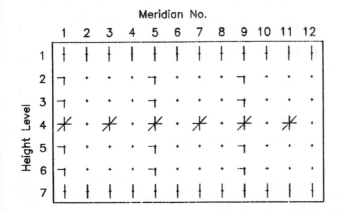

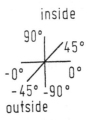

Figure 2. Map of the Strain Gauges

In addition to forces and longitudinal displacements 108 strains were measured by strain gauges at the crossings of seven circumferential lines (height levels) and twelve meridians. Figure 2 illustrates the arrangement of the strain gauges. At those nodes various types of strain gauges were applied at the inner and outer surface. Some nodes remained empty. The symbols used in Figure 2 show the three directions 0°, 45°, 90°. The upper right quadrant refers to the inner surface, the lower left quadrant to the outer surface.

During the tests an in-process computer prescribed the axial shortening in small steps by transmitting the set-point signal to the servo-hydraulic control. The measurement signals were read 20 times at each step and averaged in order to suppress the error coming from the noise of the signals down to the level of the resolution of a 12-bit AD-converter.

Test Results

The experimental buckling loads and buckling modes are summarized in Table 2, which also contains the computed classical buckling loads and buckling modes.

TABLE 2
Results of the Buckling Tests

Shell-No.	Buckling Loads in kN		Buckling Modes M/N		Remarks
	comp.	exper.	comp.*	exper.**	
Z11	288.74	228.0	16/6	2/7	
Z12	98.52	93.5	13/0	2/10	
Z14	80.30	82.8	1/7	2/8	
Z17	288.79	278.5	4/11	2/7	Some buckles were lacking
Z18	225.79	212.6	17/0	2/8	
Z21	72.21	69.3	19/0	2/10	
Z22	36.26	34.4	17/0	1/9	Partially M = 2
Z23	288.74	221.7	16/6	2/7	
Z24	98.52	90.2	13/0	1/8	
Z25	288.79	227.9	4/11	2/7	

* Buckling mode at bifurcation   **Stable postbuckling mode

M and N are the numbers of axial half waves and of circumferential waves, respectively. The computed buckling modes belong to the bifurcation, whereas the experimental ones describe stable postbuckling modes.

For exploitation of the strain measurement a comfortable graphics software has been developed, which enables a fast visualization of the measured strains on different types of diagrams. An example is given in Figure 3 which, for the shell Z24 (pessimum for $n_p = 5$), illustrates in its upper part the strains in axial direction - measured inside and outside - as function of the axial load, and in its lower part the variation of axial strains around the circumference at the height levels 1 and 7 (close

to the upper and lower edge of the shell, respectively), taken at 65 % of the experimental buckling load.

Axial Load in kN

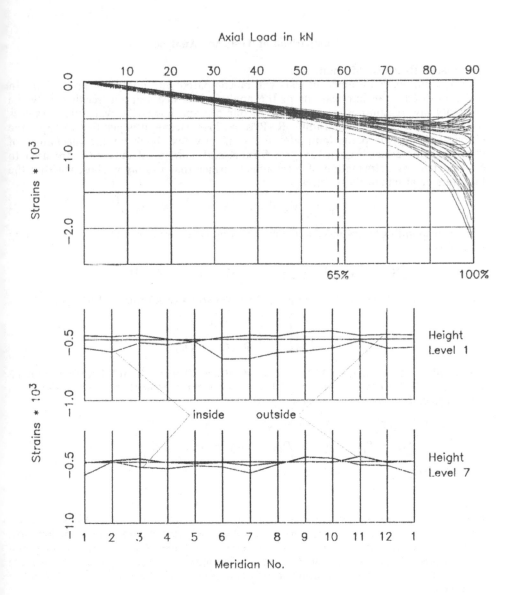

Figure 3.   Some Measured Axial Strains, Shell Z24

It is impossible to demonstrate here the exploitation of the strains. The conclusion is, that - apart from the aspects of imperfections and of realized stiffnesses - the
buckling loads are governed by the design and not by some other conditions. When

comparing the experimental loads with the computed ones in the next section, the two aspects mentioned above will be taken into account.

Comparison of Tests and Analysis

The Elastic Response of Laminates

The buckling loads obtained in tests will be compared with analytical predictions. In order to discuss the implications of the analysis it is necessary to recall briefly the elasticity law of composite laminates. In our test specimens the wall thickness h is only a small fraction of the radius R (R/h $\geq$ 200). Hence, it is justified to apply the classical theory for plane laminates found in any text book on the mechanics of composite materials, e.g. in ref. [3]. Strains and changes of curvature are used to characterize the deformation of a reference surface that usually is chosen within the laminate or at one of its surfaces,

$$\varepsilon = [\, \varepsilon_x, \, \varepsilon_y, \, \gamma_{xy} \,]^T \quad , \quad \kappa = [\, \kappa_x, \, \kappa_y, \, \kappa_{xy} \,]^T \ .$$

The corresponding stress resultants are the membrane forces and the bending/twisting moments,

$$N = [\, N_x, \, N_y, \, N_{xy} \,]^T \quad , \quad M = [\, M_x, \, M_y, \, M_{xy} \,]^T \ .$$

The lamination theory yields a relation between stress resultants and deformation,

$$\left\{ \begin{matrix} N \\ M \end{matrix} \right\} = \begin{bmatrix} A & B \\ B & D \end{bmatrix} \left\{ \begin{matrix} \varepsilon \\ \kappa \end{matrix} \right\} \ .$$

Three different symmetric (3,3)-matrices A, B and D form the total stiffness matrix. In general all those three partitions have non-vanishing components: The laminate is anisotropic. Two special cases deserve attention:

1. The orthotropic laminate for which

    $$A_{13} = A_{23} = B_{13} = B_{23} = D_{13} = D_{23} = 0 \ .$$

2. The balanced laminate for which

    $$A_{13} = A_{23} = 0 \ .$$

A balanced laminate is obtained if there are as many plies oriented $-\Theta$ as there are plies with the orientation $+\Theta$. Very often such laminates are treated like orthotropic ones assuming that the terms D_{13}, D_{23} as well as the terms B_{13}, B_{23} are small enough to be neglected. This neglection may lead to erroneous results, in particular for plate problems, cf. another contribution to this conference [4]. All test specimens were of balanced design.

Analytical Models

The analytical predictions of the buckling loads were based on the simplifying assumption of a uniform membrane state of stress in the prebuckling region. In reality the lateral expansion of the cylinders is prevented at the edges and, as a consequence, circumferential membrane stresses are produced. They vary along the generator from compression to tension, again to compression, and so on, in a decaying short-wave mode. It is supposed that they do not influence the buckling load considerably. In fact, for isotropic cylinders it was shown that they might reduce the computed buckling load by at most 15 % [5]. We shall evaluate here two methods to calculate buckling loads:

1. The <u>classical buckling load</u> which is a closed-form solution for <u>orthotropic</u> cylinders with simply supported edges which is based on the theory of <u>shallow shells</u>. More precisely, the boundary conditions implied are

$$w^* = 0 \, , \quad M_x^* = 0 \, , \quad N_x^* = 0 \, , \quad v^* = 0 \, .$$

The asterisk indicates the increase of quantities due to buckling. The formula for the classical buckling load is not reproduced here. It can be found in refs. [6, 7].

2. An <u>analytical solution for anisotropic cylinders</u> which forms the base of the computer program BACCUS, cf. ref. [8]. It allows solutions for arbitrary boundary conditions and any combination of axial load, internal or external pressure and torsion. The buckling loads given here are calculated for built-in (clamped) edges boundary conditions

$$w^* = 0 \, , \quad \frac{\partial w^*}{\partial x} = 0 \, , \quad u^* = 0 \, , \quad v^* = 0 \, .$$

These boundary conditions closely describe the support of the shells in the experiments. A <u>small strain - moderate rotation shell theory</u> is exploited in BACCUS.

Knock-Down Factors

The comparison will be based on a consideration of the knock-down factor. It relates the experimental buckling loads to the theoretical ones. The relevant information is comprised in Table 3.

The knock-down factor η ranges from 0.80 to 1.03. This result is remarkable for three reasons:

1. The knock-down factors are high,
2. their scatter is small,
3. the knock-down factor is not significantly smaller for the optimal than for the pessimal cylinders.

The cylinders No. 23, 24, 25 were duplicates, with equal nominal design, of cylinders No. 11, 12 and 17, respectively. They were fabricated and tested in order to get more information on the question of the design (optimum, pessimum or other) being decisive for the buckling load attained. Cylinder No. 17 was the optimum among all cylinders made of 5 layers, No. 12 was the pessimum, and No. 11 has a theoretical buckling load very close to the optimum, but the laminate stacking is different.

The absolute magnitude of the knock-down factor depends on the stiffness properties used in the computations. The factor might come down somewhat if higher stiffness values are used in the analysis. The strain gauge measurements taken during the tests are being evaluated presently to find improved stiffness values. This is by no means a trivial task, since the measurements give some indication of certain <u>laminate</u> stiffness coefficients, but basic input properties for the computation are the stiffnesses of the <u>unidirectional</u> plies from which the laminates are built.

TABLE 3

Comparison of Experimental and Theoretical Buckling Loads

Cyl. No.	Test		Class. Buckl. Ld.	BACCUS	η	
11 / 23	228	221.7	288.7	276.4	0.82	0.80
12 / 24 ")	93.5	90.2	98.5	99.6	0.94	0.91
14 ")	82.8	- - -	80.3	80.5	1.03	- - -
17 / 25 ")	278.5	227.9	288.8	275.0	1.01	0.83
18	212.6	- - -	225.8	226.8	0.94	- - -
21	69.3	- - -	72.2	71.6	0.97	- - -
22 ")	34.4	- - -	36.6	35.6	0.97	- - -

") : Designed as optimum
") : Designed as pessimum

$$\eta = \text{ratio} \; \frac{\text{buckling load found in test}}{\text{buckling load computed with BACCUS}} = \text{knock-down factor}$$

It is likely that the use of improved stiffnesses would not change the scatter. The fact that it is unusually small may be attributed to the extreme care obeyed during all phases of the manufacturing and testing process.

We expected that optimal cylinders would be more sensitive to imperfections than pessimal or intermediate ones. This supposition is supported by the larger span between buckling and postbuckling loads observed with the optimal shells. If we compare the knock-down factors of the four cylinders which are optimal or near optimal (cylinders No.11, 23, 17, 25) we find three of the knock-down factors slightly above 0.8, but No. 11 is an exception with $\eta = 1.01$. The pessima had knock-down factors above 0.9. Hence, the results might admit the weakly supported statement that the imperfection sensitivity is slightly greater for optimal than for pessimal cylinders. But the absolute values of the buckling loads are by a factor > 2.75 higher for the optima. This shows clearly that imperfection effects do not overwhelm the influence which the individual nominal design has on the buckling load.

CONCLUSIONS

The results presented and discussed above admit the conclusion that the small strain, moderate rotation theory provides a good basis for the prediction of buckling loads of laminated composite cylinders subjected to axial compression, even if edge effects are disregarded. The use of the classical buckling load in optimization also turns out to be justified. It implies that the large range between optimum and pessimum is confirmed.

This does not mean that we advocate the uncritical use of buckling loads computed for perfect cylindrical shells with such high knock-down factors in any design situation.

REFERENCES

1. Pappas, M., An Improved Direct Search Numerical Optimization Procedure, Computers and Structures, 1980, 11, pp. 539-557.

2. Zimmermann, R., Optimization of Axially Compressed Fiber Composite Cylindrical Shells. In Optimization. Methods and Applications, Possibilities and Limitations, ed. H.W. Bergmann, International Seminar, Deutsche Forschungsanstalt für Luft- und Raumfahrt (DLR), Bonn, Proceedings, Springer Verlag, Berlin/Heidelberg/New York, 1989, pp. 63-82.

3. Jones, R.M., Mechanics of Composite Materials, Scripta Book Company, Washington D.C., 1975.

4. Rohwer, K., Malki, G., Steck, E., Influence of Bending-Twisting Coupling on the Buckling Loads of Symmetrically Layered Curved Panels. Paper presented at this conference.

5. Fischer, G., Über den Einfluß der gelenkigen Lagerung auf die Stabilität dünnwandiger Kreiszylinderschalen unter Axiallast und Innendruck. Zeitschrift für Flugwissenschaften, Bd. 11, 1963, 3, pp. 111-119.

6. Geier, B., Das Beulverhalten versteifter Zylinderschalen. Teil 1, Zeitschrift für Flugwissenschaften, Bd. 14, 1966, 7, pp. 306-323.

7. Seggelke, P., Geier, B., Das Beulverhalten versteifter Zylinderschalen. Teil 2, Zeitschrift für Flugwissenschaften, Bd. 15, 1967, 12, pp. 477-488.

8. Geier, B., Rohwer, K., Buckling and Postbuckling of Anisotropic Shells. Proc. of an International Colloquium "Stability of Plate and Shell Structures", Ghent, Belgium, 06-08.04.1987, edited by P. Dubas and D. Vandepitte, pp.321-329.

TOWARDS RECOMMENDATIONS FOR SHELL STABILITY DESIGN BY MEANS OF NUMERICALLY DETERMINED BUCKLING LOADS

HERBERT SCMMIDT, ROLAND KRYSIK
University of Essen, FB 10 - Steel Structures,
Universitaetsstraße 15, D-4300 Essen 1

ABSTRACT

After a short description of the common 3-step-procedure in design codes for shell stability in structural engineering, four conceivable approaches how to use numerical shell analysis tools in a stability design problem are outlined. The first approach "A", based on linear bifurcation buckling analysis of the perfect elastic shell, is believed to be preferable for practising design engineers. It is explained in more detail, two versions "A1" and "A2" are discussed. Adventages and disadventages are illustrated by means of a design example (a conical vessel head structure). In a second example experimental buckling results of doubly curved steel test specimens are compared to the theoretical buckling resistances according to approach A1. Some conclusions are drawn.

1 INTRODUCTION

Design codes or design recommendations dealing with thin-walled metal shell structures usually contain some stability design rules for unstiffened or stiffened fundamental types of shells of revolution (e.g. circular cylinders, cones, spheres, torispherical caps) under fundamental loads [1-6]. The rules are, explicitly or implicitly, based on a 3-step-procedure: (1) Calculate the critical buckling resistance R_{cr} of the geometrically perfect, elastic shell by means of given approximate, but conservative formulas. (2) Apply a reduction factor α to cover the imperfection-induced "knock-down" from R_{cr} to the elastic ultimate buckling resistance $R_{u,el}$. (3) Apply, if necessary, a second reduction factor η to cover the plasticity-induced further decrease from $R_{u,el}$ to the "real" ultimate buckling resistance R_u (see figure 1):

$$R_u = \eta \cdot \alpha \cdot R_{cr} \tag{1a}$$

Both reduction factors are of empirical character and are, for each particular buckling case under consideration, experimentally calibrated. It is well-known that α may vary from ca. 0,8 down to ca. 0,15. Instead of applying the second reduction factor straightforward, the additional plasticity reduction is in some codes expressed by another type of plasticity factor Φ, relating the ultimate buckling resistance to the plastic reference resistance R_{pl} of the shell by means of a special shell slenderness parameter $\bar{\lambda}_\alpha$ (see figure 1):

$$R_u = \Phi \cdot R_{pl} \qquad (1b)$$
with $\Phi = f(\bar{\lambda}_\alpha)$ and $\bar{\lambda}_\alpha = \sqrt{R_{pl}/\alpha\, R_{cr}}$.

As a third alternative, some codes combine the two consecutive factors into **one** stability reduction factor , covering both, the imperfection-induced "knock-down" and the plasticity-induced further decrease. It relates R_u to R_{pl}, too, but depends on the general stability slenderness parameter (see figure 1):

$$R_u = \aemath \cdot R_{pl} \qquad (1c)$$
with $\ae = f(\bar{\lambda})$ and $\bar{\lambda} = \sqrt{R_{pl}/R_{cr}}$.

The three eqns. (1a) to (1c) are different formulations of virtually the same 3-step-procedure. Eqn. (1c) corresponds to the format which is favoured by Eurocode 3 [7].

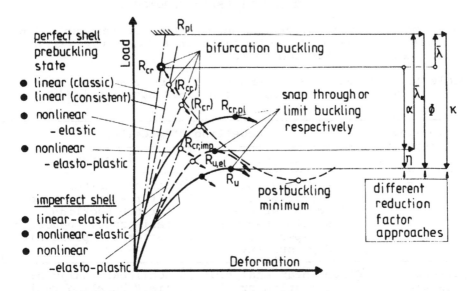

Figure 1. Buckling behavior of shells of revolution: Approaches to analysis (schematic) and design.

The design codes usually do not give much guidance, or none at all, how to handle other than the fundamental shells of revolution (e.g. with arbitrary meridional shape or with variable

thickness or with irregularly spaced stiffeners or with ela-
sticly supported edges etc.) and/or other than the fundamental
loads (e.g. wind pressure). On the other hand, powerful computer programs are available to analyse either (a) the geometrically perfect shell on the basis of some finite-ring-type discreticization or (b) the imperfect shell on the basis of a complete finite-element discreticization. The problem is not the
numerical or algorithmic capability of these programs, but the
use of their results in terms of a safe and economic design. Of
course, one may have the attitude that a scientifically educa-
ted engineer should be able to come up with a safe shell stability design without particular guidance by codes, if he is given enough time to engage in the problem. But in his every day
business, dealing with relatively ordinary structures, very
often under time pressure, he would certainly appreciate some
recommendations how to model a shell buckling problem for a
computer analysis and how to convert the numerically determined
buckling load into the "real" (characteristic) ultimate buckling resistance that he needs for his design.

The first author has discussed the task of formulating such
recommendations in several working groups. Some aspects are
presented in the following, considering only shells of revolution.

2 FOUR CONCEIVABLE APPROACHES

On principle, four different approaches are thinkable how to
use numerical shell analysis tools in a stability design problem; they are summarized in table 1.

Table 1. Conceivable approaches to shell stability design by
means of numerically determined buckling loads.

modelling of material	modelling of geometry	
	perfect	imperfect
elastic	Ⓐ linear analysis: bifurcation buckling resistance, reduction factor(s) for imperfections and plasticity	Ⓑ linear analysis: bifurcation or snap through buckling resistance, reduction factor for plasticity
elasto-plastic	Ⓒ nonlinear analysis: bifurcation or limit buckling resistance, reduction factor for imperfections	Ⓓ nonlinear analysis: limit buckling resistance as equivalant to the "real" buckling resistance

2.1   Approach A

In this approach the critical buckling resistance R_{cr} of the **geometrically perfect, elastic shell** is determined as the lowest bifurcation value obtained from an eigenvalue analysis performed on the linearly calculated prebuckling state under the applied combination F of actions. Of course, it should be ensured that the used computer program finds reliably the critical eigenmode. In some special cases it may be necessary to calculate the prebuckling state geometrically nonlinearly, but these cases are disregarded here. In detailed recommendations, some rules would have to be given how to idealize the middle surface and the boundary/continuity/support/loading conditions, how to discreticize the shell, how to choose the calculated shell portion (if not the whole shell is calculated), how to handle constructional excentricities etc.

The linear bifurcation analysis replaces the approximate formulas of step (1) in the abovementioned 3-step-procedure. The two steps (2) and (3) remain to be handled by reduction factors in the sense of eqns. (1). Two versions for this reduction (called approaches A1 and A2) are thinkable; they are outlined using eqn. (1c).

2.1.1 Approach A1

In this version of approach A the ultimate buckling resistance R_u is directly obtained by multiplying R_{pl} by **one** global reduction factor æ. The value of R_{pl} shall be determined as that value of the applied combination F of actions at which the buckling-relevant membrane stresses of the linearly calculated prebuckling state at any point of the shell firstly satisfy the v.MISES yield criterion (f_y being the yield stress):

$$\sigma_x^2 - \sigma_x \sigma_\varphi + \sigma_\varphi^2 + 3\tau^2 = f_y^2 \rightarrow R_{pl} \, . \tag{2}$$

In general, it should be sufficient to check those (maximally three) points of the shell, where
- either the maximum compressive meridional membrane stress max σ_x
- or the maximum compressive circumferential membrane stress max σ_φ
- or the maximum shear membrane stress max τ
is present, and to use the smallest of these R_{pl}-values (see figure 2).

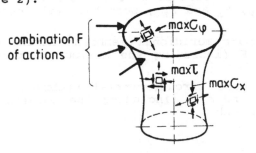

combination F of actions

Figure 2:
Buckling-relevant membrane stresses under the combination F of actions

If no specific reduction factor æ is available (e.g. from tests), it must be estimated from existing rules of the type æ = f($\bar{\lambda}$). Here the ECCS-Recommendations [5] are used (see also [8]):

$$\text{æ} = \alpha/\bar{\lambda}^2 \qquad \text{when } \bar{\lambda} \geq \sqrt{2\,\alpha}\,,$$
$$\text{æ} = 1 - 0,25\,\bar{\lambda}^2/\alpha \qquad \text{when } \bar{\lambda} \leq \sqrt{2\,\alpha}\,. \tag{3}$$

The value of α may be conservatively determined from:

$$\alpha = \frac{0,83}{\sqrt{1 + 5,40\,\bar{\lambda}^2}} \qquad \text{when } \bar{\lambda} \leq 0,626\,,$$
$$\alpha = \frac{0,70}{\sqrt{0,1 + 5,40\,\bar{\lambda}^2}} \qquad \text{when } \bar{\lambda} > 0,626\,. \tag{4}$$

The determination of α from eqn. (4) is based on the plausible assumption that no shell buckling case can be imagined to be more imperfection-sensitive than the unstiffened circular cylinder subject to uniform meridional compression which is known to be **extremely imperfection-sensitive**. This determination may in some cases be very conservative and uneconomic. If a **moderate imperfection sensitivity** is believed to hold true, a value

$$\alpha = 0,65 \tag{5}$$

could be envisaged. (Existing rules [1 - 6] offer values from α = 0,5 to α = 0,8 for moderately imperfection-sensitive shell buckling cases.)

It will be illustrated in sections 3 and 4 that using eqn. (4) may very often yield an uneconomic design, but using eqn. (5) may as often yield an unsafe design. In fact, there is a lack of experimental evidence for imperfection factors α between the two extremes of eqns. (4) and (5). One possibility of overcoming this problem would incorporate stress-specific reduction factors, as explained in the next subsection.

2.1.2 Approach A2

In this version of approach A, the combination F of actions being under consideration is decomposed into (possibly fictitious) partial actions F_x, F_φ and F_τ , such that (at least approximately)

- F_x produces only the pure meridional membrane stress field $\sigma_x(x,\varphi)$ present under F (including the maximum compressive stress max σ_x),

- F_φ produces only the pure circumferential membrane stress field $\sigma_\varphi(x,\varphi)$ present under F (including the maximum compressive stress max σ_φ) and

- F_τ produces only the pure shear membrane stress field $\tau(x,\varphi)$ present under F (including the maximum shear stress max τ).

A simple example for such a decomposition is shown in section 3.

For the partial actions, critical partial buckling resistances R_{xcr} and/or $R_{\varphi cr}$ and/or $R_{\tau cr}$ have to be determined, as described above. The maximum values of the buckling-relevant membrane stresses present under the critical partial buckling resistances would then be taken as critical buckling stresses σ_{xcr} and/or $\sigma_{\varphi cr}$ and/or τ_{cr}, and eqns. (1c) and (3) would be applied seperately to the (maximally three) stress components:

$$\sigma_{xu} = æ_x \cdot f_y$$

$$\text{with} \quad æ_x = f(\bar\lambda_x) \quad \text{and} \quad \bar\lambda_x = \sqrt{f_y/\sigma_{xcr}} \; , \tag{6}$$

$$\sigma_{\varphi u} = æ_\varphi \cdot f_y$$

$$\text{with} \quad æ_\varphi = f(\bar\lambda_\varphi) \quad \text{and} \quad \bar\lambda_\varphi = \sqrt{f_y/\sigma_{\varphi cr}} \; , \tag{7}$$

$$\tau_u = æ_\tau \cdot f_y/\sqrt{3}$$

$$\text{with} \quad æ_\tau = f(\bar\lambda_\tau) \quad \text{and} \quad \bar\lambda_\tau = \sqrt{f_y/\tau_{cr}\sqrt{3}}. \tag{8}$$

The ultimate buckling resistance R_u would in this approach be implicitly given by an interaction formula, e.g. the linear ECCS [5] interaction formula

$$\frac{\max \sigma_x (R_u)}{\sigma_{xu}} + \frac{\max \sigma_\varphi (R_u)}{\sigma_{\varphi u}} + \frac{\max \tau (R_u)}{\tau_u} = 1, \tag{9}$$

where $\max \sigma_x (R_u)$, $\max \sigma_\varphi (R_u)$ and $\max \tau (R_u)$ are those maximum membrane meridional compressive, circumferential compressive and shear stresses respectively, which are present under the applied combination F of actions (see figure 2), once F equals R_u. In detailed recommendations, some rules would have to be given how to procede if the points of maximum membrane stresses are far away from each other so that buckling interaction is physically impossible.

Of course, we end up at the question of stress-specific imperfection factors α_x, α_φ and α_τ respectively. Very vaguely, a system as sketched in table 2 could be imagined.

Table 2. Hypothetical stress-specific imperfection factors.

Shells of revolution with	α_x	α_φ	α_τ
$K = 1/(r_x \cdot r_\varphi) > 0$	eqn. (4)	eqn. (4)	eqn. (5)
$K = 1/(r_x \cdot r_\varphi) \leq 0$		eqn. (5)	

In table 2 r_x and r_φ are the two curvature radii of the shell at the point where max σ_φ is present, and K is the GAUSS-parameter.

2.2   Approaches B and

These two approaches try to eliminate one of the two empirical factors in eqns. (1) by investing more numerical input. As a consequence, new specific reduction factors are needed to take care of the remaining discrepancies between numerically determined buckling loads and "real" (characteristic) ultimate buckling resistances (seee figure 1):

For approach B:

$$R_u = \eta^* \cdot R_{cr,imp}$$ (10)

with η^* = specific new plasticity factor.

For approach C:

$$R_u = \alpha^* \cdot R_{cr,pl}$$ (11)

with α^* = specific new imperfection factor.

If no specific tests were available, the new reduction factors would have to be calibrated to the lots of known tests which have already been the basis for the traditional reduction factors α, η, Φ and $\ae$ in eqns. (1). But this would actually not improve the accuracy of the results.

Therefore, the authors suppose that approaches B and C will be confined to
- either particular design problems where specific test series are feasible,
- or shell geometries where η^* or α^* respectively may be simply estimated because of having not much influence.

2.3   Approach D

In the authors' opinion this "fully nonlinear" approach is the only real alternative to approach A. The German DASt-Working Group "Shell Buckling" has tried to come up with some draft recommendations [10] how to model the equivalent geometric imperfections, both qualitatively (imperfection pattern) and quantitatively (imperfection amplitude). The present paper does not provide enough space to discuss the ideas (and the many unsolved problems!).

3   EXAMPLE 1

Figure 3a shows the conical head of a vacuum vessel with an additional axial load on its top. The upper edge may be assumed radially restrained; the lower edge is stiffened by a ring with angle cross section; the adjacent cylindrical portion is neglected. The "real" (characteristic) ultimate buckling resistance $R_u \triangleq q_u$ shall be determined.

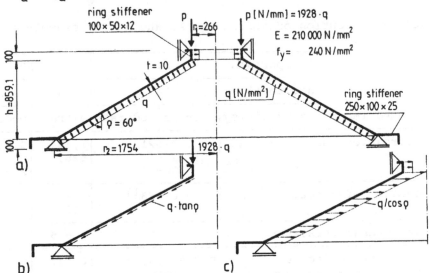

Figure 3. Example 1: Overall system (a) and decomposition into partial actions with "pure" meridional (b) and circumferential (c) membrane stress fields respectively.

The execution of approaches A1 and A2 according to the equations in section 2.1 of this paper is summarized in table 3. The numerical bifurcation analyses have been carried out with the program F04 B08 [11] using its linear prebuckling analysis option. The stress values σ_{xcr} and $\sigma_{\varphi cr}$ are the maximum elementary cone membrane stresses under the critical actions q_{cr}.

It can be seen from table 1 that

- assuming in approach A1 a moderate overall imperfection sensitivity would overestimate the safety margin considerably and possibly lead to an unsafe design,

- assuming in approach A1 an extreme overall imperfection sensitivity would underestimate the safety margin considerably and possibly lead to an uneconomic design.

4 EXAMPLE 2

Figure 4 shows the nominal dimensions of two types of doubly curved steel shells of revolution that have been tested under axial compression; the details are reported in [12]. The wall thicknesses were, because of the manufacturing process, variable in meridional direction, but fairly constant in circumferential direction. In figure 5, one of the two specimens of each type is shown after the buckling test. The experimental ultimate loads shall be compared with calculated predictions according to approach A1.

Table 3. Execution of approaches A1 and A2 for example 1.

approach	A1		A2	
acc.to fig.	3a		3b	3c
assumed imperf. sensitivity	extreme	moderate	extreme	moderate
bifurcation analysis: q_{cr}	0,521		1,364 526,1[1]	0,618
σ_{xcr}	-			-
$\sigma_{\varphi cr}$	-		-	216,8[2]
plastic reference: q_{pl}	0,663[1]		-	-
f_y	-		240	240
slenderness parameter: $\bar{\lambda}$	1,128		0,675	1,052
imperfection factor: α	0,265	0,65	0,437	0,65
stability reduction factor: æ	0,208	0,511	0,739	0,574
ult. buckling stresses: σ_{xu}	-		177,4	-
$\sigma_{\varphi u}$	-		-	137,8
ult. buckling resistance: q_u	0,138	0,339	0,212	

Dimensions: $[N/mm^2]$,
1) determining point: upper edge of cone, 2) lower edge of cone

The execution of approach A1 according to the equations in section 2.1 of this paper is summarized in table 4. The middle surfaces have been dicreticized to finite rings in order to model the variable thicknesses. The numerical bifurcation analyses have again been carried out with the linear version of F04 B08.

It can be seen from table 4 that
- assuming a moderate overall imperfection sensitivity leads for both types to an unconservative prediction,
- assuming an extreme overall imperfection sensitivity leads still to a slightly unconservative prediction for type K > 0, but to a very conservative prediction for type K < 0.

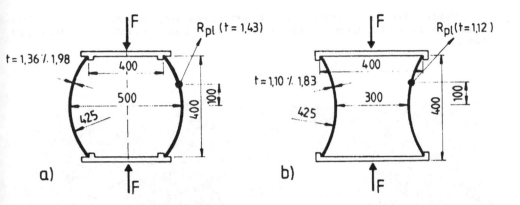

Figure 4. Example 2: Nominal dimensions of shell specimens with
K > 0 (a) and K < 0 (b) respectivelly.

Table 4. Execution of approach Al for example 2.

	acc. to fig.	type K > 0 4a		type K < 0 4b	
assumed imperf. sensitivity		extreme	moderate	extreme	moderate
material:	E f_y	182000 279		195000 415	
bifurcation analysis:	R_{cr}	1545		360,0	
plastic reference:	R_{pl}	$408,0^{1)}$		$528,0^{1)}$	
slenderness parameter:	$\bar{\lambda}$	0,514		1,211	
imperfection factor:	α	0,533	0,65	0,247	0,65
stability reduction factor:	æ	0,876	0,898	0,169	0,443
ultim. buckling resistance:	R_u	357,4	366,4	89,2	233,9
test results: No.1 No.2	R_u^{ex}	342 349		177 198	

Dimensions: [kN] and [N/mm^2]
1) determining points marked in figure 4

It should be mentioned that approaches A2, B and C have also been applied to these test specimens, but cannot be discussed here.

Figure 5.Test specimens of example 2 after buckling test.

5   CONCLUSIONS

(a) Of four conceivable approaches how to use numerical shell analysis tools in a general stability design problem, as outlined in the present paper, the linear bifurcation analysis approach is believed to be, at the time being, preferable for practising design engineers.

(b) It is easy to present a design procedure based on the beforementioned approach and yielding conservative designs.

(c) In order to come up with designs which are not only conservative but also economic, a refined stress-specific reduction factor system has to be developed.

REFERENCES

1. American Society of Mechanical Engineers: ASME Boiler and Pressure Vessel Code - Section III, Code Case N-284. New York: ASME 1980.

2. Det Norske Veritas: Rules for the Design, Construction and Inspection of Offshore Structures - Appendix C: Steel Structures. Hovik/Norway: DnV 1982.

3. British Standards Institute: Specification for Unfired Welded Pressure Vessels. London: BSI 1976.

4. Deutscher Ausschuß für Stahlbau: DASt-Richtlinie 013 – Beulsicherheitsnachweise für Schalen: Köln: DASt 1980.

5. European Convention for Constructional Steelwork: Buckling of Shells – European Recommendations, 4th edition. Bruxelles: ECCS 1988.

6. Deutsches Institut für Normung: DIN 18800 Part 4 – Steel Structures, Stability, Buckling of Shells. Köln: Beuth Verlag 1990.

7. Commission of the European Communities: EUROCODE No. 3 – Design of Steel Structures, Part 1 – General Rules and Rules for Buildings. Edited draft Nov. 1990 (unpublished).

8. Samuelson, L.A.: The ECCS Recommendations on Shell Stability, Design Philosophy and Practical Applications. Paper for the Round Table Discussion at Lyon Conference 1991.

9. Schmidt, H.: The German Code DIN 18800 Part 4, Design Philosophy and Practical Applications. Paper for the Round Table Discussion at Lyon Conference 1991.

10. DASt-Working Group "Shell Buckling": Beulsicherheitsnachweis für Schalen mit Hilfe numerisch ermittelter Beullasten. Draft Dec. 1990 (unpublished).

11. Esslinger, M./Geier, B./Wendt, U.: Berechnung der Spannungen und Deformationen von Rotationsschalen im elasto-plastischen Bereich. Stahlbau 53 (1984), 17-25.

12. Stracke, M./Düsing, H./Krysik, R./Schmidt, H.: Belastungs- und Beulversuche an axialsymmetrisch belasteten Rotationsschalen aus Metall im elast.-plast. Bereich zur Überprüfung nichtlinearer Rechenprogramme. Research Report No. 38, University Essen-FB 10, 1986 (in German, with an English summary).

THE RCC-MR DESIGN RULES AGAINST BUCKLING

Denis ACKER
Département de Mécanique et Technologie
Service d'Etudes Mécaniques et Thermiques
C.E.N.-SACLAY
91191-GIF-SUR-YVETTE-CEDEX

INTRODUCTION

Fast breeder reactor structures are designed with large thin shells (typicaly about 20 meters diameter for 20 millimeters of thickness for Super Phenix or other pool type demonstration plant projects) and loaded with a Low Level of pressure and large cyclic thermal transient during the in service operation conditions, but shall be available to support without damage the dynamic pressures due to fluid structure interaction under seismic loading or other hypothetical situations.

Such structures are susceptible to buckle and relatively large geometrical imperfections cannot be avoided.

Even if buckling is not, strictly speaking a damage since the induced deformations can be limited, its appearance may induce failure mode as elastic or elastoplastic instability and in certain cases, increases the risks of fatigue damage or progressive deformation. This latter phenomenon may in its turn increase the risks of instability. Buckling may be accelerated or amplified by geometrical imperfections permitted by the manufacturing tolerances.

The RCC-MR Design rules against buckling, edited in 1985 with the first edition, of the french design and construction code for fast breeder reactors, are under reviewing and should undergo large modification in the next times.

DESIGN PHILOSOPHY

The RCC-MR [1] objective is to provide the same safety margins against failure or damage induced by buckling than by other causes. To meet these objectives.

RCC-MR distinguishes between-Load controlled buckling, Strain controlled buckling and time dependant buckling and requieres to meet specific criterion for each type of buckling, connected with their specific risks.

Load controlled buckling

Buckling is said to be load controlled when it is the result of imposed loads which cannot be reduced by the deformation associated with buckling.

The existence of other external (imposed displacements) or internal (temperatures) loadings, act simultaneously with the imposed loads to modify the imposed loading leading to buckling.

The risk of load controlled buckling is to induce elasto plastic instability of the structure. Also RCC-MR requieres to have a safety margin on the concerned loading equal to 2.5 against this failure mode (under normal or upset in service conditions).

This factor shall be applied to all the loading components, excepted the throught wall thermal gradient wich is affected with a coeeficient equal to 1.

RCC-MR recall to the designer attention that instability load connected with the buckling phenomenon may be reduced if cyclic stress such as thermal transients exist, because their combination with maintained compressive stress load to ratchet phenomenon. In such situations, the current rules against ratchetting are not available [2] to exclude large increase of the geometrical defects and whether the allowable stress limits shall be tightened or the instability load reduction shall be appraised.

Strain controlled buckling

Buckling is said to be strain controlled only if the imposed loads, whatever their intensity, could not on their own produce it. In all other cases, buckling is said to be load controlled.

Strain controlled buckling cannot induce instability but the buckled structure shape changes modify the stress and strain fields and the risk of fatigue or excessive deformations by ratchetting.

Also, RCC-MR requieres either to verify that elasto-plastic bifurcation do not appears with a loading equal to 1.67 times the considered loading level or to take account of the effect of geometrical imperfections and non linear phenomena associated with buckling during evaluation of the quantities (strain range, stress range) used in the calculation of fatigue damage and progressive deformation evaluation [3].

Time-dependent buckling

At high temperatures, maintained loadings could cause time-dependent buckling chiefly because of the evolution of the properties of the material and the shape of the structures with time (amplification of geometrical defects).

In order to warrant constant safety margins during the in Service life of the structure, RCC-MR requieres to cheek the compliance with the above acceptability conditions of load controlled buckling risk at all times in the life of the structure, taking account of the modification of the shape due to creep produced by in service loadings. An additional sheck shall also be made that instability do not appears, taking account of defect size increasing under 1.5 times the imposed loads in normal in service conditions and assuming that they are applied throughout the life of the structure.

RECOMMANDED PRACTICES

RCC-MR provides the designers with recommended methodes of buckling risk analysis in its appendix A12, devoted to shells of revolutions subject to external pressure and cylinders under axial compression, and apppendix A7.

This appendix thus initially offers two alternative methods to avoid elastic instability or elastoplastic instability damage under monotonic loading. The first described in chapter A7.2000, uses the results of a linear elastic analysis of the structure without defects to determine its elastic buckling load E. This elastic buckling load must be affected by a knock down factor given diagrams and depending on the plasticity, the imperfection size and the stable or unstable post buckling behaviour of the structure [4] [5]

The second described in A7.3000 gives the methodology to applied an inelastic analysis [6].

When the cyclic loadings are superimposed on monotonic loadings, as mentioned before, the instability load can be reduced. Chapter A7.4000 can be used either to limit the variations of the cyclic loadings so that they do not lead to a notable reduction of the buckling load, or to calculate this reduction on the basis of an elastic analysis.

Both of these routes need to evaluate the actual stress in the structure with imperfection and use the primary effective stress concept proposed by R. ROCHE to limit or to appraise the ratchet consequences on the critical load reduction during normal in service conditions.

CURRENT DEVELOPMENT ACTIVITIES

Some problems are unsolved in the RCC-MR 85 edition and also Appendix A7 should be improved to provide :

- Guidancies to evaluate the buckling risk during earthquake, taking account for the knock down factor between elastically computed load and equivalent static loading [7].

- Guidancies for simplified analysis of imperfection growth by creep.

- Guidancies for simplified analysis of imperfection growth by ratchet during in service normal condition and their effect or critical load under accidental events.

- Condition of no interaction between thermal buckling and mechanical buckling modes.

- Guidancies to appraise the buckling risk of particular geometries structures under specific load combinations featuring the fast breeder reactor situations [8].

REFERENCES

1   **RCC-MR.** Design and construction rules for mechanical components of F.B.R. Nuclear Islands edited by AFCEN - PARIS 1985. Published by AFNOR 1985

2   **CLEMENT G., DRUBAY B.** - Progressive buckling - International Conference SMIRT 11 paper E11/3 August 18-23 - TOKYO (Japan)

3   **MOULIN D., COMBESCURE A., ACKER D.** - A review of analysis methods about thermal buckling-6 International Seminar on Inelastic Analysis and Life Prediction in High temperature environment - Paris (France) - August 24-25-1987

524

4   **AUTRUSSON B., ACKER D., DEVOS J.** - Design rules to
     prevent elastoplastic buckling. International Conference
     SMIRT 8 paper E7/4 - Brussel (Belgium) August 19-24-1985

5   **AUTRUSSON B., ACKER D., HOFFMANN A.** - Discussion and
     validation of a simplified analysis against buckling -
     Nucl. Eng. and Design 98 (1987) 379-393

6   **COMBESCURE A.** - Upon the different theories of plastic
     buckling - Element for a chance - International
     Conference SMIRT 10 - paper E11/6 - August 14-18-1989 -
     Anaheim (U.S.A.).

7   **COMBESCURE A., QUEVAL J.C., ALLIOT P., HERMANN R.** -
     Dynamic buckling of Liquid Storage Towks - International
     Conference SMIRT 9 - Vol. K - p.781 - LAUSANNE
     (Switzerland) - August 17-21-1987

8   **COMBESCURE A., BROCHARD J.** - Recent advances on Thermal
     buckling at CEA-Saclay - International Conference on
     SMIRT 11 - paper EII/1 - TOKYO (Japan)- August 18-
     23 1991

525

INDEX OF CONTRIBUTORS

Milton Keynes UK
Ingram Content Group UK Ltd.
UKHW020006071024
449327UK00031B/2677